Advanced Polymers and its Selected Applications

Advanced Polymers and its Selected Applications

Edited by **Jan Cooper**

NY RESEARCH PRESS

New York

Published by NY Research Press,
23 West, 55th Street, Suite 816,
New York, NY 10019, USA
www.nyresearchpress.com

Advanced Polymers and its Selected Applications
Edited by Jan Cooper

International Standard Book Number: 978-1-63238-018-0 (Hardback)

Printed in China.

Contents

Preface

The advanced polymers along with its selected applications are comprehensively illustrated in this extensive book. This book is a compilation of contributions by reputed authors from across the globe engaged in various fields of polymer physics. It encompasses various topics like applications of conducting polymers; thin film polymers; photopolymers for holographic media; hydrogel films on optical fiber core; microwave absorption and EMI shielding behavior of nanocomposites based on naturally conducting polymers and graphene as well as carbon nanotubes; electro reductive polymerization; nitrogen-rich and radiation cross-linking poly(urethane-imide) polymers as applicants for energetic applications; oxidative polymerization of aniline; polysilanes with organized sequences; development of ruthenium complexes to new functional nanocomposites. The motive of this book is to inspire readers in pursuing research and study in this extensive field.

All of the data presented henceforth, was collaborated in the wake of recent advancements in the field. The aim of this book is to present the diversified developments from across the globe in a comprehensible manner. The opinions expressed in each chapter belong solely to the contributing authors. Their interpretations of the topics are the integral part of this book, which I have carefully compiled for a better understanding of the readers.

At the end, I would like to thank all those who dedicated their time and efforts for the successful completion of this book. I also wish to convey my gratitude towards my friends and family who supported me at every step.

Editor

Section 1

Polymer Physics

Chapter 1

Microwave Absorption and EMI Shielding Behavior of Nanocomposites Based on Intrinsically Conducting Polymers, Graphene and Carbon Nanotubes

Parveen Saini and Manju Arora

Additional information is available at the end of the chapter

1. Introduction

Electromagnetic interference (EMI) is an undesirable and uncontrolled off-shoot of explosive growth of electronics and widespread use of transient power sources. Conducting polymers nanocomposites represent a novel class of materials that possess unique combination of electrical, thermal, dielectric, magnetic and/or mechanical properties which are useful for suppression of electromagnetic noises. Now it is possible to incorporate various dielectric or magnetic fillers within conducting polymer matrices to form multifunctional nanocomposites. The first section of this chapter gives a brief overview of fundamentals of EMI shielding & microwave absorption, theoretical aspects of shielding, governing equations, various techniques for measurement of shielding effectiveness and different strategies for controlling EMI. In the next section, a comprehensive account of potential materials for handling of EMI are described with special reference to nanocomposites based on intrinsically conducting polymer matrix filled with conducting [e.g. metals, graphite, carbon back, carbon nanotubes, graphene], dielectric (e.g. $BaTiO_3$ or TiO_2) or magnetic (e.g. γ-Fe_2O_3, Fe_3O_4, $BaFe_{12}O_{19}$) inclusions.

2. Electromagnetic Interference (EMI) shielding

Electromagnetic interference shielding (EMI) is an undesired electromagnetic (EM) induction triggered by extensive use of alternating current/Voltage which tries to produce corresponding induced signals (Voltage and current) in the nearby electronic circuitry, thereby trying to spoil its performance. The mutual interference among electronic gadgets, business machines, process equipments, measuring instruments and appliances lead to disturbance or complete breakdown of normal performance of appliances. The EM disturbances across communication channels, automation, and process control may lead to loss of time, energy, resources and also adversely affect human health. Due to these reasons only, use of mobile phone is restricted inside robotic operation theatres or during onboard/flight which may trigger series of electronic failures and a crash in worst case/scenario. Therefore, some shielding mechanism must be provided to ensure undisturbed functioning of devices even in the presence of external electromagnetic (EM) noises. For efficient shielding action, shield should possess either mobile charge carriers (electrons or holes) or electric and/or magnetic dipoles which interact with the electric (E) and magnetic (H) vectors of the incident EM radiation. Therefore, in the recent past, a wide variety of materials (Abbas et al, 2007; Colaneri et al, 1992; Joo & Epstein, 1994; Ott, 2009; Paul, 2004; Saini et al, 2009a, 2010, 2011; Schulz et al, 1988; Singh et al, 1999a, 2000b) have been used for EMI shielding with a broad range of electrical conductivity (σ), good electromagnetic attributes such as permittivity (ε) or permeability (μ) and engineered geometries. The designing a EMI shielding with a certain level of attenuation, meeting a set of physical criteria, maintaining economics and regulating the involved shielding mechanism is not a straight forward task and involves complex interplay of intrinsic properties (σ, ε and μ) of shield material and logical selection of extrinsic parameters. Therefore, to touch the theoretically predicted shielding performance of a materials and to satisfy stringent design criteria, elementary knowledge of shielding theory, set of governing theoretical equations, important design parameters and relevant measurement technique becomes a prime prerequisite.

3. Shielding definitions and phenomenon

EMI shield is essentially a barrier to regulate the transmission of the electromagnetic EM wave across its bulk. In power electronics, term shield usually refers to an enclosure that completely encloses an electronic product or a portion of that product and prevents the EM emission from an outside source to deteriorate its electronic performance.

Conversely, it may also be used to prevent an external susceptible (electronic items or living organisms) from internal emissions of an instrument's electronic circuitry. Shielding is the process by which a certain level of attenuation is extended using a strategically designed EM shield. The shielding efficiency is generally measured in terms of reduction in magnitude of incident power/field upon transition across the shield. Mathematically shielding effectiveness (SE_T) can be expressed in logarithmic scale as per expressions (Saini et al 2009a, 2011):

$$SE_T\,(\mathrm{dB}) = SE_R + SE_A + SE_M = 10\log_{10}\left(\frac{P_T}{P_I}\right) = 20\log_{10}\left(\frac{E_T}{E_I}\right) = 20\log_{10}\left(\frac{H_T}{H_I}\right) \tag{1}$$

where P_I (E_I or H_I) and P_T (E_T or H_T) are the power (electric or magnetic field intensity) of incident and transmitted EM waves respectively. As shown in Fig. 1, three different mechanisms namely reflection (R), absorption (A) and multiple internal reflections (MIRs) contribute towards overall attenuation with SE_R, SE_A and SE_M as corresponding shielding effectiveness components due to reflection, absorption and multiple reflections respectively.

3.1. Theoretical shielding effectiveness

Before starting the shielding analysis, it is necessary to understand the various electromagnetic terminologies (Ott, 2009).

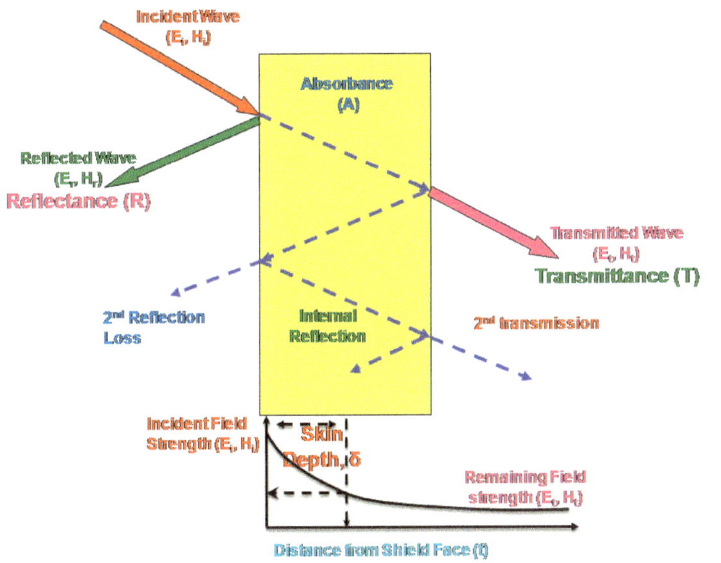

Figure 1. Schematic representation of EMI shielding mechanism

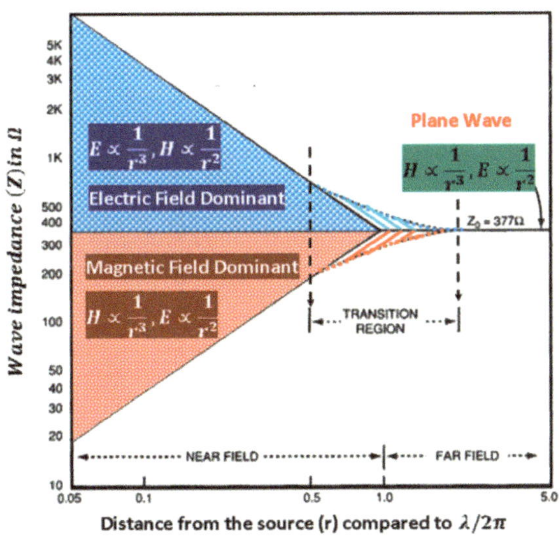

Figure 2. Dependence of wave impedance on distance from source normalized to $\lambda/2\pi$

According to the distance r between the radiating source and the observation point, an electromagnetic radiative region can be divided into three parts (Fig. 2) relative total wavelength λ of the electromagnetic wave. The region within the distance $r < \lambda/2\pi$ is the near field while the distance $r > \lambda/2\pi$ is the far field. Between the two regions, as the distance $r \approx \lambda/2\pi$, is the transition region. For designing a material for particular shielding application, it is imperative to have in-depth knowledge of both intrinsic & extrinsic parameters on which shielding effectiveness depend alongwith suitable theoretical relations correlating them with reflection, absorption and multiple-reflection loss components.

3.1.1. *Shielding theory*

This section presents the shielding basics based on the transmission line theory (Schelkunoff, 1943) and the plane wave shielding theory (Schulz et al,1988). Assume a uniform plane wave characteristic by E and H that vary within a plane only with x direction as showed in Fig. 3. The Maxwell's curl equations give:

$$\frac{dE}{dx} = -j\omega\mu H \text{ and } \frac{dH}{dx} = -(\sigma + j\omega\varepsilon)E \tag{2}$$

where μ is the permeability of the material and $\mu = \mu_0\mu_r$. μ_0 and μ_r are the permeabilities of air (*or free space*) and shield material respectively, σ is the conductivity of material in S/m. where ε is the permittivity of the material and $\varepsilon = \varepsilon_0\varepsilon_r$. ε_0 and ε_r are the permittivities of air (or free space) and shield material respectively, $\omega = 2\pi f$. $\omega(f)$ is angular frequency (linear frequency) in Hz.

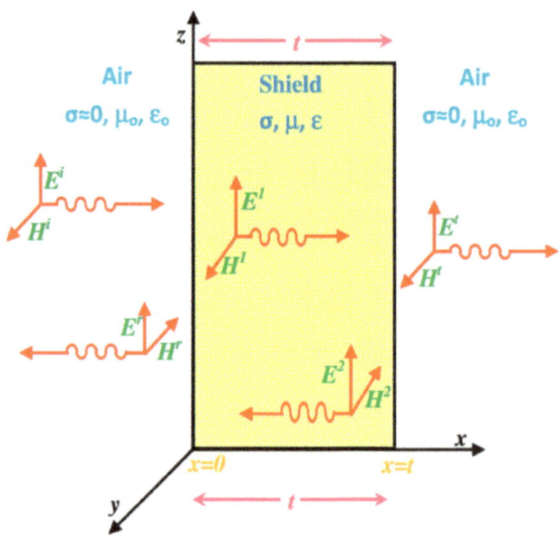

Figure 3. Propagation of electromagnetic waves and its interaction with the shield material

All homogenous materials are characterized by a quantity known as the intrinsic impedance:

$$\eta = \sqrt{\frac{j\omega\mu}{\sigma + j\omega\varepsilon}} \quad (3)$$

When an electromagnetic wave propagates through the material, the wave impedance approaches the intrinsic impedance of the material. For dielectric material, the conductivity is extremely small ($\sigma \ll \omega\varepsilon$) and the intrinsic impedance of Eq. (3) becomes:

$$\eta = \sqrt{\frac{\mu}{\varepsilon}} \quad (4)$$

For a conductor used below optical frequencies defined by $\sigma \gg \omega\varepsilon$, the intrinsic impedance of Eq. (3) can be written as:

$$\eta = \sqrt{\frac{j\omega\mu}{\sigma}} = (1+j)\sqrt{\frac{\pi\mu f}{\sigma}} \quad (5)$$

It is customary to define propagation constant (γ) in the media such that:

$$\gamma = (\alpha + j\beta) = \sqrt{j\omega\mu(\sigma + j\omega\varepsilon)}. \quad (6)$$

where α is attenuation constant and β is phase constant. A good conductor is a medium for which $\sigma / \omega\varepsilon \gg 1$. Under this condition the Eq. (6) becomes:

$$\gamma = \sqrt{j\omega\mu} = (1+j)\sqrt{\pi\mu f\sigma} \quad (7)$$

Therefore, we can write $\alpha = \beta = 1/\delta = \sqrt{\pi\mu f\sigma}$, where quantity δ represents skin depth which is defined as the distance required by the wave to be attenuated to $1/e$ or 37% of its original strength. For a dielectric plane sheet $\sigma / \omega\varepsilon \ll 1$ and Eq. (6) becomes:

$$\gamma = \sqrt{-\omega^2\mu\varepsilon} = j\omega\sqrt{\mu\varepsilon} \quad (8)$$

The impedance of a homogenous barrier of thickness t is

$$Z = \eta\frac{Z(t)\cosh(\gamma t) + \eta\sinh(\gamma t)}{\eta\cosh(\gamma t) + Z(t)\sinh(\gamma t)} \quad (9)$$

$$H(t) = \frac{\eta}{\eta\cosh(\gamma t) + Z(t)\sinh(\gamma t)}H(0) \quad (10)$$

$$E(t) = \frac{Z(t)}{Z(t)\cosh(\gamma t) + \eta\sinh(\gamma t)}E(0) \quad (11)$$

where $Z(0)$ is the impedance at interface 0 looking into the plane and $H(t)$ is the impedance at interface t looking into the right of the plane at $x = t$. If $Z(t) \neq \eta$, reflection occurs at the boundary $x=t$. Let E^i and H^i are the incident electric and magnetic fields, E^r and H^r the reflected fields, and E^t and H^t the transmitted fields as shown in Fig. 3. With the continuity of the tangential field component at a boundary we can write:

$$E^i + E^r = E^t \quad \text{and} \quad H^i + H^r = H^t \tag{12}$$

The electric and magnetic fields of a plane wave are related by the intrinsic impedance of the medium

$$E^i = \eta H^i, \; E^r = -\eta H^r \text{ and } E^t = Z(l)H^i \tag{13}$$

Solving the above equations, the expression of reflection coefficients can be written as:

$$q_E = \frac{E^r}{E^i} = \frac{Z(t)-\eta}{Z(t)+\eta} \tag{14}$$

$$q_H = \frac{H^r}{H^i} = \frac{\eta - Z(t)}{\eta + Z(t)} = -q_E \tag{15}$$

The corresponding transmission coefficients can be written as:

$$p_E = \frac{E^t}{E^i} = \frac{2Z(t)}{\eta + Z(t)} = 1 + q_E \tag{16}$$

$$p_H = \frac{H^t}{H^i} = \frac{2\eta}{\eta + Z(t)} = 1 + q_H \tag{17}$$

When two mismatched interfaces must be considered in the same plane, the net transmission coefficients is the product of the transmission coefficient across the two boundaries i.e.:

$$p = p_E = p_H = p_E(0)p_E(t) = p_H(0)p_H(t) \tag{18}$$

Considering the re-reflection effect, the transmission coefficients across the plane are:

$$T_H = \frac{H(t)}{H^i} = \frac{H(t)}{H(0)} \cdot \frac{H(0)}{H^i} \tag{19}$$

$$T_E = \frac{E(t)}{E^i} = \frac{Z(t)}{Z_w} \cdot \frac{H(t)}{H^i} = \frac{Z(t)}{Z_w} T_H \tag{20}$$

where $E(0)$, $E(t)$, $H(0)$ and $H(t)$ are the actual values at interfaces i.e. at $x=0$ and $x=t$. Z_w is the impedance of the incident wave. Using equations (9), (10) and (11) for the plane of the thickness 0 and t

$$\frac{H(t)}{H(0)} = \frac{\eta}{\eta\cosh(\gamma t) + Z(t)\sinh(\gamma t)} \tag{21}$$

$$\frac{E(t)}{E(0)} = \frac{Z(t)}{Z(t)\cosh(\gamma t) + \eta \sinh(\gamma t)} \quad (22)$$

From equations (14) and (15) we may write:

$$\frac{H(t)}{H^i} = \frac{2Z_w}{Z_w + Z(0)} \quad (23)$$

$$\frac{E(t)}{E^i} = \frac{2Z(0)}{Z_w + Z(0)} \quad (24)$$

where $Z(0)$ is the impedance at interface $x = 0$ looking into the plane. By substituting (23) and (24) into equations (19) and (20) we get:

$$T = T_E = p_H\left(1 - q_H e^{-2\gamma t}\right)e^{-\gamma t} \quad (25)$$

where

$$p_H = \frac{4Z_w \eta}{\left(Z_w + \eta\right)\left(Z(t) + \eta\right)} \quad (26)$$

$$q_H = \frac{\left(Z_w - \eta\right)\left(Z(t) - \eta\right)}{\left(Z_w + \eta\right)\left(Z(t) + \eta\right)} \quad (27)$$

when $Z(t) = Z_w$, taking $k = Z_w / \eta$ we can write:

$$p = p_H = \frac{4k}{\left(k + 1\right)^2} \quad (28)$$

$$q = q_H = \frac{\left(k - 1\right)^2}{\left(k + 1\right)^2} \quad (29)$$

$$T_E = T_H = T = p\left(1 - qe^{-2\gamma t}\right)e^{-\gamma t} \quad (30)$$

By definition total shielding effectiveness is:

$$SE_T = 20\log_{10}|T| = 20\log_{10}\left|p\left(1 - qe^{-2\gamma t}\right)e^{-\gamma t}\right| \quad (31)$$

$$SE_T = \underbrace{20\log_{10}|p|}_{SE_R} + \underbrace{20\log_{10}\left|e^{-\gamma t}\right|}_{SE_A} + \underbrace{20\log_{10}\left|\left(1 - qe^{-2\gamma t}\right)\right|}_{SE_M} \quad (32)$$

Therefore, after careful comparison of δ with shield thickness (t) two situations can be visualized:

a. **When (t << δ):** which occurs either at low frequencies or in case of electrically thin sample where actual shield thickness is much less than skin depth. In such a case the absorption which is a bulk (or thickness) related phenomenon, can be neglected and attenuation occurs almost exclusively by reflection. The total shielding becomes frequency independent and can be expressed in terms of free space impedance (Z_o=377 Ω) as:

$$SE\,(dB) = -20\log_{10}\left[1+\frac{Z_O}{2}t\sigma_T\right] \tag{33}$$

b. **When (t >> δ):** which is valid in our case and generally occurs at higher frequencies where skin depth becomes much less as compared to actual shield thickness i.e. in case of electrically thick samples. In such regime, attenuations due reflection, absorption and multiple internal sub-phenomenon becomes a straight forward exercise after making good conductor approximation i.e. $\sigma_T / \omega\varepsilon \gg 1$ [or $k \ll 1(i.e.\, Z_w \gg \eta)$].

3.1.2. Reflection loss

The reflection loss (SE_R) is related to the relative impedance mismatch between the shield's surface and propagating wave. The magnitude of reflection loss under plane wave (far field conditions) can be expressed as (Saini et al, 2011):

$$SE_R(dB) = -10\log_{10}\left(\frac{\sigma_T}{16\omega\varepsilon_o\mu_r}\right) \tag{34}$$

where σ_T is the total conductivity, f is the frequency in Hz, μ_r is the relative permeability referred to free space; The above expression shows that SE_R is a function of the ratio of conductivity (σ_T) and permeability (μ_r) of the shield material i.e. quantity (σ_T/μ_r). Further, for a given material (i.e. fixed σ_T and μ_r) SE_R decreases with increase in frequency.

3.1.3. Absorption loss

As shown in Fig. 1, when an electromagnetic wave pass through a medium its amplitude decreases exponentially. This decay or absorption loss occurs because currents induced in the medium produce ohmic losses and heating of the material, and E_t and H_t can be expressed as $E_t = E_i e^{-t/\delta}$ and $H_t = H_i e^{-t/\delta}$ (Ott, 2009). Therefore, the magnitude of absorption term (SE_A) in decibel (dB) can be expressed by following equation:

$$SE_A(dB) = -20\frac{t}{\delta}\log_{10} e = -8.68\left(\frac{t}{\delta}\right) = -8.68t\left(\frac{\sigma_T\omega\mu_r}{2}\right)^{\frac{1}{2}} \tag{35}$$

where t is shield thickness in inches and f is frequency in Hertz. The above expression revealed that SE_A is proportional to the square root of the product of the permeability (μ_r)

times the conductivity (σ_T) of the shield material i.e. quantity $(\sigma_T.\mu_r)^{1/2}$ (Saini et al, 2009a, 2011). Further, for a given material, absorption loss increases with increase in frequency. Therefore, a good absorbing material should possess high conductivity and high permeability, and sufficient thickness to achieve the required number of skin depths even at the lowest frequency of concern.

3.1.4. Multiple Internal Reflections (MIRs)

If the shield is thin, the reflected wave from the second boundary is re-reflected from the first boundary and returns to the second boundary to be reflected again and again as shown in Fig. 1. The attenuation due these multiple internal reflections i.e. SE_M can be mathematically expressed as (Ott, 2009, Saini et al, 2011):

$$SE_M = 20\log_{10}(1-e^{-2t/\delta}) = 20\log_{10}\left|\left(1-10^{-\frac{SE_A}{10}}\right)\right| \tag{36}$$

Therefore, it can be seen from the above expression that SE_M is closely related to absorption loss (SE_A). SE_M is also important for porous structures and for certain type of filled composites or for certain design geometries. It can be neglected in the case of a thick absorbing shield due high value of SE_A so that by the time the wave reaches the second boundary, it is of negligible amplitude. For practical purposes, when SE_A is ≥ 10 dB (Saini, et al 2009a, 2011) SE_M can be safely neglected. Usually SE_M is important only when metals are thin and are used at very low frequencies (i.e. ~kHz range). However, for highly absorbing materials or at very high frequencies (~GHz or high), condition $|SE_A| \geq 10dB$ gets satisfied and re-reflections can be safely ignored i.e. $SE_M \approx 0$.

3.2. Experimental shielding effectiveness

Experimentally, shielding is measured using instruments called network analyzer. Scalar network analyzer (SNA) measures only the amplitude of signals whereas vector network analyzer (VNA) measures magnitude as well as phases of various signals. Consequently, SNA can not be used to measure complex signals (e.g. complex permittivity or permeability) and therefore, despite its higher cost VNA is the most widely used instrument.

The incident and transmitted waves in a two port VNA (Fig. 4) can be mathematically represented by complex scattering parameters (or S-parameters) i.e. S_{11} (or S_{22}) and S_{12} (or S_{21}) respectively which in-turn can be conveniently correlated with reflectance (R) and transmittance (T) i.e. $T = |E_T/E_I|^2 = |S_{12}|^2 = |S_{21}|^2$, $R = |E_R/E_I|^2 = |S_{11}|^2 = |S_{22}|^2$, giving absorbance (A) as: A = (1-R-T). When SE_A is greater than 10 dB, SE_M becomes negligible (~ -1.0 dB) and can be neglected (Saini et al, 2011) so that SE_T can be expressed as: $SE_T = SE_R + SE_A$. In addition, the intensity of the EM wave inside the shield after primary reflection is based on quantity (1-R), which can be subsequently used for normalization of absorbance (A) to yield effective absorbance $\{A_{eff}=[(1-R-T)/(1-R)]\}$ so that experimental reflection and absorption losses can be expressed as (Hong et al, 2003; Saini et al, 2009a, 2011):

$$SE_R = 10\log_{10}(1 - R) \quad (37)$$

$$SE_A = 10\log_{10}(1 - A_{eff}) = 10\log_{10}\left[\frac{T}{(1 - R)}\right] \quad (38)$$

Therefore, from the knowledge of reflected and transmitted signals i.e. R and T, VNA can easily compute reflection and absorption loss components of total shielding.

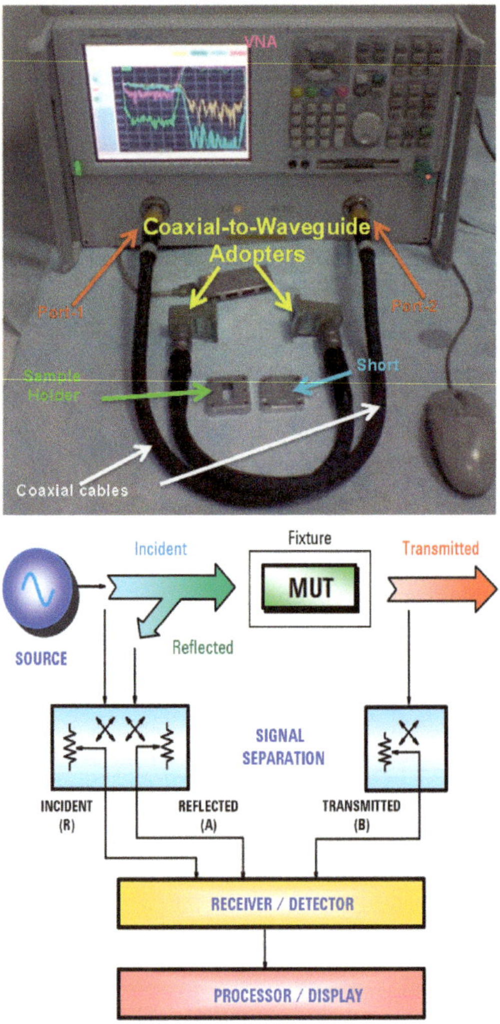

Figure 4. A two port VNA (left) and its internal block diagram (right)

3.3. Estimation of electromagnetic attributes

The attenuation of EM radiation by intended shield material is critically dependent on its electromagnetic attributes like complex permittivity [$\varepsilon^*=(\varepsilon' \;\; j\varepsilon'')$] and complex permeability [$\mu^*=(\mu' \;\; j\mu'')$], and their estimation is of paramount importance. Both complex dielectric permittivity and magnetic permeability consists of real part and imaginary parts as shown in Fig. 5

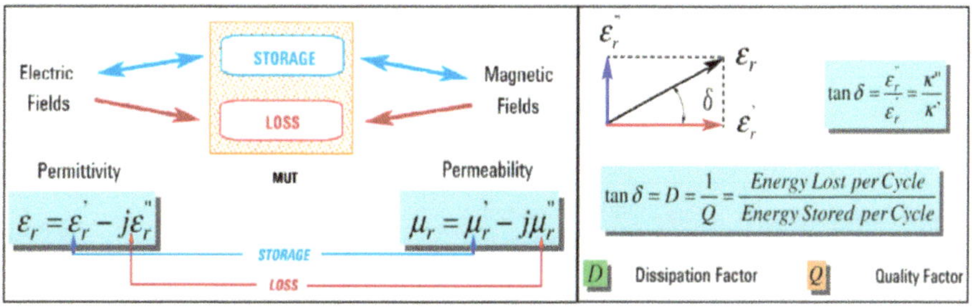

Figure 5. Complex electromagnetic attributes of a shield

Parameter ε' or ε_r (real permittivity) represents the charge storage (or dielectric constant) whereas ε'' (imaginary permittivity) is a measure of dielectric dissipation or losses. Similarly, μ' (or ε_r) and μ'' represents magnetic storage and losses respectively. The extent of losses can be assessed by calculating dielectric/magnetic loss tangent (tan δ) (Colaneri et al, 1992; Joo. et al, 1994; Saini. et al, 2009a, 2011) which is the ratio of imaginary and real permittivity/permeability.

3.3.1. *Measurement and conversion techniques*

While designing a shield, all the above parameters must be taken into consideration. The incident and transmitted travelling waves inside a VNA can be represented by complex scattering parameters (or S-parameters) i.e. S_{11} (or S_{22}) and S_{12} (or S_{21}) respectively, which are in-turn closely related to the electromagnetic (EM) attributes (Nicolson & Ross, 1970; Ott, 2009; Paul, 2004; Weir, 1974). There are many techniques developed for measuring these S-parameters like Transmission/Reflection method, Open ended coaxial probe technique, Free space technique, Resonant cavity method and Parallel plate technique (Ott, 2009; Tong, 2009). Among these techniques Transmission/Reflection method is the most popular as it simultaneously measures of all four S-parameters and gives complex permittivity as well as magnetic permeability by using suitable algorithms or models developed for obtaining the permittivity and permeability from the recorded S-parameters. Table 1 gives an overview of some the conversion techniques, S-parameters & to and their evaluation capability for output attributes.

Each of the above conversion technique has different advantages and limitations. The selection of the technique depends on several factors such as the measured S-parameters, sample length, desired output properties, speed of conversion and accuracies in the converted results. Among above-mentioned procedures, Nicholson-Ross-Weir (NRW) technique is the most widely used regressive/iterative analysis as it provides direct calculation of both the permittivity (ε^*) and permeability (μ^*) from the input S-parameters.

Conversion technique	Input S-parameters	Output attributes
Nicolson-Ross-Weir (NRW)	S_{11}, S_{21}, S_{12} and S_{22} or and S_{11} and S_{21} (or S_{22} and S_{12})	ε_r and μ_r
NIST iterative	S_{11}, S_{21}, S_{12} and S_{22} or pair S_{11} and S_{21} (or S_{22} and S_{12})	ε_r and $\mu_r = 1$
New non-iterative	S_{11}, S_{21}, S_{12} and S_{22} or and S_{11} and S_{21} (or S_{22} and S_{12})	ε_r and $\mu_r = 1$
Short circuit line (SCL)	S_{11} or S_{22}	ε_r

Table 1. Conversion techniques, input S-parameters & output attributes

3.3.2. Nicholson-Ross-Weir (NRW) technique

Nicholson-Ross-Weir (NRW) technique (Nicolson & Ross, 1970; weir, 1974) provides direct calculation of both the permittivity (ε^*) and permeability (μ^*) from the input S-parameters. It is the most commonly used technique for performing such conversions where the measurement of reflection (Γ) and transmission (T) coefficient requires all four (S_{11}, S_{21}, S_{12}, S_{22}) or a pair (S_{11}, S_{21}) of S-parameters of the material under test to be measured. The procedure proposed by NRW method is deduced from the following set of equations:

$$S_{11} = \frac{\Gamma\left(1-T^2\right)}{\left(1-\Gamma^2 T^2\right)} \quad \text{and} \quad S_{21} = \frac{T\left(1-\Gamma^2\right)}{\left(1-\Gamma^2 T^2\right)} \tag{39}$$

Once these S-parameters are extracted from the network analyzer, simultaneous solving of equation set (39) gives the reflection coefficient as:

$$\Gamma = X \pm \sqrt{X^2 - 1} \tag{40}$$

The condition [$|\Gamma| < 1$] is used for finding the correct root of the quadratic equation so that parameter X can be expressed as:

$$X = \frac{S_{11}^2 - S_{21}^2 + 1}{2S_{11}} \tag{41}$$

Therefore, the transmission coefficient can be written as:

$$T = \frac{S_{11} + S_{21} - \Gamma}{1-\left(S_{11} - S_{21}\right)\Gamma} \tag{42}$$

The permeability is then given by:

$$\mu^* = \frac{1+\Gamma}{\Lambda(1-\Gamma)\sqrt{\frac{1}{\lambda_o^2}-\frac{1}{\lambda_c^2}}} \tag{43}$$

where λ_o and λ_c are free space and cutoff wavelength respectively and Λ is given by (Tong, 2009):

$$\frac{1}{\Lambda^2} = \left(\frac{\varepsilon^* \mu^*}{\lambda_o^2} - \frac{1}{\lambda_c^2}\right) = -\left[\frac{1}{2\pi L}\ln\left(\frac{1}{T}\right)\right]^2 \tag{44}$$

Therefore, the permittivity can be written as:

$$\varepsilon^* = \frac{\lambda_o^2}{\mu^*}\left(\frac{1}{\lambda_c^2} - \left[\frac{1}{2\pi L}\ln\left(\frac{1}{T}\right)\right]^2\right) \tag{45}$$

Equations (44) & (45) have infinite number of roots since the imaginary part of the term ln(1/T) is equal to $i(\theta + 2\pi n)$ where n= 0, ± 1, ± 2…, i.e. the integral multiples of ratio L/λ_g, where L is sample length and λ_g is wavelength inside the sample. This brings phase ambiguity and the correct value of 'n' can be determined by either of two methods:

a. **The analysis of group delay:**

The calculated group delay for n[th] solution can be determined from:

$$\tau_{cal,n} = L\frac{d}{df}\left(\sqrt{\frac{\varepsilon_r\mu_r f^2}{c^2} - \frac{1}{\lambda_c^2}}\right) = \frac{\left(f\varepsilon_r\mu_r + \frac{f^2}{2}\frac{d(\varepsilon_r\mu_r)}{df}\right)}{c^2\sqrt{\frac{\varepsilon_r\mu_r f^2}{c^2} - \frac{1}{\lambda_c^2}}}L \tag{46}$$

The group can also be directly measured by network analyzer by measuring the slope of the plot between phase (ϕ) of the transmission coefficient versus frequency as:

$$\tau_{meas} = -\frac{1}{2\pi}\frac{d\phi}{df} \tag{47}$$

The correct root (n=k) should satisfy the condition $\tau_{cal,k} - \tau_{meas} = 0$

b. **Phase unwrapping method:**

By estimating n from λ_g using initial guess values of ε and μ for the sample, we get:

$$\varepsilon_r = \mu_r\frac{(1-\Gamma)^2}{(1+\Gamma)^2}\left(1-\frac{\lambda_o^2}{\lambda_c^2}\right) + \frac{\lambda_o^2}{\mu_r\lambda_c^2} \tag{48}$$

where ε_r and μ_r are initially guessed permittivity and permeability respectively, γ is propagation constant of material, c is velocity of light and f is frequency of incident EM radiation.

3.4. Shielding material and design considerations

The careful analysis of theoretical shielding expressions revealed that in order to meet design requirements and for extending efficient shielding action, shield should possess a balanced combination of electrical conductivity (σ), dielectric permittivity (ε) and magnetic permeability (μ) and physical geometry (Chung, 2001; Joo & Epstein, 1994; Saini, 2009a). Further, as shown in Fig. 1, the primary mechanism of EMI shielding is reflection from the front face of the shield, for which the shield must possess mobile charge carriers (electrons or holes) that can interact with the electromagnetic fields to cause ohmic (heating) losses in the shield. As a result, the shield needs to be electrically conducting, although only moderate conductivity (10^{-3} to 1.0 S/cm) is sufficient (Olmedo et al, 1997; Saini, et al 2011). The secondary EMI shielding mechanism is absorption for which shield should possess electric and/or magnetic dipoles which can interact with the electromagnetic fields in the radiation.

Metals are by far the most common materials for EMI shielding (Ott 2009; Paul 2004; Schulz et al, 1988) owing to their high electrical conductivity. In principle, for a highly conducting material (e.g. metals like Cu, Ag or Ni), only conductivity (σ) and magnetic permeability (μ) are important, such that the reflection loss (SE_R) is dependent upon their ratio (i.e. σ/μ) whereas the absorption loss (SE_A) is a function of their product (i.e. $\sigma.\mu$) [Chung, 2001; Joo & Epstein, 1994; Ott, 2009, Saini et al, 2011). However, in the case of moderately conducting materials permittivity (ε) also plays a significant role (besides σ and μ) in deciding absolute values of SE_R and SE_A. Such compounds are capable of displaying dynamic dielectric and/or magnetic loss, upon impingement by incident electromagnetic waves (Abbas et al, 2005, 2006; Joo & Epstein, 1994; Olmedo et al, 1997). Nevertheless, metal based compositions are suffered from problems (Ott, 2009; Paul, 2004; Saini et al, 2009a, 2009b) such as high reflectivity, corrosion susceptibility, weight penalty and uneconomic processing. Among other alternatives, carbon based materials (graphite, expanded graphite, carbon black, carbon nanotubes and graphene) have also been widely explored for possible applications in EMI shielding (Chung 2000, 2001, Gupta & Choudhary, 2011; Huang et al, 2006; Joo et al 1999, Makeiff & Huber, 2006; Pandey et al, 2009; Saini et al 2007, 2009a, 2009b, 2010, 2011; Singh et al, 2011; Yang, 2005a, 2005b). However, graphite exhibit poor dispersibility and high percolation threshold (Friend, 1993; Olmedo, 1997; Saini, 2009a). Similarly, CNTs are economically non-viable, difficult to produce at bulk scale and often require purification, auxiliary treatment and functionalization steps (Bal, 2007; Olmedo, 1997; Saini, 2009a, 2011). In this regard, intrinsically conducting polymers (ICPs) with tunable electrical conductivity/dielectric properties, facile processing and compatibility with other polymeric matrices can offer an attractive solution over other conducting fillers (Chandrasekhar, 1999; Ellis, 1986; Olmedo, 1997; Skotheim, 1986; Trivedi, 1997). Interestingly, due to their inherent electrical conductivity and dielectric properties, these ICPs can be used either as conducting filler for various insulating matrices or as an electrically conducting matrix with incorporated conducting/dielectric/magnetic inclusions.

3.4.1. Intrinsically conducting polymers (ICPs)

Intrinsically conducting polymers (ICPs) combine moderate conductivity, good compatibility and ease of processability (as compared with carbons) with low density (~ 1.1–1.3 g/cm^3 compared to metals e.g. ~9.0 g/cm^3 for copper) and corrosion resistance (compared to metals) (Baeriswyl, 1992; Chandrasekhar, 1999; Ellis, 1986; Freund & Deore, 2007; Heeger, 2001a, 2001b; Joo & Epstein, 1994; MacDiarmid, 2001; Nalwa, 1997; Olmedo, 1997; Saini, 2011, Shirakawa, 2001; Skotheim, 1986; Trivedi, 1997). They possess unique shielding mechanism of reflection plus absorption rather than dominated reflection for metals and carbons. The ability to regulate their electrical conductivity by controlling parameters such as oxidation state, doping level, morphology and chemical structure, makes them powerful candidate for various techno-commercial applications. Fig. 6 shows the structure of some of the well known conducting polymers in their undoped forms.

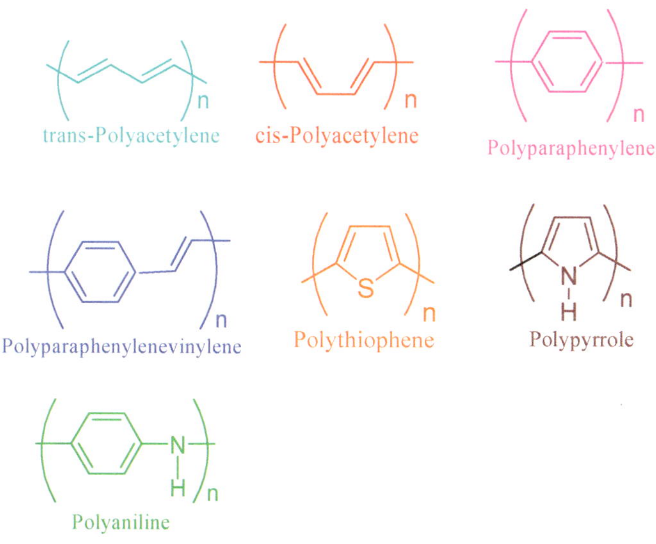

Figure 6. Chemical structures of some undoped conjugated polymers

Since the first ICP, polyacetylene (PA), was successfully synthesized by Shirakawa et al. (1977) and Chiang et al (1977, 1978a, 1978b) with conductivity as high as 10^6 S/cm in doped form, great interest has been aroused and a series of ICPs such as polyaniline (PANI), polypyrrole (PPY), polythiophene (PTH), poly(p-phenylene-vinylene) (PPV) etc have been developed (Carter et al, 1985; Rahman et al, 1989; Saxman et al, 1985; Snow, 1981; Soga et al, 1983; Thomas et al, 1988; Yamamoto et al, 1988). These undoped polymers display poor conducting properties and lies in insulating or semiconducting range (10^{-10} to 10^{-5} S/cm) as shown in (Fig. 7). However, the controlled doping can transforms the poorly conducting undoped material into a system which displays semiconducting or metallic conductivity (10^{-6} to 10^{5} S/cm). The predicted theoretical value for highly doped PA is about 2×10^{7} S/cm, which is even higher than that of copper (Chiang et al, 1978a, 1978b). However, the highest experimentally recorded conductivity for PA (in highly oriented thin films form) was greater than 10^{5} S/cm, which is still the highest value that has been reported for any conducting polymer till date. In contrast, conductivity of other conjugated polymers reaches a maximum value ~10^{3} S/cm (Baeriswyl, 1992; Cao et al, 1992, 1995; Chaing et al, 1978a, 1978b; Chandrasekhar, 1999; Ellis, 1986; Heeger, 2001a; Nalwa, 1997; MacDiarmid, 2001; Shirakawa, 2001; Skotheim, 1986).

The display of metal like electrical and optical properties by the highly doped forms of these ICPs (synthetic polymers) also entitled them to be called synthetic metals (Freund & Deore, 2007; Heeger, 2001a, 2001b; Nalwa, 1997; MacDiarmid, 2001; Shirakawa, 2001; Skotheim, 1986). The intrinsic conductivity of conjugated polymers in the field of microwave (100 MHz – 20 GHz) makes them a viable shielding material. In particular, dependence of their conductivity on frequency, has inspired many scientific ideas to adopt these phenomenon to microwave applications (Coleman & Petanck, 1986; Karasz et al, 1985; Natta et al, 1958; Olmedo, 1995, 1997; Saini et al, 2009a, 2011).

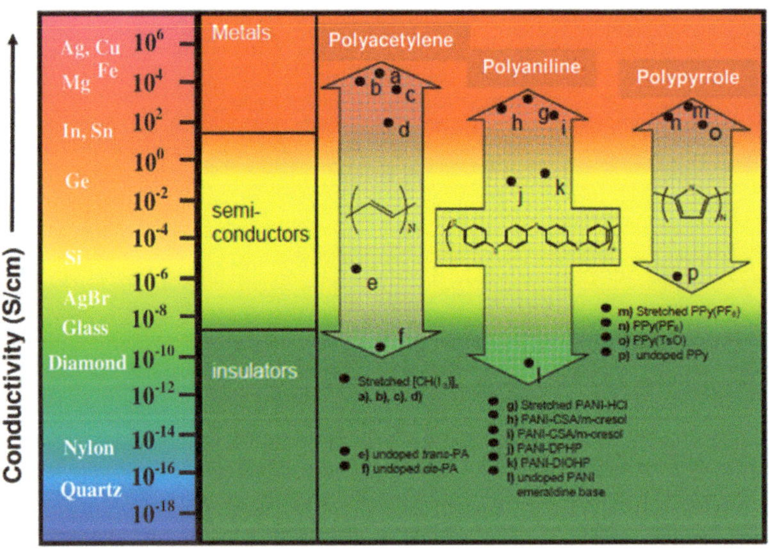

Figure 7. Conductivity of some conjugated polymers in comparison to typical metals, semiconductors or insulators.

The unique properties like tunable conductivity (between insulating and metallic limits), adjustable permittivity/permeability via synthetic means, low density, non-corrosiveness, nominal cost, facile processing (melt or solution), and controllable electromagnetic attributes, further strengthen their candidature as futuristic shielding material for various techno-commercial applications. Their utility can also be extended to high-tech areas like space, defense (military), or navigation/communication control or as a radar absorbing material (RAM) in the stealth technology (Ellis, 1986; Knott et al, 1993; Olmedo, 1997; Nalwa, 1977). Especially, conducting polymers appear to be one of the few materials capable of displaying dynamic (switchable) microwave absorption behavior, which are called "intelligent stealth materials", due to the reversible electrical properties of conducting polymers affected by redox doping/de-doping processes.

A careful comparison of properties of large number of available shielding materials revealed that no single phase material can take care of all the aspects of shield (e.g. absorption coefficient, thickness, volume, broadband response) to give desired level of performance under different environments and applications. Therefore, several attempts have also been made to exploit the worthy property of above materials by making strategic combinations e.g. admixtures, blends and composites (Ajayan et al, 2000; Cao et al, 1992, 1995; Chung, 2000, 2001; Colaneri et al, 1992; Dhawan, 2003; Gangopadhyay et al, 2001; Grimes, 1994; Gupta & Choudhary, 2011; Huang et al, 2000; In et al, 2010; Joo et al, 1999; Koul et al, 2000; Liang et al, 2009; Pomposo, 1999; Ramanathan et al, 2008; Saini et al, 2009a, 2009b, 2011; Sanjai et al, 1997; Shacklette et al, 1992; Shi & Liang, 2008; Singh et al, 2011; Stankovich et al, 2006; Taka, 1991; Varrla, 2011; Wang & Jing, 2005; Wessling, 1999; Wojkiewicz et al, 2003; Zhang et al, 2011). Among these options, composites based on various organic/inorganic filler (guests) loaded ICP matrices (hosts) as well as ICP (guest) loaded insulating matrices (hosts) have captured maximum attention due to fascinating properties and wealth of prevalent applications (Chandrasekhar, 1999; Ellis, 1986; Freund & Deore, 2007; Heeger, 2001a, 2001b; MacDiarmid, 2001; Nalwa, 1997; Shirakawa, 2001; Skotheim, 1986). Recently, the discovery of various nanomaterials (NMs) and ability to design and tailor their electrical and electromagnetic properties has lead to scientific surge to identify the best materials for shielding and other applications (Ajayan et al, 1994; Alexandre, 2000; Baughman et al, 2002; Geim & Novoselov, 2007; Geim, 2009; Ijima, 1991; Meyer et al, 2007; Moniruzzaman & Winey, 2006; Rozenberga & Tenn, 2008; Stankovich et al, 2006, 2007; Thostenson et al, 2005). Especially, nanocomposites have attracted enormous scientific attention due to distinguished set of properties as well as promising applications.

3.4.2. ICP based nanocomposites

Nature has the astonishing ability to form self-organized functional nanomaterials with perfect structures and unusual properties e.g. bacteria, viruses, proteins, cells etc. which ordinarily falls in the size range of 1-100 nm (1 nm = 10^{-9} m). In fact, nature is considered as maestro nanotechnologist who has created one of the best known nanocomposites such as bones, hairs, shells, and wood. Therefore, in quest of making perfect nanocomposites, researches are trying to learn and mimic the natural material synthesis principles. However, though high quality bulk composites (e.g. straw reinforced mud, concrete, carbon/glass fiber reinforced polymers) were already realized by researchers, formation of perfect nanocomposite remained a biggest scientific challenge. In case of nanocomposites, fillers possess nanoscale dimensions (~10^4 times finer than a human hair) and extend ultra-high interfacial area per volume to host polymeric matrices. Consequently, marked differences in the properties of nanocomposites are observed compared to their bulk counterparts e.g. enhanced strength, better optical or electrical properties etc. (Ajayan et al, 1994; Alexandre, 2000; Choudhary & Gupta, 2011; Mathur et al, 2010; Moniruzzaman & Winey, 2006; Ramasubramaniam, 2003; Rozenberga & Tenn, 2008; Thostenson et al, 2005) even at the lower loadings. Polymer is a versatile choice as a matrix material due to advantages like low density, mechanical flexibility, facile processing and corrosion resistance. Interestingly, most polymeric matrices possess poor electrical, dielectric or magnetic properties and are transparent to electromagnetic radiations (Saini et al, 2009a, 2011). Therefore, most of the electrical and electromagnetic properties of the conventional nanocomposites are mainly contributed by the nanofillers (nature and concentration) and matrix simply plays the role of holding the filler particles. In this consideration, utilization of ICPs as host matrix can offer an attractive solution over conventional (insulating) polymer based matrices (Ellis, 1986; Nalwa, 1997; Olmdo, 1995, 1997; Saini et al, 2011) primarily due to microwave non-transparency and design flexibility. However, the incorporation of nanofillers within polymeric matrices is not a straightforward task because of the ultrahigh surface area and agglomeration tendencies. These often resulted in failure to efficiently translate the nanoscopic properties of these fillers into macroscopic properties of resultant nanocomposites, thereby inability to utilize their full potential. Hence, handling and dispersion of nanofiller is the biggest challenge for nanocomposite science and technology.

3.4.3. Synthesis of ICP based nanocomposites

i. ICP as filler

As already mentioned in the previous section, inherent electric conductivity/dielectric properties (i.e. without any added conducting additive e.g. metals, graphite or carbon nanotubes), design flexibility and good compatibility with various insulating polymer matrices (e.g. thermoplastic/thermoset/rubber/elastomer/fiber/fabric etc.), ICPs can be used as filler to form composites.

As shown in Fig. 8, such composites are formed either by solution processing or by melt phase mixing/blending (Pud et al, 2003; Cao et al, 1992, 1995; Colaneri & Shacklette, 1992; Taka, 1991; Shacklette et al, 1992; Saini, et al, 2011; Wessling, 1999). In the former case both ICP and matrix polymer are dissolved/dispersed in a common solvent and stirred/sonicated to achieve the final mixing followed by casting (shaping) and drying/curing. In contrast, melt blending involves mixing of filler particles with molten matrix polymer followed by molding (shaping) and cooling/curing. In some case e.g. thermosts, ICPs are mixed with pre-polymer (resin) by solution blending technique. Finally, cross-linkers (curing agents) are added and curing is achieved by a combination of heat (not required for room temperature cross-linkers) and pressure (not required when no volatiles are expelled during curing process).

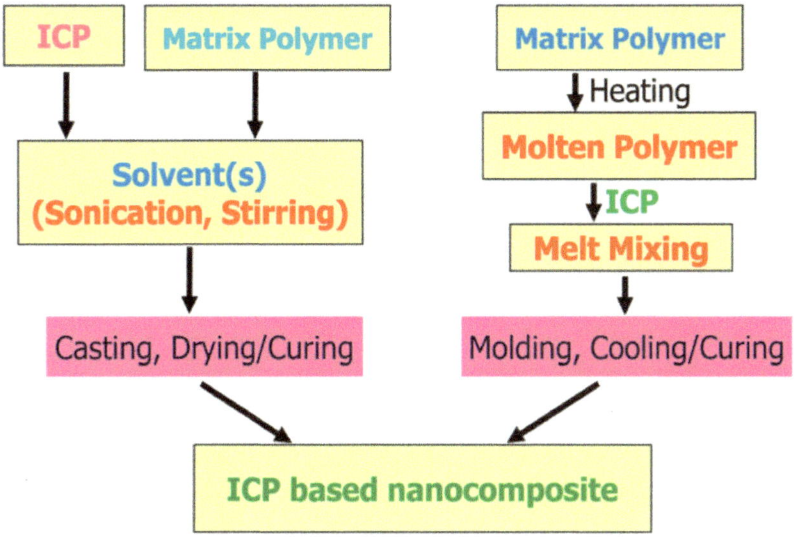

Figure 8. Schematic representation of steps involved in the formation of ICP (as conducting filler) loaded insulating polymer matrix by solution and melt processing techniques

ii. ICP as matrix polymer

The utilization of different ICPs as nanocomposite matrix can be attributed to advantages as design flexibility, good filler incorporation-ability, specific interactions with fillers and microwave non-transparency.

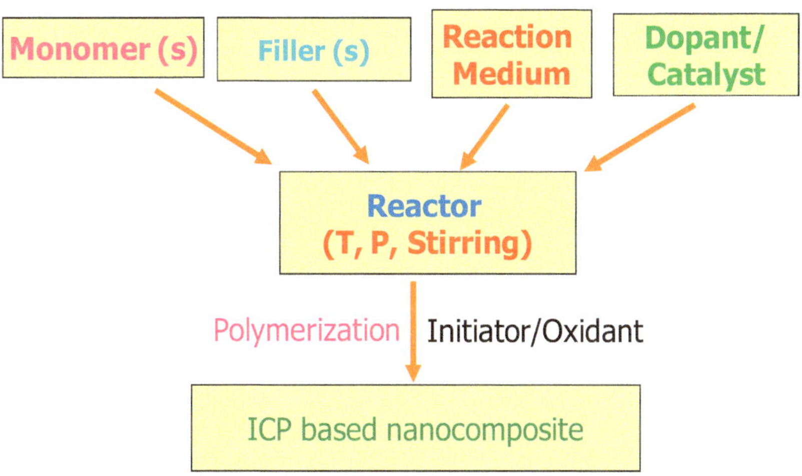

Figure 9. Schematic representation of formation of mechanism of ICP matrix based nano-composites by in-situ polymerization route.

The incorporation of various conducting, dielectric or magnetic nanoparticles within conducting polymer matrices can be achieved either by ex-situ physical mixing processes or by in-situ polymerization (Abbas et al, 2005, 2006; Das & Mandal, 2012; Dong et al, 2008; Fang et al, 2006; Joo et al, 1999; Moniruzzaman & Das, 2010; Pant et al, 2006; Phang et al, 2007, 2008; Saini et al, 2007, 2009a, 2009b, 2010; Yang et al, 2010, 2011). However, ex-situ mixing leads to poor dispersion of filler particles and failure to overcome their agglomeration tendencies that results in inferior and non-reproducible electrical and electromagnetic attributes. In contrast, the electronic properties of such synthetic metals can be strictly controlled by following in-situ incorporation (Fig. 9) approach i.e. carrying out the polymerization under the controlled conditions and in the presence of specific dopants and fillers (Bredas et al, 1998; Chandrasekhar et al, 2002, Mattoso et al, 1994; Nalwa et al, 1997; Saini et al, 2007, 2009a, 2009b; Savitha et al, 2005; Skotheim, 1986). In a typical reaction, monomer(s), filler and dopant or catalyst are charged into a suitably designed reactor to maintain required temperatre (T), pressure (P) and agitation (stirring) conditions. During such pre-polymerization process, monomers are generally adsorbed over dispersed nano-filler particles. The polymerization was initiated by addition of specific initiator/oxidant and allowed to proceed till reaction gets completed leading to formation of ICP based nanocomposite.

3.4.4. Electrical properties of ICP based nanocomposites

As already mentioned and shown in Fig. 1, the primary shielding mechanism is reflection for which shield should possess free charge carriers (electrons/holes) that can interact with incident EM field. But the organic conjugated polymers are insulators in their undoped forms e.g. room temperature electrical conductivity (σ_{dc}) of emeraldine base (EB) is ~10^{-9} S/cm (Fig. 10, Gupta et al, 2005). However, controlled doping leads to enhancement of conductivity due to formation of charge carriers (Fig. 11) i.e. polarons/ bipolarons (Saini et al, 2008; Stafstrom et al, 1987; Trivedi, 1997; Zuo et al, 1989) that can move under the influence of external potential and in the Coulmbic field of counter-ions distributed along the chain.

Therefore, increasing dopant concentration leads to increase in concentration and mobility of proto-generated charge carriers resulting in enhancement of conductivity. Furthermore, such a conductivity enhancement in conductivity is strongly dependent on nature and concentration of dopant and in some case conductivity well exceeds the required limit (Olmedo et al, 1997; Saini et al, 2011) for exhibiting good shielding effectiveness.

The addition of ICPs particles (guests) as a conducting filler within insulating polymer matrices (hosts) leads to establishment of electrical conductivity (in resultant nanocomposites) due to formation of percolation networks (Colaneri & Shacklette, 1992; Hsieh, 2012; Lakshmi et al, 2009; Shacklette et al, 1992; Taka, 1991; Wessling, 1999). At percolation threshold, ICP particles form a 3D conductive network within host matrix, which can be easily estimated by plotting the electrical conductivity as a function of the reduced volume fraction of filler (Fig. 9) and performing data fitting with a power law function (Saini et al, 2011):

$$\sigma = \sigma_0 \left(v - v_c\right)^t \tag{49}$$

where σ is the electrical conductivity of the composite, σ_0 is characteristic conductivity, v is the volume fraction of filler, v_c is volume fraction at the percolation threshold and t is the critical exponent. The log (σ) versus log ($v - v_c$) plot (Fig. 12) gives a straight line according to eqn. 10. The values of scaling law parameters i.e. v_c and t can be subsequently obtained by least-square analysis of the above double logarithmic plots. When the densities of the host polymer and the filler are similar, mass fraction (m) becomes same as volume fraction (v) and can be used in above calculations.

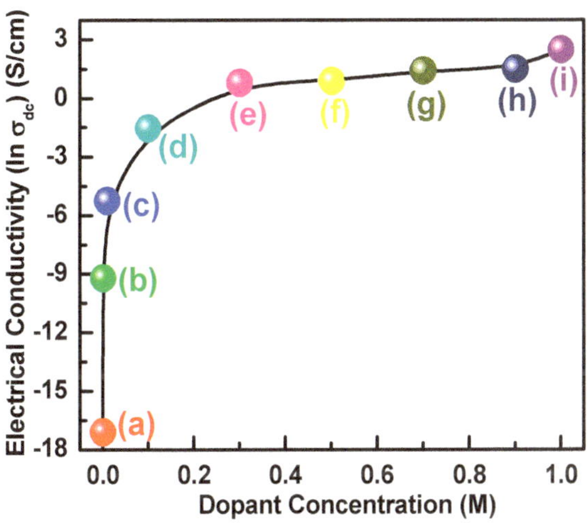

Figure 10. Variation of electrical conductivity (ln σ_{dc}) of hydrochloric acid (HCl) doped Emeraldine base (EB) samples as a function of dopant (HCl) concentration (a) 0.0 M (b) 0.001 M, (c) 0.01 M, (d) 0.1 M, (e) 0.3 M, (f) 0.5 M, (g) 0.7 M, (h) 0.9 M and (i) 1.0 M.

However, it has been observed that formation of such networks and percolation thresholds (minimum loading level at which first continuous network of conducting particles is formed) critically depend on nature of ICP, its intrinsic conductivity, particle shape, morphology, aspect ratio, its concentration, degree of dispersion and extent of compatibility with host matrix.

Nevertheless, at percolation conductivity (σ_p) remained too low to exhibit any acceptable shielding action and generally higher loadings (>30 wt. %) are required though in most cases, σ_p is sufficient to extend antistatic action. Interestingly, when ICPs are combined with other conducting fillers (e.g. Polyaniline with MWCNT, Saini et al, 2011) significant reduction in percolation threshold, higher conductivity and better shielding performance is observed as compared to pristine (unfilled) ICPs.

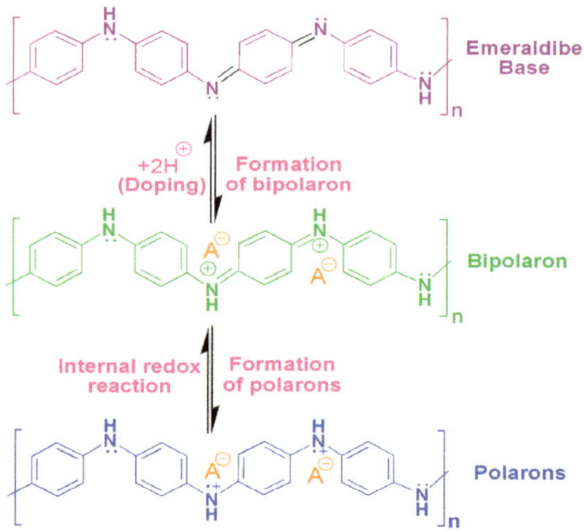

Figure 11. Protonic acid doping of polyaniline leading to formation of charge carriers polarons (radical cations) and bipolarons (dications)

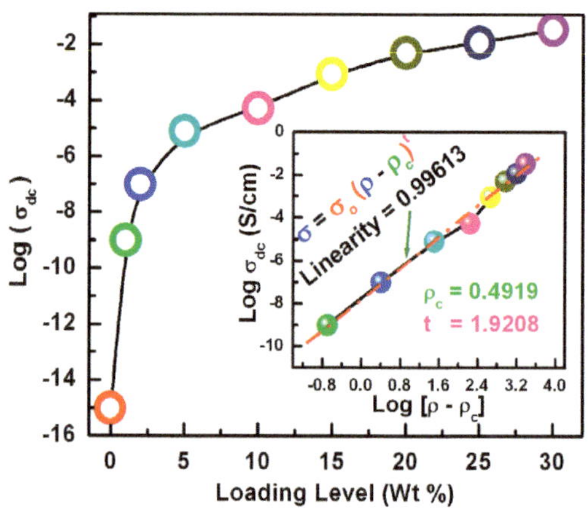

Figure 12. Variation of conductivity (σ_{dc}) of PANI-MWCNT nanofiller loaded polystyrene solution blends. Inset shows the percolation and scaling details

In many cases conjugated polymers are used as matrix instead of conventional insulating polymers. When conducting fillers (e.g. metal particles, carbon black, graphite or CNTs) are incorporated within undoped (poorly conducting) ICP matrices, electrical conductivity increases and follows a typical percolation behavior. In contrast, the loading of above conducting fillers within microwave non-transparent doped (intrinsically conducting) ICP matrices lead to further enhancement (Fig. 13) of electrical conductivity. Such improvement can be explained on the basis of granular metal/inhomogeneous doping model (Sheng &

Klafter, 1983) which considered that ICPs consists of highly conducting metallic islands dispersed within low conductivity amorphous matrix. Therefore, above improvement in conductivity can be attributed to bridging of these metallic islands (Saini et al, 2009a) by the metallic filler particles facilitating enhanced inter-particle transport. The increase in conductivity is strongly dependent on nature, concentration and aspect ratio of filler particles as well as type and morphology of host ICP matrix.

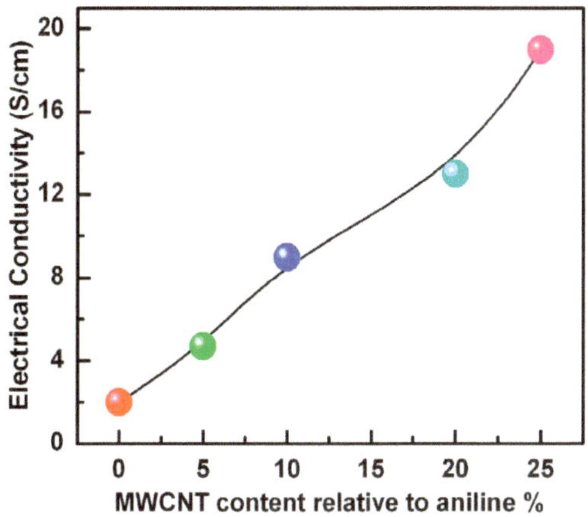

Figure 13. Dependence of electrical conductivity of in-situ polymerized PANI-MWCNT nanocomposites on MWCNT content

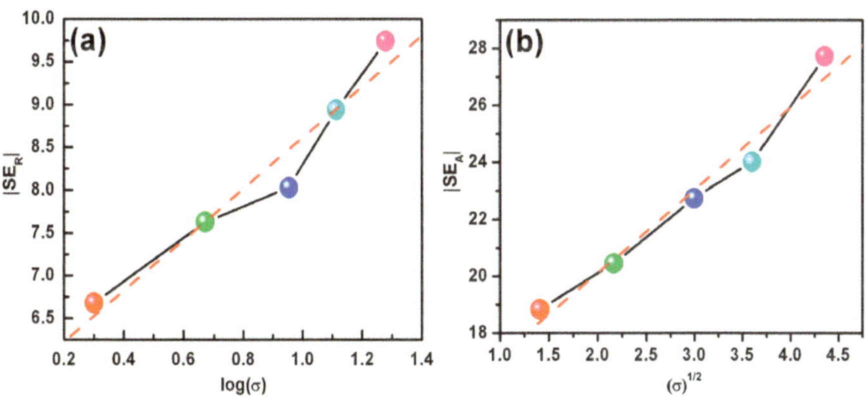

Figure 14. Correlation between electrical conductivity (σ) and shielding effectiveness (SE) showing linear dependences of (a) reflection loss (SE_R) on log σ and (b) absorption loss (SE_A) on $\sigma^{1/2}$.

Nevertheless, the establishment and enhancement of electrical conductivity is of paramount importance because it leads to parallel enhancement of reflection and absorption loss components (Fig. 14, Saini et al, 2009a) leading to enhancement of overall shielding effectiveness. Interestingly, absorption loss (SE_A) increases by much larger magnitude (with conductivity) compared to corresponding reflection loss (SE_R) component. For a non-magnetic material, this can be explained on the basis of logarithmic [i.e. log(σ)] and square root [i.e. $(\sigma)^{1/2}$] conductivity dependence of SE_R and SE_A respectively as shown in Fig. 14.

3.4.5. Dielectric and magnetic properties of ICP based nanocomposites

A secondary mechanism of shielding is absorption for which shield should possess electric or magnetic dipoles. These dipoles can interact with transverse electric (E) and magnetic (H) vectors of the incident EM waves to introduce losses into the system. It is interesting to note that pure (without any external filler loading) conjugated polymers in their undoped (base) forms possess poor dielectric and magnetic properties. However, controlled doping leads to marked improvement (Fig. 15) in dielectric properties (e.g. dielectric constant/real-permittivity, dielectric losses/imaginary-permittivity), although even after doping magnetic properties (e.g. real and imaginary magnetic permeability) remained poor.

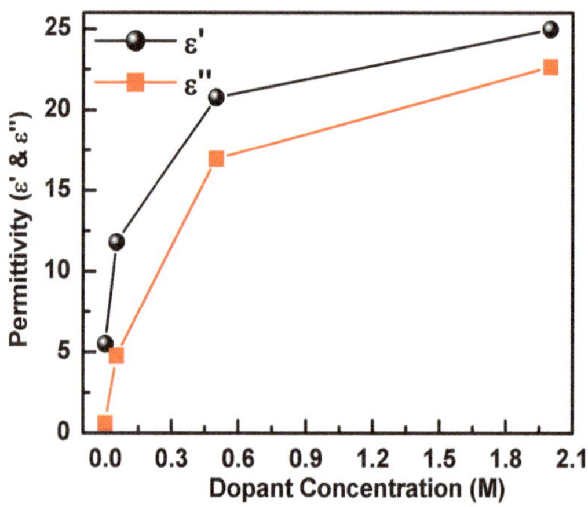

Figure 15. Dependence of (a) real permittivity or dielectric constant (ε') and (b) imaginary permittivity or dielectric loss (ε'') on the dopant concentration for acrylic acid (AA) doped emeraldine base (EB) samples

As already discussed, doping of ICPs leads to formation of polarons/bipolarons (Fig. 11) that produces pronounced polarization/relaxation effects (Olmedo et al, 1995, 1997; Saini et al, 2008, 2009a, 2011; Stafstrom et al, 1987). Therefore, observed improvement of dielectric properties with doping level can be attributed formation and increase in concentration of above localized carriers. The correlation between dielectric properties and shielding response for various ICPs is presented in Fig.16 which clearly shows that the total shielding efficiency (SE_T) increases as the absolute value of complex dielectric constant increases.

The increase of both real and imaginary parts of dielectric permittivity contributes (Joo & Epstein, 1994) towards enhancement of SE_T. Furthermore, the complex dielectric constant dependence of absorption loss (SE_A) component (inset of Fig. 16) was found to be much stronger compared to that due to reflection loss (SE_R). In some cases, especially for highly doped and stretch oriented ICPs; dielectric constant becomes negative (Javadi et al, 1989; Joo & Epstein, 1994; Joo et al, 1994; Hsieh et al, 2012; Wang et al, 1991) and ultra-high attenuation is observed which suggests the possibility of ICP based left handed materials (LHMs) or meta-materials.

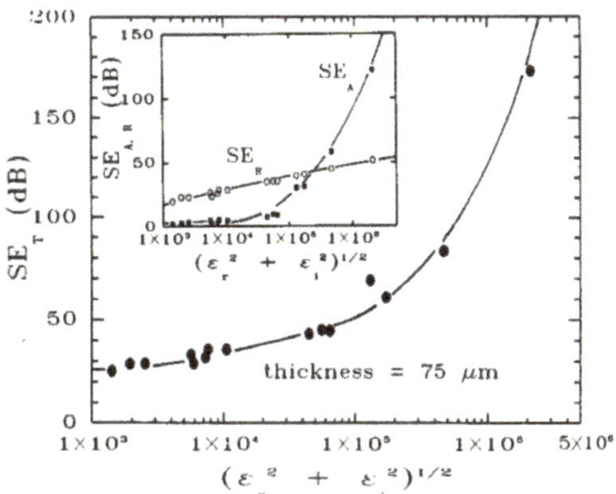

Figure 16. Total shielding efficiency (SE_T) vs absolute value of complex dielectric constant $[(\varepsilon_r^2 + \varepsilon_i^2)^{1/2}]$ of various conducting polymers. Inset: comparison of reflection (◦) and absorption (▪) shielding efficiency as a function of absolute value of complex dielectric constant. Solid lines are guides to the eye. Reprinted with permission from [J. Joo and A. J. Epstein, Appl. Phys. Lett. 65 (18), 2278-2280, 1994]. Copyright [1994], American Institute of Physics.

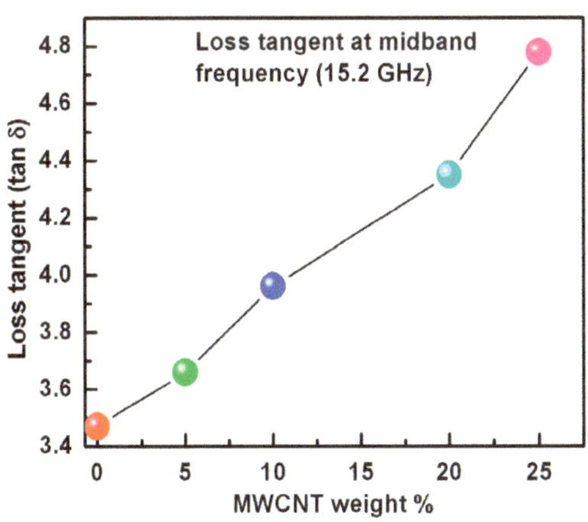

Figure 17. Loss tangent (tan δ) of in-situ synthesized PANI-MWCNT nanocomposites as a function of MWCNT loading

Interestingly, for ICPs, besides doping induced polarization filler induced interfacial polarization may also contribute towards dielectric properties. For example, when conducting fillers like metal particles, graphite or carbon nanotubes are introduced into ICP matrices; further improvement of dielectric properties was observed. Such a polarization occurs due to electrical conductivity differences between ICP and metallic fillers leading to charge localization at interfaces via Maxwell-Wagner-Sillars (Kremer & Schönhals, 2003; Riande & Diaz-Calleja, 2004; Sillars, 1937; Wagner, 1914) interfacial polarization phenomenon. Such polarization and related relaxation phenomenon contribute towards energy storage and losses. The actual losses can be computed by normalization of these losses with storage terms [i.e. by ratio of dielectric losses/imaginary permittivity (ε'') with dielectric constant/real permittivity (ε')] to quantity loss tangent (tan δ).

In case of in-situ formed MWCNT-polyaniline nanocomposites, improvement of dielectric properties leads to high value of loss tangent (Fig. 17) which further increases (Saini et al, 2009a) with increase in MWCNT loading. However, though for a given thickness, total shielding is dominated by absorption, reflection loss component becomes too high form the viewpoint stealth technology. Nevertheless, despite good dielectric properties, magnetic properties of ICPs remained poor to extend any significant contribution towards EMI regulation. In principle, for highly conducting materials, only conductivity (σ) and magnetic permeability (μ) are important, such that the reflection loss (SE_R) is dependent upon their ratio (i.e. σ/μ) whereas the absorption loss (SE_A) is a function of their product (i.e. $\sigma.\mu$) (Saini et al, 2011). In contrast, for moderately conducting materials (e.g. ICPs) permittivity (ε) also plays a significant role (besides σ and μ) in deciding absolute values of SE_R and SE_A (Joo & Epstein, 1994). As most ICPs are non magnetic in nature ($\mu_r \approx \mu_i \approx 0$), observed attenuations are mainly governed by σ and ε only. Therefore, it is expected that any improvement in magnetic properties will lead to definite improvement of absorption loss alongwith parallel reduction of reflection loss. In addition, the incorporation of high dielectric constant materials like $BaTiO_3$, ZnO, TiO_2 etc. within ICP matrices are expected to further improve the microwave absorption response. Consequently, in recent years, lot of work has been carried out to formulate composites of polyaniline with the dielectric or magnetic filled inclusions, either by in-situ polymerization or by ex-situ physical mixing processes (Abbas et al, 2005, 2006, 2007, 2008). Such composites possess moderate polarization or/and magnetization alongwith good microwave conductivity so as to introduce absorbing properties into the material. They display dynamic dielectric and/or magnetic losses, upon impingement by incident electromagnetic waves. As electromagnetic wave consists of an pulsating (orthogonal to each other) electric (E) and the magnetic (H) fields; therefore, above multi-component composites are expected to yield good attenuation efficiencies, primarily due to interaction of conducting/dielectric and conducting/magnetic phases with E and H vectors of the incident EM waves (Fig. 5). Furthermore, most insulating polymer matrices possess poor electrical, dielectric or magnetic properties and are transparent to radio frequency (RF) or microwave (MW) electromagnetic radiations (EMRs). Therefore, only fillers contribute towards shielding and leakage of radiation from EMR transparent regions tends to degrade shielding effectiveness. However, microwave non-transparency (Olmedo et al, 1995, 1997; Saini et al, 2011) of ICPs compared to conventional polymers is an added advantage as both filler and matrix contribute towards shielding. Moreover, the dominant shielding characteristic of absorption for above nanocomposites materials other than that of reflection for metals render ICPs more useful in applications requiring not only high EMI SE but also shielding by absorption, such as in stealth technology.

When ICPs are exploited as microwave non-transparent matrices, the added dielectric or magnetic filler particles result in establishment of properties (e.g. dielectric/magnetic character, thermal conductivity etc.), that are extrinsic to these intrinsically conducting polymers. Therefore, combination of dielectric or magnetic nanoparticles with conducting polymer leads to formation of multi-component composite possessing unique combination of electrical, dielectric and magnetic properties useful for suppression of electromagnetic noises and reduction of radar signatures (Abbas et al, 2005, 2007; Chan, 1999; Chandrasekhar, 1999; Cho & Kim, 1999; Dong et al, 2008; Ellis, 1986; Gairola et al, 2010; Huang, 1990; Knott et al, 1993; Kurlyandskaya et al, 2007; Meshram et al, 2004; Nalwa, 1997; Ngoma et al, 1990; Pant et al, 2006; Phang et al, 2007, 2008, 2009, 2010; Xiaoling et al, 2006; Xu et al, 2007; Yang et al, 2009). The incorporation of magnetic fillers (e.g. ferrites like γ-Fe_2O_3 or Fe_3O_4) within ICP matrices leads to improvement of magnetic properties (Fig. 18) without much loss of conductivity. Such a combination is expected to display additional magnetic loss leading to enhanced absorption.

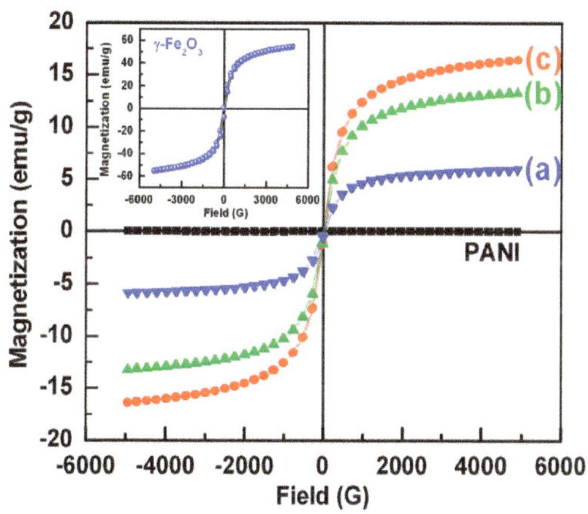

Figure 18. Magnetization of polyaniline (PANI), γ–Fe_2O_3 nanoparticles and nanocomposites formed by in-situ polymerization taking different weight ratio of aniline: γ–Fe_2O_3 (a) 2:1, (b) 1:1 & (c) 1:2

The magnetization plots (Fig. 18) of polyaniline/γ-Fe_2O_3 composites revealed that pure γ-Fe_2O_3 nanoparticles display pronounced magnetic signatures with narrow hysteresis loop. The saturation magnetization (M_s) value of of these particles was found to be 59.3 emu/g (at 5.0 kG) alongwith very small retentivity (M_r~ 4.3 emu/g) and coercivity (H_c~ 83.8 G), which indicate the super-paramagnetic (SPM) nature of these particles. The SPM character imparts fast relaxation behaviour and originates due to small size of the ferrite particles i.e. approaching towards the single domain limit (Qiao et al, 2009). However, PANI possesses weak ferromagnetic behaviour and with increase in ferrite content, enhancement of M_s was observed with as parallel reduction of coercivity (H_c). The initial permeability (μ_i) of ferromagnetic materials can be expressed as (Stonier, 1991):

$$\mu_i = \left(\frac{M_s^{\,2}}{akH_c M_s + b\lambda\xi} \right) \tag{50}$$

where a and b are two constants determined by the material composition, λ is the magneto-striction constant, ξ is elastic strain parameter of crystal, and k is a proportionality coefficient. The above equation shows that permeability can be enhanced either by enhancing M_s or by reducing H_C. In the present system, the incorporation of ferrite within PANI matrix is expected to affects the surface electron density of γ-Fe_2O_3 nanoparticles and hence the spin-spin or spin-lattice interactions. The results show that M_s value increases (5.94 to 16.4 emu/g) with increasing ferrite content (plots a-c) whereas H_c shows a simultaneous decrement (35.7 to 57.8 G). Therefore, It can be seen from the eqn. (50) that both higher M_s and lower H_c values are favorable to the improvement of μ_i value, which in turn is expected to enhance the microwave absorption capability.

In many cases, highly doped ICP particles are used as conducting fillers (in place of metal or carbon based materials) for various insulating polymer host matrices. This not only leads to establishment and improvement of electrical conductivity but also contribute towards improvement of both real as well as imaginary permittivity (Abbas et al, 2005; Colaneri & Shacklette, 1992; Joo et al, 1994; Saini et al, 2011; Shacklette et al, 1992; Taka, 1991; Wessling, 1999).

Again the magnetic properties remained poor due to non-magnetic nature of most ICPs. However, when magnetic filler loaded ICPs is use hybrid filler, improvement in magnetic properties has also been observed besides regular improvement of dielectric attributes. For example use of PANI-MWCNT hybrid filler within polystyrene matrix leads to formation of composites with magnetic properties due to MWCNT core (containing entrapped ferromagnetic iron catalyst phase) and electrical conductivity/dielectric properties due to ICP and MWCNTs. As the concentration of PANI-MWCNT filler increases, real and imaginary parts of both permittivity and permeability increases as shown in Fig. 19. Most importantly, losses due to reflection (SE_R) and absorption (SE_A) follows the permittivity and permeability trends and exhibit corresponding increase. However, SE_A was more sensitive towards electromagnetic attributes compared to SE_R which may be attributed to their square root and logarithmic dependences. Furthermore, two most important parameters that decide the relative magnitudes of SE_R and SE_A are microwave conductivity (σ_T) and skin depth (δ). The σ_T can be related to imaginary permittivity (ε" or ε_i) as (Saini et al, 2011):

$$\sigma_T = (\sigma_{ac} + \sigma_{dc}) = \omega\varepsilon_o\varepsilon'' \tag{51}$$

where σ_{ac} and σ_{dc} are frequency dependent (ac) and independent (dc) components of σ_T respectively, ω is angular frequency and ε_o is permittivity of free space (8.85×10^{-12} F/m). Higher the value of σ_T more will be reflection for a given absorption. Further, the skin depth (δ) of the shield is defined as depth of penetration at which strength of incident EM signal is reduced to ~37% of its original magnitude. For a good conductor (i.e. $\sigma_T \gg \omega\varepsilon$), it (δ) can be expressed in terms of σ_T real permeability (μ' or μ_i) and ω as (Joo. et al, 1994):

$$\delta = \left(\frac{2}{\sigma_T \omega \mu'} \right)^{\frac{1}{2}} \tag{52}$$

Now shallower is the skin depth, higher will be absorption loss for a given thickness of material. Fig. 20 shows that as the loading level of PANI-MWCNT filler within PS matrix increases, σ_T increases whereas δ value decreases.

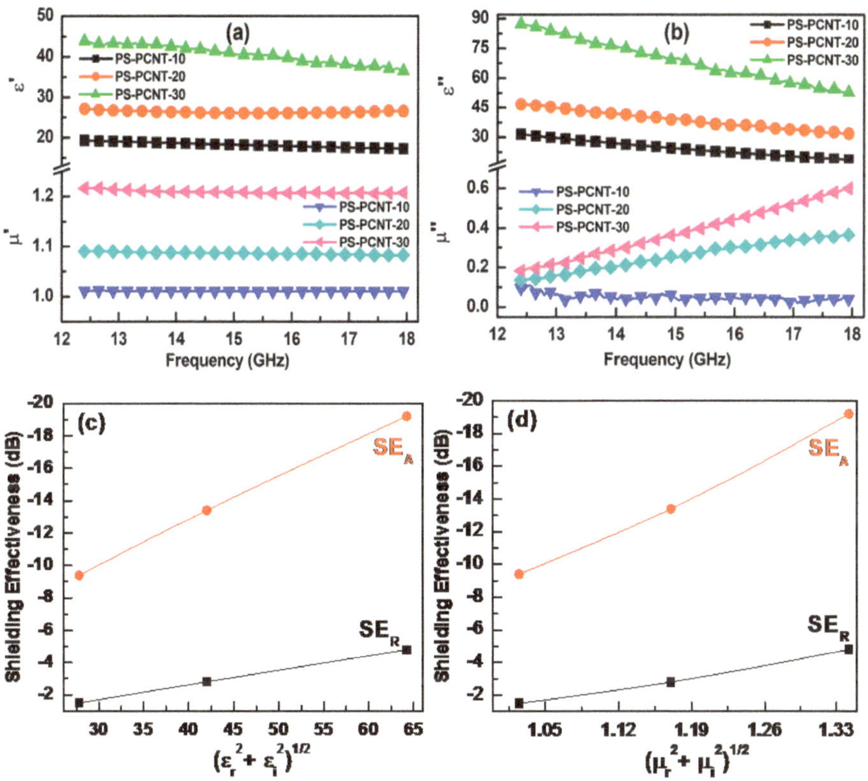

Figure 19. Frequency dependence of (a) dielectric constant (ε') & real permeability (μ'), (b) dielectric loss (ε'') & magnetic loss (μ'') of PANI-MWCNT/Polystyrene nanocomposites with increasing loading (10, 20 & 30 weight %) of PANI-MWCNT filler. Dependences of losses due to absorption (SE_A) and reflection (SE_R) of above composites as a function of absolute value of (c) complex permittivity [($\varepsilon_r^2 + \varepsilon_i^2$)$^{1/2}$] , (d) complex permeability [($\mu_r^2 + \mu_i^2$)$^{1/2}$].

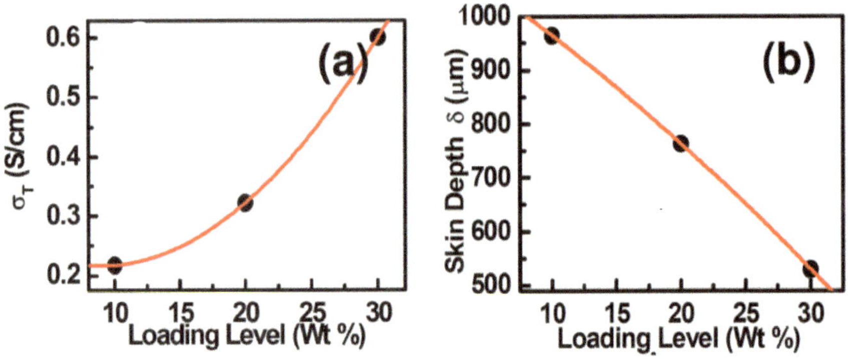

Figure 20. Dependence of (a) σ_T and (b) δ value of PANI-MWCNT loaded polystyrene composites on the loading of PANI-MWCNT

3.4.6. EMI shielding performance of ICP based nanocomposites

In the previous section we learned about the importance of parameters such as electrical conductivity and dielectric/magnetic attributes in regulating the shielding effectiveness and their correlation with reflection and absorption loss components. This section is devoted to measurement and interpretation of shielding effectiveness alongwith detailed analysis of reflection and absorption sub-components.

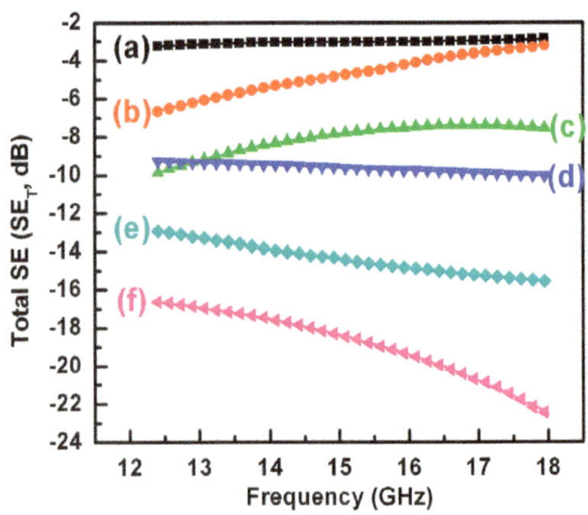

Figure 21. Frequency and dopant concentration dependence of total shielding effectiveness (SE_T) value of samples prepared by doping of emeraldine base (EB) with different concentrations of acrylic acid (AA) viz. (a) 0.0 M, (b) 0.05 M, (c) 0.1 M, (d) 0.5 M, (e)1.0 M and (f) 2.0 M

As already discussed doping produces localized defects (polarons/bipolarons) that are responsible for polarization and electrical conductivity. With the increase in dopant concentration, achieved doping level increases leading to enhancement of polaronic concentration as well as related conductivity/permittivity, which ultimately leads to improvement of shielding effectiveness (Fig. 21). Nevertheless, though based on microwave dielectric constant and electrical conductivity values many speculations were made about the shielding properties of ICPs, the first direct evidence of shielding response of ICP based composites alongwith actual shielding effectiveness values was presented by Taka (1991). He prepared poly(3-octyl thiophene) composites by melt mixing chemically synthesized poly(3-octyl thiophene) with PS, PVC, and EVA in and tested for EMI shielding at frequency range from 100 kHz to 1GHz. EMI SE of these composites (3 mm thick) increased with the polymer loadings and -45 dB (from 100 kHz to 10 MHz) was achieved with high (i.e. 20%) loading in the PVC that was still lower than that of a nickel painted sample (-80 dB). The measurements showed that P3OT blends behave as pseudo-homogenous metals (PHM). A PHM has no intentional holes or slits but lacks homogeneity. The shielding efficiency depends strongly on the amount of conducting polymer mixed in the blends due to regulation of conductivity. The authors concluded that composites with 20% or less loading of poly(3-octyl thiophene) were not readily applicable as EMI shielding.

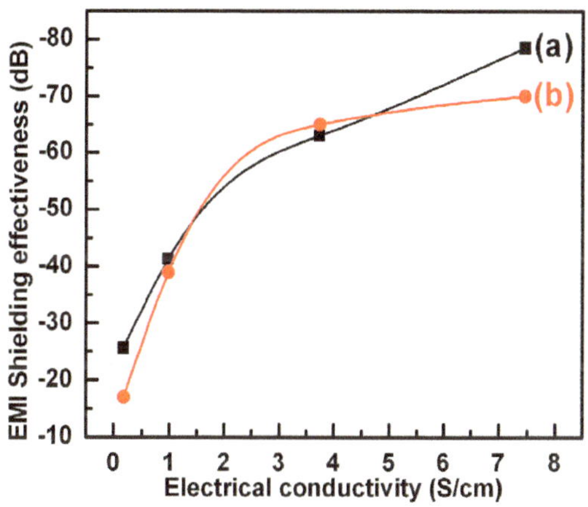

Figure 22. EMI shielding effectiveness (at 1.0 GHz frequency) of ICP loaded PVC blends as a function of DC electrical conductivity under (a) far field and (b) near field regimes

Later, systematic study of EMI shielding behavior of conducting polymer (PANI) based thermoplastic blends with polyvinyl chloride (PVC) or Nylon was reported (Colaneri & Shacklette, 1992; Shacklette et al, 1992). The EMI SE values of these highly conducting blends (~0.1-20 S/cm) were measured over a frequency range of 1 MHz to 3 GHz and also calculated theoretically under both near and far field regimes. The results are graphically presented in Fig. 22 which showed that both near and far filed SE followed the DC electrical conductivity and exhibit rapid initial rise followed by slow increment at higher conductivity. Far field SE of -70 dB was obtained for the melt blend of polyaniline (again at higher loading level of 30 wt. %) with PVC which agreed well with the theoretical calculations as per expressions derived by authors. Pomposo et al (1999) have prepared PPY based conducting hot melt adhesives by melt mixing appropriate amounts of ethylene-co-vinyl acetate (EVA) copolymer and PPY, which was synthesized with oxidant of $FeCl_3$. Both near and far field EMI shielding properties of the adhesives were measured at room temperature and found to increase with the loading of PPY. Near field SE in excess of -80 dB was determined at 1MHz and above -30 dB at 300 MHz, though a decrease with increase of frequency. Far field SE values of -22, -27 and -30 dB were determined (in the 1 to 300 MHz frequency range) for PPY loadings of 15, 20 and 25% respectively. Similarly, Wessling (1999) prepared highly conductive blends of PANI with PVC, polymethylmethacrylate (PMMA) or polyester at Ormecon Chemie, with conductivities of ca. 20 S/cm and in some cases up to 100 S/cm. These blends exhibited EMI SE of -40 to -75 dB for both near and far field conditions. However, mechanical properties were not encouraging and demanded considerable improvement. In addition the higher necessary thicknesses of 2-3 mm of these blends were found to be higher than technically acceptable thickness of 0.5–0.8 mm for practical uses.

Naishadham & Kadaba (1991), Naishadham & Chandrasekhar (1998) and Chandrasekhar & Naishadham (1998) reported the cumulative broadband (4-18 GHz) measurements and computations of all microwave parameters (e.g. conductivity, absorption, complex permittivity, shielding and reflection) of sulfonate doped PANI. It was found that the total SE of -35 to -15 dB was obtained with return loss of -5 to -1 dB and nominal absorption of -5 dB for PANI samples of conductivity 1-7 S/cm. Authors also demonstrated that better SE value upto -50 dB can be realized by stacking several polymeric sheets of different thicknesses or by sandwiching a lossy dielectric between two sheets of the same thickness.

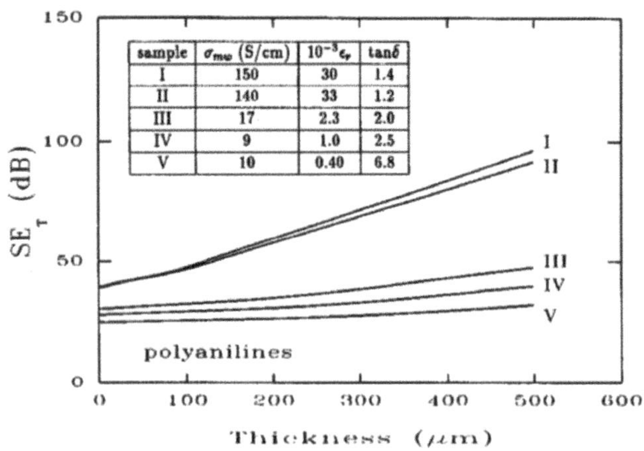

Figure 23. Thickness dependence of total shielding efficiency (SE_T) of various "crosslinked" polyaniline (XPANI-ES) samples. Sample I: highly XPANI-ES [3.5 times (×) stretched, parallel (| |)], sample II: intermediate XPANI-ES [3.5×,| |), sample III: highly XPANI-ES (3.5×, ⊥), sample IV: highly XPANI-ES (unstretched), and sample V: non-XPANI-ES (12.5×, ⊥). Inset: comparison of σ_{mw} , ε_r, and tan δ. Reprinted with permission from [J. Joo and A. J. Epstein, Appl. Phys. Lett. 65 (18), 2278-2280, 1994]. Copyright [1994], American Institute of Physics.

It has been found that electrical conductivity is not the sole scientific criteria for exhibiting high shielding effectiveness (Joo & Epstein, 1994) and good attenuations were also extended by moderate conductors with good dielectric properties. In fact as mentioned in the previous section, it has now been established that shielding effectiveness increases (as shown previously in Fig. 16) with absolute value of complex dielectric permittivity. Furthermore, absorption loss was found to be more sensitive towards permittivity (inset Fig. 16) than corresponding reflection loss. Figures 23 and 24 show the microwave SE of pure ICPs including PANI and PPY (in thin film forms) as a function of their intrinsic properties (insets of Fig. 23 and Fig. 24) such as microwave conductivity (σ_{mw}) and dielectric constant (ε_r) alongwith its dependence on extrinsic parameters like thickness (t) and temperature (T). The role of parameters like degree of crosslinking and parallel (| |) or perpendicular ($\perp$) stretch orientation which tends to affect σ_{mw}, ε_r or loss tangent (tanδ) has been clear from the table data (above insets).

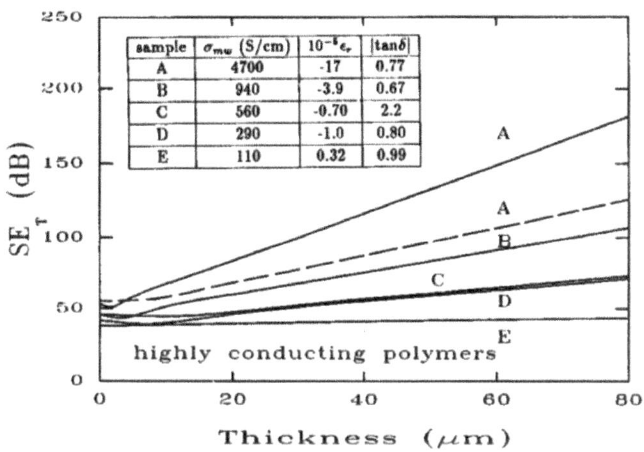

Figure 24. Thickness dependence of total shielding efficiency (SE_T) of highly conducting polymers. Sample A: stretched heavily iodine doped Tsukamoto polyacetylene (dotted line is obtained by using approximated α and n), sample B: unstretched heavily iodine doped Tsukamoto polyacetylene, sample C: camphor sulfonic acid doped polyaniline in m-cresol solvent, sample D: PF_6 doped polypyrrole, and sample E: TsO doped polypyrrole. Inset: comparison of σ_{mw}, ε_r, and tan δ. Reprinted with permission from [J. Joo and A. J. Epstein, Appl. Phys. Lett. 65 (18), 2278-2280, 1994]. Copyright [1994], American Institute of Physics.

It can be concluded that SE of PANI and PPY films show weak temperature dependence. However, pronounced thickness effects were observed with attenuation level of -30 to -90 dB depending on thickness and conductivity. Different types of shielding mechanisms i.e. reflection, absorption and multiple reflections were discussed and corresponding theoretical equations were also presented. It has been found that absolute value of tanδ plays a critical role in determining the shielding effectiveness. When tanδ >> 1 (e.g. for heavily doped and highly conducting ICPs or metals), shielding is solely determined by σ. However, when tanδ ~ 1, both σ and ε_r must be considered when calculating absorption coefficient (α) and complex index of refraction (n) which decide overall shielding effectiveness. Therefore, one can expect higher shielding efficiency for materials with higher σ and ε_r.

It has been observed that magnetic properties also play a vital role in improving shielding response. Kathirgamanathan et al (1993) have demonstrated that PPY impregnation microporous membranes such as polyurethane, polyethylene, poly(ethylene terephthalate) (PET), poly(propylene) etc., showed higher SE (~ -10 to -50 dB) in the 10 kHz to 1000 MHz frequency range as compared to metal (e.g. Al) based membranes. The authors pointed out the higher relative magnetic permeability (μ_r>1) due to the incorporation of paramagnetic Fe (III) during the synthesis process provided extra shielding by absorption as compared with the μ_r~1 for aluminum. Furthermore, the microscopic orientation of ICPs is expected to improve SE as showed by the fact that higher SE was exhibited by the PPY impregnated polyethylene membranes (-40 to -45 dB) than that of the impregnated polyurethane membranes (-20 to -25 dB), despite the much lowered thickness (1/5th) which was due to the more oriented PPY produced in polyethylene than that in polyurethane.

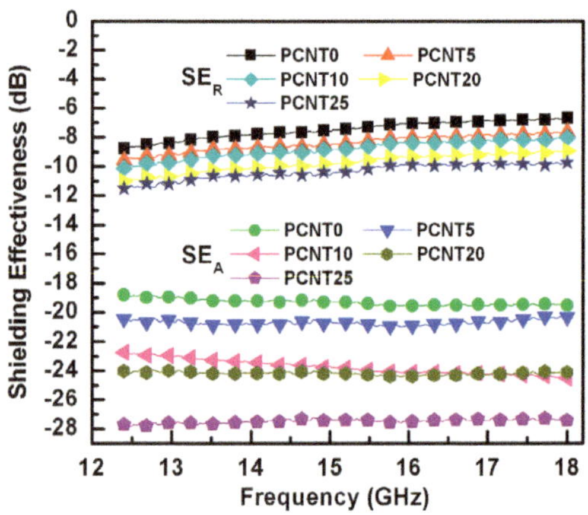

Figure 25. Frequency dependence of losses due to reflection (SE_R) and absorption (SE_A) for MWCNT loaded PANI nanocomposites having different loadings of MWCNT relative to aniline monomer viz. PCNT0 (0.0 wt. %),PCNT5 (5.0 wt. %), PCNT10 (10 wt. %), PCNT20 (20 wt. %) and PCNT25 (25 wt. %).

Synergistic coupling of fillers can give unique combination of properties (Saini et al 2009a) like enhanced conductivity, better dielectric/magnetic traits and improved processability/thermal conductivity that can not be achieved by individual fillers. This ultimately gets resulted in superior shielding performance (Fig. 25) so that reflection loss (SE_R) increases slightly from -8.0 to -12.0 dB whereas absorption loss (SE_A) exhibited rapid enhancement from -18.5 to -28.0 dB with the increase in CNT loading. This may be ascribed to increase in the conductivity (as well as capacitive coupling effects) of composites leading to proportional decrease in skin depth which may be helpful in designing thinner EMI shields. The increased conductivity may manifest itself as increase in both long range charge transport as well as number of possible relaxation modes, leading to enhanced ohmic losses.

The well-dispersed PANI NPs within insulating epoxy matrix provides continuous conducting networks with higher level of charge delocalization which leads to huge negative permittivity (Hsieh et al, 2012) which is a signature of left handed materials (LHM). The observed EMI SE in an electric field at low frequency (100–1000 MHz) range was found to be -30 to -60 dB.

A straight forward solution for handling low conductivity and poor processability (or agglomeration tendency) of ICPs and CNTs respectively is combining these two fillers in composites. Saini et al (2009a) prepared polyaniline (PANI) coated multiwall carbon nanotubes (MWCNTs) which inherit dielectric and magnetic attributes (ferromagnetism due to entrapped iron phase) from PANI and MWCNT respectively.

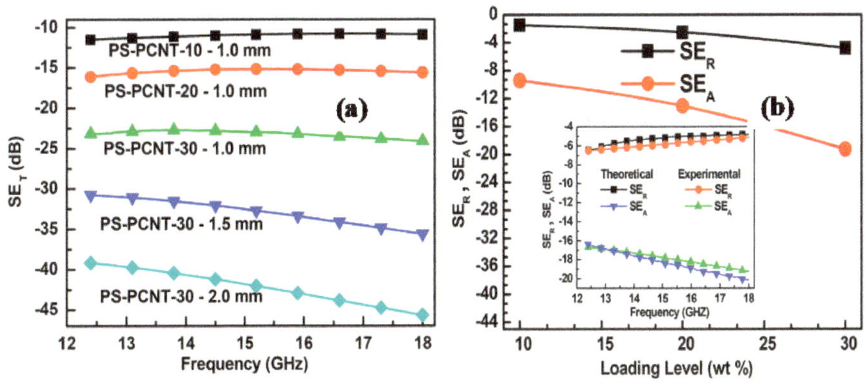

Figure 26. (a) Frequency dependence of SE_T and (b) variation of SE_R and SE_A with loading of PANI-MWCNT. Inset shows the theoretical and experimental SE_R and SE_A value of the composite (PCNT30) in the 12.4 - 18.0 GHz frequency band.

This PANI-MWCNT hybrid filler was solution blended with polystyrene (PS) matrix (10-30 wt % loading) resulting in absorption dominated total shielding effectiveness (SE_T) of -45.7 dB (Fig. 26a) in the 12.4–18.0 GHz range and at a sample thickness of ~2.0 mm. The SE_T was found to exhibit strong dependence on shield thickness as well as loading level of hybrid filler (PANI-MWCNT).

The enhanced SE_T was ascribed to optimization of conductivity, skin-depth, complex permittivity and permeability leading to nominal reflection and high absorption (Fig. 26b). A good agreement between theoretical and experimental shielding measurements (inset of Fig. 26b & Fig. 27) was also observed. Besides, role of highly reflecting planes of PANI-MWCNTs separated with less conducting matrix regions was also explained to introduce multiple reflections resulting in enhancement of absorption loss.

The above studies suggests that ICPs based nanocomposites may give SE value as high as -70 to -80 dB depending on nature of ICP, its loading level and presence of co-fillers. However, high loadings (>30-40%) are required which leads to phase segregation and extreme disturbance of physical properties of host matrices and consequently poor mechanical properties in most cases. Nevertheless, combination of strategies like thin film/membrane technologies, porous structures, negative permittivity materials (or left handed materials), multilayered structures and hybrid fillers based on broad range of ICP-filler combination are expected to provide an effective solution to realize a lightweight, mechanically strong, processable and economically viable shielding material suitable for commercial and defence sectors.

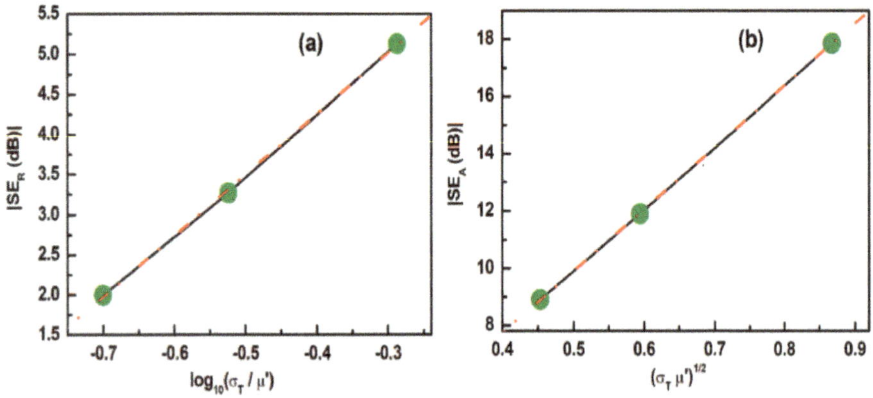

Figure 27. Variation of (a) SE$_R$ as function of $\log_{10}\left(\frac{\sigma_T}{\mu'}\right)$ and (b) SE$_A$ as function of $\left(\sigma_T\mu'\right)^{\frac{1}{2}}$ for PANI-MWCNT filled polystyrene composites

Recently, nanoscale materials based on 2-D graphene sheets have attracted much attention recently due to unusual properties (Geim, 2009; Geim & Novoselov, 2007; Meyer et al, 2007). Like CNTs, here again it is expected that the use of graphene, with large aspect ratio and high conductivity would provide a high EMI SE. Although many studies (In et al, 2010; Liang et al, 2009; Ramanathan et al, 2008; Stankovich et al, 2006; Varrla et al, 2011, Zhang et al, 2011) about the EMI shielding properties of graphene loaded insulating polymer matrix composite systems are available, Basavaraja et al (2011) presented the first EMI shielding results on ICP/oxidized graphene based nanocomposites i.e. Polyaniline/gold-nanoparticles/graphene-oxide (PANI-GNP-GO) based composites in the 2.0-12.0 GHz frequency range. According to authors, the SE values observed for GO and PANI-GNP and PANI-GNP-GO composites were in the ranges -20 to -33 dB, -45 to -69 dB and -90 to -120 dB respectively. However, considering the fact that GO is a poor conductor, conclusion from the presented thickness dependences of above composites and from our own experience, the results seem to be far from realistic. Nevertheless, the graphene nanocomposites research is still at very early stage of evolution especially from the view point of EMI shielding material development.

For many applications e.g. radar absorbers or stealth technology, the sample should reflect as low energy as possible. However, conducting filler loaded composites gives significant reflection (primary shielding mechanism) alongwith absorption which is secondary EMI shielding mechanism. For reduction of reflection loss and significant absorption of the radiation, the shield should have electric and/or magnetic dipoles which interact with the electromagnetic fields in the incident radiation. Therefore, numerous attempts have also been made to introduce dielectric ($BaTiO_3$, TiO_2 etc.) or magnetic (γ-Fe_2O_3, Fe_3O_4, $BaFe_{12}O_{19}$ etc.) materials within various ICP matrices as filled inclusions (Abbas et al, 2005, 2007; Chan, 1999; Chandrasekhar, 1999; Cho & Kim, 1999; Dong et al, 2008; Ellis, 1986; Gairola et al, 2010; Huang, 1990; Knott et al, 1993; Kurlyandskaya et al, 2007; Meshram et al, 2004; Nalwa, 1997; Ngoma et al, 1990; Pant et al, 2006; Phang et al, 2007, 2008, 2009, 2010; Xiaoling et al, 2006; Xu et al, 2007; Yang et al, 2010, 2011). It has been observed that thickness is an extrinsic parameter that can be adjusted to regulate the shielding offered by a shielding with given permittivity or permeability which can be tuned by nature and concentration of filler. An optimized dielectric particulates filled composite sample based on $BaTiO_3$ and polyaniline in polyurethane matrix (Abbas et al 2005) exhibited a maximum reflection loss of -15 dB (>99% power absorption) at 10 GHz with a bandwidth of 3.0 GHz for a 2.98 mm thick sample. Again the role of thickness and dielectric attributes to modulate absorption was demonstrated by theoretical calculations and experimental results. Similarly, they have also prepared polyaniline-$BaTiO_3$-carbon based composites (Abbas et al 2006) with maximum reflection loss of -25 dB (2.5 mm thick sample) at 11.2 GHz and bandwidth of 2.7 GHz. Many attempts were also made to introduce magnetic losses into the system for example, Yang et al (2009) produced PANI-Fe_3O_4 composites with reflection loss of -2 dB at 14.6 GHz for 3 mm thick sample. Gairola and coworkers (2010) prepared PANI with $Mn_{0.2}Ni_{0.4}Zn_{0.4}Fe_2O_4$ ferrite nanocomposites by mechanical blending with absorption loss of -49.2 dB in the 8.2-12.4 GHz range. Dong et al (2008) synthesized PANI-Ni core shell composites with reflection loss of less than -10 dB in the 4.2–18 GHz range. Phang and coworkers (2009) formulated PANI-HA based nanocomposites containing TiO_2 and Fe_3O_4 nanoparticles as dielectric filler and magnetic filler, respectively. The resultant composites show good microwave absorption response with attenuation of -48.9 dB. Phang et al (2007, 2008) produced PANI nanocomposites containing combination of dielectric (TiO_2) and conducting (CNTs) fillers possessing moderate conductivity and dielectric property with maximum reflection loss of -31 dB (for PANI-TiO_2) at 10 GHz and -21.7 dB (PANI-TiO_2-CNT) at 6GHz. In above composites, use of conducting fillers such as CNTs is expected to improve thermal conductivity (e.g. 0.19 W/mK for PANI and 0.3-0.6 W/mK for PANI-CNT composites) besides extending enhanced shielding performance. Such an improvement in thermal conductivity is beneficial for fast dissipation of heat which is generated due to interaction of shield with high frequency (GHz range) microwave radiations.

4. Conclusions

Although, much work has been done to introduce electrical conductivity in various polymer matrices but high percolation threshold and lower aspect ratios of ICPs compared to metals or carbon based fillers remained a challenging issue. Therefore, considerable work is still needed to improve further the SE as well as mechanical properties of conducting polymer based composites. Synthesis of hybrid filler materials based on various combination of conducting polymers, carbon based materials and dielectric/magnetic nanoparticles seem to be a possible solution. Nevertheless, in the light of current scenario it may be stated that there is a lot be done to attain a shielding material that can satisfy all the techno commercial specification and maintain the process economics at the same time.

Author details

Parveen Saini and Manju Arora
National Physical Laboratory, New Delhi, India

5. References

Abbas, S.M., Dixit, A.K., Chatterjee, R. and Goel, T.C. 2005, *Mater. Sci. Eng. B*, 123, 167.

Abbas, S.M., Chandra, M., Verma, A., Chatterjee, R. and Goel, T.C. 2006, *Composites: Part A*, 37, 2148.

Abbas, S.M., Chatterjee, R., Dixit, A.K., Kumar, A.V.R. and Goel, T.C. 2007, *J. Appl. Phys.*, 101, 074105.

Ajayan, P.M., Schadler, L.S., Giannaris, C. and Rubio, A. 2000, *Adv. Mater.*, 12, 750.

Ajayan, P.M., Stephan, O., Colliex, C. and Trauth, D. 1994, *Science*, 265, 1212.

Alexandre, M. and Dubois, P. 2000, *Materials Science and Engineering: R: Reports*, 28, 1.

Bal, S. and Samal, S.S. 2007, *Bull. Mater. Sci.*, 30, 379.

Baeriswyl, D., Campbell, D.K. and Mazumdar, S. 1992, *Conjugated Conducting Polymers*, Edited by Keiss, H.G. Springer-Verlag, Berlin, 7.

Basavaraja, C.,Kim, W.J., Kim, Y.D. and Huh, D.S. 2011, *Mater. Lett.*, 65, 3120.

Baughman, R.H., Zakhidov, A.A. and Heer, W.A. de 2002, *Science*, 297, 787.

Bredas, J.L., Cornil, K., Meyers, F. and Beljonne, D.1998, *Handbook of Conducting Polymers*, Ed.: Skotheim, T.A., Elsenbaumer, R.L., Reynolds, J.R. and Dekker, M. Inc., New York.

Cao, Y., Smith, P. and Heeger, A.J. 1992, *Synth. Met.*, 48, 91.

Cao, Y., Qiu, J., Smith, P. 1995, *Synth. Met.*, 69, 187.

Carter, G.M., Thakur, M.K., Chen, Y.J. and Hryniewicz, J.V. 1985, *Appl. Phys. Lett.*, 47, 457.

Chan, H.L.W., Cheung, M.C. and Choy, C.L. 1999, *Ferroelec.*, 224, 113.

Chandrasekhar, P. 1999, *Conducting Polymers: fundamental and Applications, A Practical Approach*, Kluwer Academic Publishers.

Chandrasekhar, P. and Naishadham, K. 1999, *Synth. Met.*, 105, 115.

Chandrasekhar, P., Zay, B.J., Birur, G.C., Rawal, S., Pierson, E.A., Kauder, L. and Swanson, T. 2002, *Adv. Funct. Mater.*, 12, 2137.

Chiang, C.K., Fincher, C.R.J., Park, Jr., Y.W., Heeger, A.J., Shirakawa, H., Louis, E.J., Gau, S.C. and MacDiarmid, A.G. 1977, *Phys. Rev. Lett.*, 39, 1098.

(a) Chiang, C.K., Gau, S.C., Fincher, C.R.J., Park, Y.W., MacDiarmid, A.G. and Heeger, A.J. 1978, *Appl. Phys. Lett.*, 33, 18.

(b) Chiang, C.K., Druy, M.A., Gau, S.C., Heeger, A.J., Louis, E.J., MacDiarmid, A.G., Park, Y.W. and Shirakawa, H. 1978, *J. Amer. Chem. Soc.*, 100, 1013.

Cho, H.S. and Kim, S.S. 1999, *IEEE Trans. Magn.*, 35, 3151.

Choudhary,V. and Gupta, A. 2011, *Carbon Nanotubes-Polymer Nanocomposites*, Ed. Siva Yellampalli, InTech.

Chung, D.D.L. 2000, *J. Mater. Eng. Perform.*, 9, 350.

Chung, D.D.L. 2001, Carbon, 39, 279.

Colaneri, N.F. and Shacklette, L.W. 1992, *IEEE Trans. Instrum. Meas.*, 41, 291.

Coleman, M.M. and Petanck, R.J. 1986, *J. Polym. Sci.*, 16, 821.

Das, C. K. and Mandal, A. 2012, *J. Mater. Sci. Res.*, *1*, 45.

Dhawan S.K., Singh, N. and Rodrigues, D. 2003 *Sci. Technol. Adv. Mater.*, 4,105.

Dong, X.L., Zhang, X.F., Huang, H. and Zuo, F. 2008, *Appl. Phys. Lett.*, 92, 013127.

Ellis, J.R. 1986, *Handbook of Conducting Polymers*, 1st Edition; T.A. Skotheim, Marcel Dekker: New York.

Fang, F.F., Kim, J.H. and Choi, H.J. 2006, *Macromol. Symp.* 2006, *242*, 49.

Freund, M.S. and Deore, B. 2007, *Self-Doped Conducting Polymers*, John Wiley & Sons Ltd., Chichester.

Friend, R.H. 1993, *Rapra Rev. Repo.*, 6

Gairola, S.P., Verma, V., Kumar, L., Dar, M.A., Annapoorni, S. and Kotnala, R.K. 2010, *Synth. Met.*, 160, 2315.

Gangopadhyay, R., De, A. and Ghosh, G. 2001, *Synth. Met.*,123, 21.

Geim, A.K. 2009, *Science*, 324,1530.

Geim, A.K. and Novoselov, K.S. 2007, *Nature Mater*, 6, 183.

Grimes, C.A., 1994, *Aerospace Applications Conference*, 1994. *Proceedings.*, 1994 *IEEE*, 211, Doi: 10.1109/AERO.1994.291194.

Gupta, A. and Choudhary, V. 2011, *Comp. Sci. Technol.*, 71, 1563.

Gupta, S.K., Dhawan, S.K., Arora, M. and Arya S.K. 2005, *Proc. IWPSD*, II, 1318.

(a) Heeger, A.J. 2001, *Angew. Chem. Intern. Ed.*, 40, 2591.

(b) Heeger, A.J. 2001, *Rev. Mod. Phys.*, 73, 681.

Hsieh, C.-H., Lee,A.-H, Liu, C.-D., Han, J.-L., Hsieh, K.-H. and Lee, S.-N. 2012, *AIP Adv.*, 2, 012127.

Hong, Y.K., Lee, C.Y., Jeong, C.K., Lee, D.E, Kim, K., Joo, J., 2003, *Rev. Sci. Instrum.*, 74, 1098.

Huang, Chi-Yuan and Wu, Chang-Cheng 2000, *European Polymer Journal*, 36, 2729.

Huang, Li, N., Du, Y., He, F., Lin, X., Gao, X., Ma, H., Li, Y., Chen, F., Y. and Eklund, P.C. 2006, *Nano letters*, 6, 1141.

Huang, X.X., Chen, Z.F., Zou, W.Q., Liu, Y.S. and Li, J.D. 1990, *Ferroelec.*, 101, 111.

Iijima, S. 1991, *Nature (London)*, 354, 56.

In, K.M., I.K., Lee, J., Ruoff, R.S. and Lee, H. 2010, *Nature Communications*,1, 1067.

Javadi, H.H.S., Cromack, K.R., MacDiarmid, A.G. and Epstein, A.J. 1989, *Phys. Rev. B*, 39, 3579.

Joo, J. and Epstein, A.J. 1994, *Appl. Phys. Lett.*, 65, 2278.

Joo, J., Song, H. G., Jang, K. S., and Oh, E. J. 1999, *Synth. Met.*, 102, 1349.

Joo, J., Oblakowski, Z., Du, G., Pouget, J.P., Oh, E.J., Weisinger, J.M., Min, Y. , MacDiarmid, A.G. and Epstein, A.J. 1994, *Phys. Rev. B*, 49, 2977.

Karasz, F.E., Capistran, J.D., Gagnon, D.R. and Lenz, R.W. 1985, *Mol. Cryst. Liq. Cryst.*, 118, 327.

Kathirgamanathan, P. 1993, Adv. Mater., 5, 281.

Knott, E.F. and Schaeffer, J.F. 1993, *M.T. Radar. Cross Section Handbook*, Artech House: New York.

Koul, S., Chandra, R. and Dhawan S.K. 2000, *Polymer*, 41, 9305.

Kremer F. and Schönhals A. 2003, *Broadband Dielectric Spectroscopy*, Springer Verlag,

Lakshmi, K., John, H., Mathew, K.T., Joseph, R. and George, K.E. 2009, *Acta Materialia*, 57, 371.

Kurlyandskaya, G.V., Cunanan, J., Bhagat, S.M., Aphesteguy, J.C. and Jacobo, S.E. 2007, *J. Phys. Chem. Solids*, 68, 1527.

Liang, J., Wang, Y., Huang, Y., Ma, Y., Liu, Z., Cai, J., Zhang, C., Gao, H. and Chen, Y. 2009, *Carbon*, 47, 922.

MacDiarmid, A.G. 2001, *Angew. Chem. Intern. Ed.*, 40, 2581

Makeiff, D.A. and Huber, T. 2006, *Synth. Met.*, 156, No. 7-8, 497.

Mathur, R.B., Singh, B.P. and Pandey, S. 2010, *Polymer nanotubes nanocomposites, Synthesis properties and applications*, Ed. Vikas Mittal, Wiley Scrivener.

Mattosso, L.H.C., Faria, R.M., Bulhoes, L.O.S., MacDiarmid, A.G. and Epstein, A.J. 1994, *J. Polym. Sci. Part A: Polym. Chem.*, 32, 2147.

Meshram, M.R., Agrawal, N.K., Sinha, B. and Misra, P.S. 2004, *J. Magn. Magn. Mater.*, 271, 207.

Meyer, J.C., Geim, A.K., Katsnelson, M.I., Novoselov, K.S., Booth, T.J. and Roth, S. 2007, *Nature*, 446, 60.

Moniruzzaman, M. and Das, C. K. 2010, *Macromol. Symp., 298*, 34.

Moniruzzaman, M. and Winey, K.I. 2006, *Macromolecules*, 39, 5194.

Moon, I.K., Lee, J., Ruoff, R.S. and Lee, H. 2010, *Nature Communications*, 1,1067.

Naishadham, K. and Chandrasekhar, P. 1998, *IEEE Trans. Microw. Theo. Technol.*, 3, 1353.

Naishadham, K. and Kadaba, P.K. 1991, *IEEE Trans. Microw.*, 39, 1158.

Ngoma, J.B., Cavaille, J.Y., Paletto, J., Perez, J. and Macchi, F. 1990, *Ferroelec.*, 109, 205.

Nalwa, H.S. 1997, *Handbook of Organic Conductive Molecules and Polymers* (four volumes), Wiley, New York.

Natta, G., Mazzanti, G. and Corradini, P. 1958 *Atti Accad. Naz Linceicl. Sci. Fis. Mat. Nat. Rend.*, 2, 25.

Nicolson, A.M. and Ross, G. F. 1970, *IEEE Trans. Instrum. Meas.*, 19, 377.

Olmedo L., Hourquebie, P. and Jousse, F. 1997, *Handbook of Organic Conductive Molecules and Polymers*, vol. 2; John Wiley & Sons Ltd, Chichester.

Olmedo, L., Hourquebie, P. and Jousse, F. 1995, *Synth. Met.*, 69, 205.

Ott, H.W. 2009, *Electromagnetic Compatibility Engineering*, New Jersey, John Wiley & Sons.

Pandey, S., Singh, B.P., Mathur, R.B., Dhami, T.L., Saini, P. and Dhawan, S.K. 2009, *Nanoscale Research Letters*, 4, 327.

Pant, H.C., Patra, M.K., Verma, A. , Vadera S.R. and Kumar, N. 2006, *Acta Materialia*, 54, 3163.

Phang, S.W., Hino, T., Abdullah, M.H. and Karamoto, N. 2007, *Mater. Chem. Phys.*, 104, 327.

Phang, S.W., Tadokoro, M., Watanabe, J. and Kuramoto, N. 2008, *Synth. Met.*, 158, 251.

Phang, S.W., Tadokoro, M., Watanabe, J. and Kuramoto, N. 2009, *Poly. Adv. Technol.*, 20, 550.

Phang, S.W. and Kuramoto, N. 2010, *Poly. Comp.*, 31, 513.

Pomposo, J.A., Rodriguez, J. and Grande, H. 1999, *Synth. Met.*, 104, 107.

Pud, A.,Ogurtsov, N., Korzhenko, A. and Shapoval, G. 2003, *Prog. Poly. Sci.*, 28, 1701.

Qiao, R., Yang, C. and Gao, M. 2009, *J. Mater. Chem.*, 19, 6274.

Rahman, S., Mahapatra, M., Maiti, M.M. and Maiti, S. 1989, *J. Poly. Mater.*, 6, 135.

Ramanathan T, Abdala, A.A., Stankovich, S., Dikin, D.A., Herrera-Alonso, M., Piner, R.D., Adamson, D.H., Schniepp, H.C., Chen, X., Ruoff, R.S., Nguyen, S.T., Aksay, I.A., Prud'Homme, R.K. and Brinson L.C. 2008, *Nature Nanotechnol*, 3, 31.

Ramasubramaniam, R., Chen, J. and Liu, H. 2003, *Appl. Phys. Lett.*, 83, 2928.

Riande, E. and Diaz-Calleja, R. 2004, *Electrical Properties of Polymers*, CRC Press ISBN: 978-1-4200-3047-1.

Rozenberga, B.A., and Tenn, R. 2008, *Prog. Polym. Sci.*, 33, 40.

Saini, P., Choudhary, V. and Dhawan, S. K., 2007, *Ind. J. Engn. Mater. Sci.*, 14, 436.

Saini, P., Choudhary, V. and Dhawan, S.K. 2010, *Poly. Adv. Technol.*, 21, 1.

Saini, P., Jalan, R. and Dhawan, S.K. 2008, *J. Appl. Polym. Sci.*, 108, 1437.

(a) Saini, P., Choudhary, V., Singh, B.P., Mathur, R.B. and Dhawan, S.K. 2009, *Mater. Chem. Phys.*, 113, 919,

(b) Saini, P., Choudhary, V., Sood, K.N. and Dhawan, S.K. 2009, *J. Appl. Poly. Sci.*, 113, 3146.

Saini, P., Choudhary, V., Singh, B.P., Mathur, R.B. and Dhawan, S.K. 2011, *Synth. Met.*, 161, 1522.

Sanjai, B., Raghunathan, A., Natarajan, T.S., Rangarajan, G., Thomas, S. Prabhakaran, P.V. and Venkatachalam, S. 1997, *Phys. Rev. B*, 55, 10734.

Savitha, P., Rao, P.S. and Sathyanarayana, D.N. 2005, *Polym. Int.*, 54, 1243.

Saxman, A.M., Liepins, R. and Aldissi, M. 1985, *Prog. Polym. Sci.*, 11, 57.

Schelkunoff, S.A. 1943, *Electromagnetic Waves*, Van Nostrand, New Jersey.

Sheng, P. and Klafter, J. 1983, *Phys. Rev. B.*, 27, 2583.

Schulz, R.B., Plantz, V.C. and Brush, D.R. 1988, *IEEE Trans.*, 30, 187.

Shacklette, L.W., Colaneri, N.F., Kulkarni, V.G. and Wessling, B. 1992, *J. Vinyl Technol.*, 14, 118.

Shi, Sui-Lin and Liang, Ji, 2008 *Nanotechnology*, 19, 255707.

Shirakawa, H. 2001, *Angew. Chem. Intern. Ed.*, 40, 2575.

Shirakawa, H., Louis, E.J., MacDiarmid, A.G., Chiang, C.K. and Heeger, A.J. 1977, *Chem. Comm.*, 578.

Skotheim, T.A. 1986, *Handbook of Conducting Polymers*, 2nd Edition, CRC.

Sillars, R.W. 1937, J. Inst. Elect. Eng., 80, 378

Singh, B.P., Prabha, Saini, P., Gupta, T., Garg, P., Kumar, G., Pandey, I., Pandey, S., Seth, R.K., Dhawan, S.K. and Mathur, R.B. 2011, *J. Nanopart. Res.*, 1.

Singh, P., Babbar, V.K., Razdan, A., Srivastava, S.L. and Goel, T.C. 2000, *Mater. Sci. Eng. B*, 78, 70.

Singh, P., Babbar, V.K., Razdan, A., Srivastava, S.L. and Puri, R.K. 1999, *Mater. Sci. Eng. B*, 67, 132.

Snow, A.W. 1981, *Nature*, 292, 40.

Soga, K., Nakamura, M., Kobayashi, Y. and Ikeda, S. 1983, *Synth.. Met.*, 6, 275.

Stafstrom, S., Bredas, J.L., Epstein, A.J., Woo, H.S., Tanner, D.B., Huang, W.S. and MacDiarmid, A.G. 1987, *Phys. Rev. Lett.*, 59, 1464.

Stankovich, S., Dmitriy, A.D., Geoffrey, H.B.D., Kevin, M.K., Zimney, E.J., Stach, E.A., Piner, R.D., Nguyen, S.T. and Ruoff, R.S. 2006, *Nature*, 442, 20.

Stankovich, S., Dikin, D.A., Piner, R.D., Kohlhaas, K.M., Kleinhammes, A, Jia, Y., Wu, Y., SonBinh, T. N., Rodney, S.R. 2007, *Carbon*, 45, 1558.

Stonier, R.A. 1991, *Sampe J.*, 27, 9.

Taka, T. 1991, *Synth. Met.*, 41, 1177.

Thomas, B., Pillai, M.G.K. and Jayalakshmi, S. 1988, *J. Phys. D: Appl. Phys.*, 21, 503.

Thostenson, E.T., Li, C. and Chou, T.-W. 2005, *Nanocomposites in context, Composites Science and Technology*, 65, 491.

Tong, X.C., 2009, *Advanced Mater. And Design For Electromagnetic Interference Shielding*, CRC Press Taylor and Francis Book, London, New York.

Trivedi, D.C. 1997, *Handbook of Organic Conductive Molecules and Polymers*, 2, John Wiley & Sons Ltd.: Chichester

Varrla, E., Venkataraman, S. and Sundara R. 2011, *Macromol. Mater. Eng.*, 296, 894.

Wang, Y. and Jing, X., 2005, *Polym. Adv. Technol.*,16, 344.

Wang, Z.H., Li, C., Scherr, E.M., MacDiarmid, A.G. and Epstein, A.J. 1991, *Phys. Rev. Lett.*, 66, 1745.

Wagner, K.W . 1914, *Arch Elektrotech*, 2, 371.

Weir, W.B. 1974, *Proc. IEEE*, 62, 33.

Wessling, B. 1999, *Synth. Met.*, 102, 1396.

Wojkiewicz, J.L., Fauveaux, S. and Miane, J.L. 2003, *Synth. Met.*,135–136, 127

Xiaoling, Y., Gang, L., Duanming, Z. and Huahui, H. 2006, *Mater. Design*, 27, 700.

Xu, P., Han, X.J., Jiang, J.J., Wang, X.H., Li, X.D. and Wen, A.H. 2007, *J. Phys. Chem. C*, 111, 12603.

Yamamoto, T., Hayashi, Y. and Yamamoto, A. 1978, *Bull. Chem. Soc. Japan*, 51, 2091.

Yang, C., Du, J., Peng, Q., Qiao, R., Chen, W., Xu, C., Shuai, Z. and Gao, M. 2009, *J. Phys. Chem. B*, 113, 5052.

Yang, C.C., Gung, Y.J., Hung, W.C., Ting, T.H. and Wub, K.H. 2010, *Composites Science and Technology*, 70, 466.

Yang, C.C., Gung, Y.J., Shih, C.C., Hung, W.C. and Wub, K.H. 2011, *J. Magn. Magn. Mater.*, 323, 933.

(a) Yang, Y., Gupta, M.C., Dudley, K.L. and Lawrence, R.W. 2005, *Nano Lett.*, 5, 2131.

(b) Yang, Y., Gupta, M.C., Dudley, K.L. and Lawrence, R.W. 2005, *Adv. Mater.*, 17, 1999.

Zhang, H.-B., Yan, Q., Zheng, W.-G., He, Z. and Yu, Z.-Z.T. 2011, *ACS Appl. Mater. Interfaces*, 3, 918.

Zuo, F., Angelopoulos, M., MacDiarmid, A.G. and Epstein, A.J. 1989, *Phys. Rev. B*, 39, 3570.

Chapter 2

Polymerization of Thin Film Polymers

Markus Woehrmann and Michael Toepper

Additional information is available at the end of the chapter

1. Introduction

Polymeric coatings such as polyurethanes, acrylic, epoxies and silicones have been used for over 40 years to protect printed wiring boards (PWB) from moisture, handling and environmental influences. Special semiconductor grade polymers have been developed for chip passivation layers. Polyimide became the standard passivation layer for memory chips and other devices needing surface protection for handling and testing procedures. Photosensitive resins have been developed to reduce processing costs. Thin film polymers are used widely in the area of electronic packaging and as an interlayer dielectric (ILD) in nearly every electronic device. Typical layer thickness ranges from 5 μm up to 15 μm. These polymers films should be temperature stable up to range of 150°C (permanent), and for a short time up to 250°C and higher, depending on the soldering process of the components. This is the reason to use thermoset polymers in most cases. The coating process is the spin-coating technology. Spray coating and other techniques are only niche processes. The polymer supplied by the chemical company is a so-called pre-cursor consisting of a partially polymerized polymer in an organic solvent. Nearly all of the thin film polymers need a polymerization step, which is done in most cases by a thermal process after the deposition on the wafer. Polymerization changes the pre-polymer into a long-term stable and much more inherent polymerized dielectric layer. The polymerization process is generally called the *cure* or *cure step* of the process. But the definition "cure" does not mean in every case a fully finished polymerization reaction. Also partly cured polymer films are possible and becoming important if multiple layers have to be deposited. Partly polymerized layers have better adhesion to the following layer compared to fully cured ones. To get something more precise the following definitions have been used: partly cured or soft cured polymers for a 60 – 80 degree of polymerization, and full cured or hard cured for a complete polymerized layer.

There are a huge range of different polymer materials which are used as interlayer dielectric such as polyimides, polybenzoxazole and benzocyclobutene [1]. An example of such a polymer layer on a CMOS-wafer is shown in figure 1:

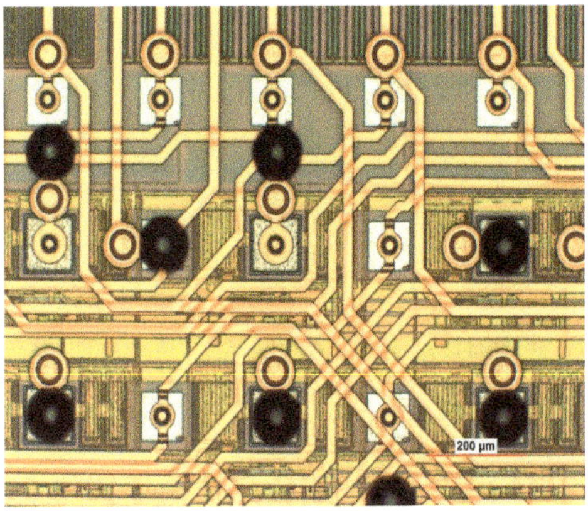

Figure 1. A thin film polymer (BCB from Dow Chemical) as protection layer on a CMOS-wafer

In the last few years, the development of new polymer materials for thin film applications has increased more and more. One main driver for this is the 3-D integration technology which demands optimized material behaviors of the polymers, such as a low coefficient of thermal expansion (CTE) or higher tensile strength. The 3-D technology also demands more and more the low temperature processing. The material combinations get more temperature critical. The reliability will be increased at the same time with low cure temperature, because of a decreased thermal stress in the material stack. In contrast to the demand of low cure temperature, a tough polymer layer is desired with a high decomposition temperature. Analyzing and engineering the polymerization process will help enormous to reach these goals.

This chapter will focus on the benzocyclobutene (BCB) from The Dow Chemical Company being a well-known material for thin-film applications. The full cured BCB has a decomposition temperature well above 350°C. The cure temperature of 250° has been reported in dozens of publications. The analyzing and modeling of the BCB polymerization process will be discussed here in detail. Understanding the behavior of the polymerization process in different states and their kinetic modeling shows the potential for decreasing the cure temperature for a BCB layer which has nearly the same material stability and decomposition temperature as standard processed BCB layers.

2. The relevance of thin film polymers in micro electronics

The Dow Chemical Co. (Midland, MI) has developed a variety of low-dielectric (low-k) polymers which are based on different benzocyclobutene monomers (BCB, also known as biscyclo[4.2.0]octa-1,3,5-triene or 1,2-dihydrobenzocyclobutene). Based on these studies, the Dow Chemical Co. commercializes a BCB-Polymer under the registered trademark CYCLOTENE. CYCLOTENE is a family of thermosetting polymers prepared from 1,3 divinyl-1,1,3,3-teramethyldisiloxane-bis-benzocyclobutene (DVS-bis-BCB) monomer. The structure of the DVS-bis-BCB monomer is illustrated in figure 2. Today CYCLOTENE is just called BCB in most of the literature. BCB has been developed for electronic coating applications. Based on the low-k characteristic of BCB, the material is very attractive as an interlayer dielectric for RF-components [2] [3]. The BCB has been widely used for multilayer re-distribution on wafer level (wafer level packaging-WLP). Since the early 90's, there are hundreds of application notes for BCB (e.g. bumping and redistributing chips, and for planarization and isolation in flat-panel displays, MEMS hermetization, wafer bonding, passivation, gap filling). The wide field of application is based on the material properties such as good planarization, no outgassing, and low copper migration. The Dow Chemical Co. also releases the chemical structure of the monomer and the final polymer, which is not common for the most commercialized thin film polymers. Knowledge of the structure allows for a better understanding of material behavior and material analyzing, which leads to the fact that BCB is used very often for academic research. Most of the newer polymers are years away from such a comprehensive database. BCB is available as photosensitive (CYCLOTENE 4000 series) and non-photosensitive (CYCLOTENE 3000 series) in the form of a spin-on solution. CYCLOTENE can be purchased in different viscosities to obtain a wide film thickness range of <1 μm up to 26 μm (see table 1). CYCLOTENE is shipped in a partly cross-linked state in mesitylene solvent (trimethylbenzene). The non-photosensitive CYCLOTENE can be structured by dry etch processes or laser ablation. On account of this, it is often called dry-etch BCB. The initial extent of polymerization is slightly different when comparing dry etch BCB with photo BCB. The dry etch version is stable at room temperature, in contrast to the photosensitive parts of the 4000 series, which have a shelf life of a week at room temperature, due to the photo-sensitizer. In a frozen state, the durability increases up to a couple of years.

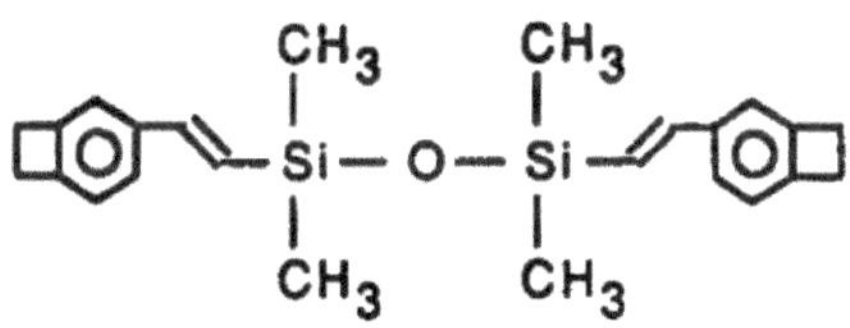

Figure 2. Structure of a DVS-bis-BCB monomer [4]

In contrast to polyimides, which are also often used as thin film layers, BCB has a couple of advantages. BCB is highly cross-linked, based on the small monomer, which leads to higher thermal stability in comparison to the cure temperature [4]. It also leads to a complete isotropic behavior of the polymer film. On the other hand, the long-chain characteristic of the polyimide leads to a higher elongation at break, which is an important factor in reliability [1]. The polymerization reaction of the BCB does not emit any volatile by-products, resulting in low cure shrinkage (less than 5 %). The non-polar chemical structure of BCB leads to low dielectric constant (ε_r=2.65) and a high breakdown voltage (530 V/μm) in combination with a low water uptake (less than 0.2 %). The good planarization properties also make BCB attractive for 3-D and adhesive bonding [5] [6].

Notation	Viscosity (cST@25°C)	Thickness (μm)	Solid content (%)	b-stage (%)
3022-35	14	1.0 – 2.4	35	35
3022-46	52	2.4-5.8	46	35
3022-57	259	5.7-15.6	57	35
3022-63	870	9.5-26.0	63	35
4022-35	192	2.5-5.0	35	47
4024-40	350	3.5-7.5	40	47
4026-46	1100	7.0-14.0	46	47

Table 1. CYCLOTENE with different solid content

3. Overview of the polymerization process

3.1. General polymerization behavior of thermosets

BCB is a thermoset like most other ILD polymers. The polymerization is the process which forms the small molecules or monomers into large polymers structures. The chemical reaction (polymerization) of the thermoset during the processing allows an easy processing of thin film layers. In most cases, a liquid solvent is deposited by spin coating on a substrate, and after the polymerization a tough solid state with a high glass transition temperature is formed. In contrast to thermosets, thermoplastics need a physical condition change for the polymerization. The disadvantage of thermoplastics is the low glass transition temperature ranging from below room temperature up to 180°C, which is a result of the low Van der Waals forces between the polymer chains. For the following processes, such as sputtering, a high thermal stability above 200°C is necessary. A softening of the polymer based on low T_g during sputtering could lead to buckling of the deposited metal layer.

The result of the polymerization of thermosets is a more or less strong 3-D cross-linking of the polymer chains with strong covalent connections in combination with solidification of the polymer layer, which leads to an insolubility of the material. In most cases, thermal energy is used for the polymerization reaction, but there are also other methods like radiation or hardener materials which work like a catalyzer being used for other polymers.

The polymerization is not a monotonic reaction, because it progresses through different stages which change the characteristics of the reaction significantly. The reaction starting point is an ideal solvent in which there are only monomer structures in the solvent. The material is in a "liquid phase" (also called an a-stage). With the supply of enough energy in the system, the cross-linking of the monomers starts and polymer chains build up. The degree of cure describes the rate between reacted parts and unreacted parts (reactants and products). In literature there is a different notation which means more or less the same thing, like extent of cure, extent of reaction, extent of conversion and fraction transformation. The reaction rate is nearly constant in a viscous liquid stage up to the gelation point, which described the abrupt and irreversible transformation of the material into an elastic gel or rubber, and the material loses its ability to flow. At this point, there are infinitesimal networks which are more or less cross-linked to each other (b-stage). The gelation leads only to a small change in the reaction rate, but could be estimated with rheological measurement. The point of gelation is for a thermoset on a fixed degree of cure with the conclusion, that it is independent of the polymerization method (radiation, convection oven or microwave). Because of that, the measurement of the gelation point allows for calculating the activation energy of the polymerization reaction [7].

With the increase in the polymer network, the material becomes more and more rigid. The movement of the monomer reactants slows down abruptly when the glass transition point (T_g) of the polymer overwhelms the process temperature. The material vitrificates as a solid state, for polymers often called the glassy state, based on their more or less amorphous structure. The reaction rate drops down by an order of two or three magnitudes, and is significantly dependent on the mobility of the reactive monomers. In gel state, the polymerization is controlled by a chemical reaction rate, but in glassy formation, the reaction is controlled by the diffusion of the reactants. The glass transition temperature increases with the polymerization degree. Based on that, the glass transition temperature is highly dependent on the cure process. The vitrification is, in contrast to the gelation, reversible by heating. When the process temperature is again above the T_g devitrification occurs. A fully cured thermoset (c-stage) has a glass transition temperature near the dissociation temperature, which makes the measurement of such a T_g in most cases impossible. Based on that it can be postulated that a fully polymerized thermoset does not have glass transition temperature anymore [8].

3.2. The polymerization reaction of BCB

The fully polymerized BCB has a high 3-D cross-linked network, which leads to isotropic properties. The polymerization reaction consists of two steps. The first step is a thermal activated BCB four-membered ring opening reaction. This reactive intermediate readily undergoes a [2+4] Diels-Alder reaction (see figure 3). The Diels-Alder reaction is a reaction where a conjugated diene will be added on a double bond, building a six terms ring. Both reaction partners are named dien and dinophile. The reaction is also called a [4+2] cycloaddition.

After ring opening, the BCB monomer has two diens and dinophiles, which allow a highly cross-linked polymer. The ring opening is a first order reaction, and depends on the concentration of the pre-polymer. The reaction rate of the ring opening and the Diels-Alder reaction was measured by ROTH et. al. The ring opening has a 10 times higher reaction rate constant [9] [10].

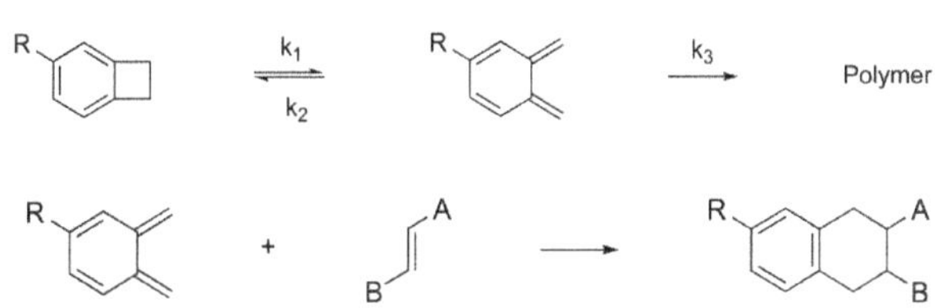

Figure 3. Ring opening reaction and Diels-Alder reaction of BCB

The polymerization reaction could be described by a pseudo first-order reaction equation, because the Diels-Alder reaction is the reaction rate limited factor. The thermal activated polymerization process of BCB or the reaction rate depends on temperature, time and extent of cure. Figure 4 show a Time-Temperature-Transformation diagram (TTT-diagram) for the polymerization reaction of BCB. The numbers in the diagram describes the extent of cure which could be measured by different methods.

The reactions rate is strongly influenced by the mobility of the molecule chains which generally increase with the temperature, but also decrease with the extent of cure, because increased cross-linking decreases the mobility. During the cure, the reaction kinetic runs through different stages, from liquid pre-polymer through a gelation to a vitrification.

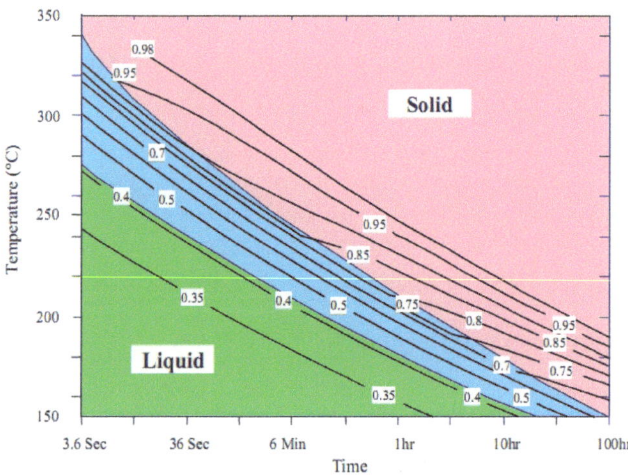

Figure 4. Extent of cure of photo BCB in relation to soak temperature and soak time. The colors describe the material conditions: liquid state (green), gel state (blue) and solid state (red) [11]

The gelation point has no significant effect on the polymerization reaction. The change into the solid state impacts the reaction rate based on the change from a chemical controlled to a diffusion controlled regime. That leads to a significant drop in molecule mobility, which could decrease the reaction rate down to nearly zero. The slowdown factor of BCB is relatively weak because of the small monomers.

The commercially available CYCLOTENE (see table 1) has been pre-cured (b-stage resin). Especially for structured photo BCB layers, a viscous liquid state during processing is avoided by a higher molecular weight of the b-stage state. After the evaporation of the solvent, the polymer layer lost its flow ability and the layer could be structured. A flowing of polymer during the cure would limit the critical structure dimension. The reaction rate is nearly constant during the entire gelation phase. The fully cured BCB has a T_g near the decomposition temperature above 350°C. But the T_g of an uncured CYCLOTENE will exceed approximately 60°C, and will rise dependent upon the degree of cure (figure 5) [6].

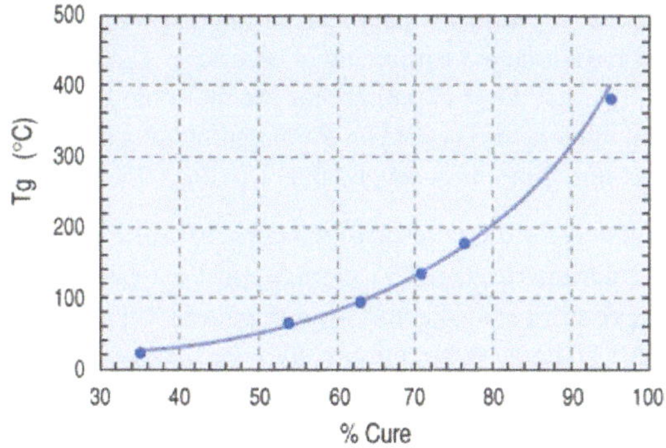

Figure 5. Glass transition temperature of photo BCB [6]

It is shown for some other polymers that the degree of cure is nearly fixed after vitrification without a temperature rise, because of the extremely slow reaction rate [12]. This means for polymerization process optimization a full cure is not possible if the temperature is decreased. A lower processing temperature leads to the glass transition temperature occurring earlier, which can be seen in figure 4. The polymerization process stops at a lower degree of cure, because the reaction changed earlier from a reaction controlled to diffusion controlled characteristic.

The slowdown of the reaction rate of BCB after vitrification has been described in previous publications [13] [14]. However, the reaction does not stop and proceeds to the solid state. In contrast to many other polymers, the diffusion controlled reaction rate is high enough to continue the polymerization process. DIBBS et. al. specifies only that above 95 % the diffusion controlled reaction rate will slow down so that it could be seen as a fixed state [2]. A degree of cure of about 95 % can be accepted as nearly fully cured. There is no significant difference measurable because of the approximation failures of measuring methods for the degree of cure.

4. Measurement of the polymerization degree of BCB

The datasheet of CYCLOTENE describes a cure at 250°C for 90 minutes to get a full cured film. Any change in the cure process or the creation of a kinetic model needs knowledge about the extent of cure. There are a couple of methods to monitor it, like Fourier transformed infrared-spectroscopy (FT-IR spectroscopy), change of the refractive index or the difference scanning calorimetric (DSC).

The determination of the reaction rate, and the extent of reaction by time and temperature are investigated further to develop a kinetic reaction model. Real-time measurements during the cure process are necessary to monitor the degree of cure. DSC measurement was used for kinetic model creation. Based on the kinetic model, cure processes were set-up and the expected extent of cure was controlled by FT-IR spectroscopy. The FT-IR spectroscopy allows the measurement on product wafers, but for the creation of a kinetic model, a special oven with integrated measurement equipment is necessary. The discrete IR wavelength gets into resonance with the rotation or vibration of some molecule groups. The rate of reactants and products of the polymerization could be estimated by the peak height [15]. The samples were measured before and after the curing by the "FT-IR 2000 System" from Perkin-Elmer.

The broadly used method for the investigation of thermal processes like polymerization is difference scanning calorimetric. The DSC is preferred because of an accurate temperature regulation and good recording of the energy dissipation [12] [16]. The DSC-analysis was done with 7 mg CYCLOTENE 4024-40 in an inert atmosphere using a "Q2000" machine from the TA Instruments Company. The samples were dried and exposed before they were analyzed. The DSC analyzer could perform high heat-up and cool-down rates, up to 300°C/min. Higher heat-up rates leads to more inaccuracy based on a stronger overshoot of the thermal system.

The DSC records the heat flow to a sample during a tempering in comparison to a reference cell. The integration of the heat flow over time gives the heat dissipation of the sample. The polymerization reaction is an exothermic reaction and occurs in DSC plot by a dip/peak of the heat flow curve [14]. The integration of the heat flow peak over time allows the estimation of the heat transformation (enthalpy change) of the reaction.

The DSC experiments for the estimation of the BCB polymerization has been done in isothermal and non-isothermal mode with a constant heat rate. In figure 6 a non-isothermal signal of a BCB DSC sample is shown. The heat of the reaction of 294.4 J/g is calculated by a linear baseline. A more accurate calculation can be done by using a blind curve, which means a second measurement of the fully polymerized sample. The difference between the two curves gives the energy dissipation of the polymerization process. The blind curve also compensates the parasitic endothermic signal which came from heating up the aluminum sample pan [8].

The heat of the reaction could be set in relation to the extent of reaction. The reaction rate $d\alpha/dt$ at any state of the polymerization can be estimated at any stage of the reaction by the following relation:

$$\frac{d\alpha}{dt} = \frac{dH / dt}{\Delta H_{total}} \tag{1}$$

The total heat of reaction ΔH_{total} describes the integral of heat flow from the initiation of reaction up until the full polymerized material form. The constant heat rate β allows for applying the extent of reaction over the heat rate, instead of the extent of reaction over time, which is measured with a higher inaccuracy:

$$\frac{d\alpha}{dT}\beta = \frac{dH / dt}{\Delta H_{total}} \tag{2}$$

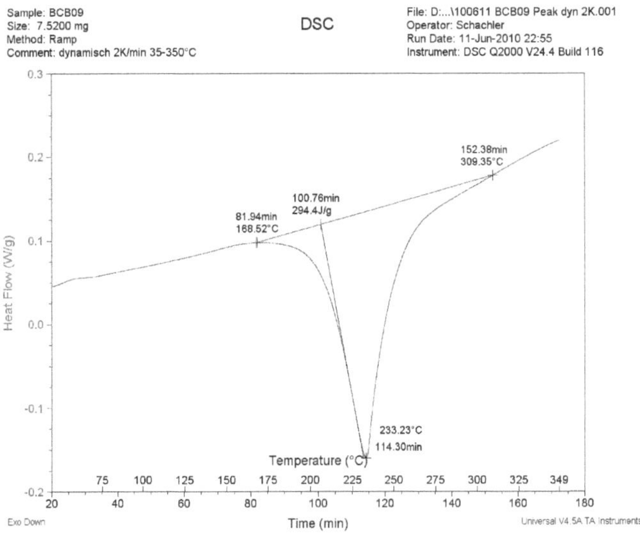

Figure 6. DSC plot of the polymerization of a photo BCB sample during a non-isothermal experiment with a heat rate of 2 K/min

This is the reason why non-isothermal experiments are often used to analyze the reaction. The degree of cure at any state could be estimated by the rate of the heat of reaction ΔH to the total reaction heat ΔH_{total} of a fully cured layer:

$$\alpha = \frac{\Delta H}{\Delta H_{total}} + \alpha_i = \frac{1}{\Delta H_{total}} \int \frac{dH}{dt} dt \tag{3}$$

where α_i is the initial degree of cure of the b-stage resin. The estimation has included an inaccuracy by the calculation of the total heat reaction of the complete polymerization. Linear heating experiments at different heating rates are performed. The ΔH_{total} of the non-isothermal experiments with different heat rates are printed in table 2. The blind curve corrected DSC-curves are shown in figure 7. By increasing the heat rate, the reaction is shifted more and more into higher temperatures. The average total heat of reaction of the fully cured BCB film is about 281.8 ± 5.3 J/g. The results are in good correlation to the measurement of CHAN, who measured a heat of reaction for CYCLOTENE 4024-40 of 291.3 ± 8.8 J/g [17]. The measurement of the heat reaction is strongly dependent on base line settings. The estimation of the total heat reaction has in general an inaccuracy of 3 to 5%. It should be noted that the photosensitive CYCLOTENE will be shipped out with an initial degree of cure of about 47 %. Based on this, the true total heat of the reaction is around 531.7 J/g. The value is confirmed by BAIR et. al. which shows a value of 515 ± 10 J/g for the complete polymerization of a monomer BCB resin [18]. T An increased heat rate leads to more intensive reaction peak but with less time resolution, which increases the inaccuracy for a kinetic model development.

Heat rate [K/min]	Reaction heat ΔH [J/g]	T_{Peak} [°C]
2	289,6	233,35
5	283,6	245,24
10	276,4	255,52
20	281,8	266,00
40	277,4	277,12
Average ΔH_{total}	281,8 ± 5,3	

Table 2. Heat reaction and the peak temperature of photo BCB

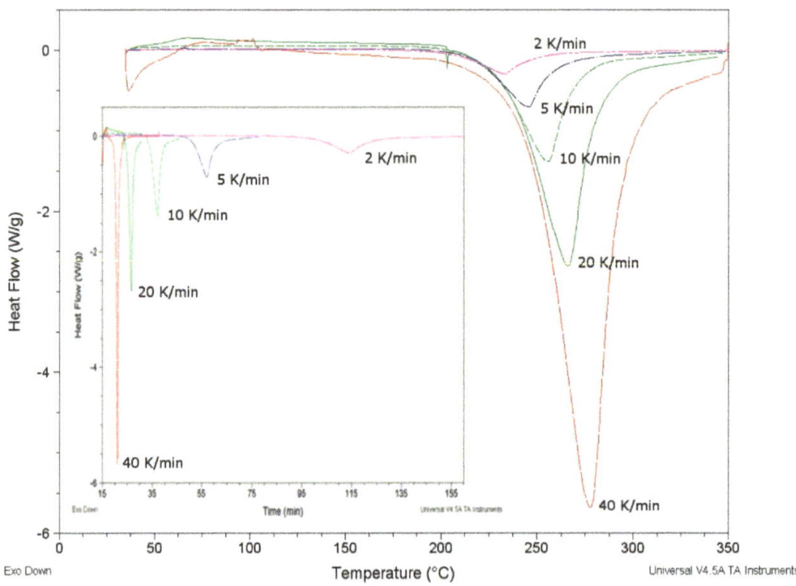

Figure 7. Non-isothermal DSC measurement of CYCLOTENE 4026-46 versus temperature and time

The state change at the glass transition temperature caused more or less a peak in the DSC curve. This peak increases significantly with higher heat rates [19]. The uncured coated BCB has a T_g of around 60°C. In figure 7 the parasitic peak which is caused by the overwhelming of the glass transition temperature (gelation) could be seen for the two heat rates at 20K/min and 40K/min in the temperature range from 60°C up to 100°C.

The vitrification of the polymer is the important factor for the reaction kinetic caused by the gradual decreasing of the rate. When the material returns to a glassy state, the same peak is generated, but the high slope of the reaction signal absorbs the additional peak. The vitrification could only be observed by the gradual change of the reaction rate.

The gradual change is in the case of non-isothermal experiments softened by the permanent increase in the temperature. The heat signal of the state change is calculated into the total heat of reaction.

Based on the knowledge of the T_g (figure 5), the influence of the vitrification could be estimated for the different heat rate experiments. The theoretical development of the T_g for the non-isothermal experiments is plotted in figure 8. The influence of the diffusion controlled part decreases with higher heat rates and has shifted the vitrification to higher temperatures.

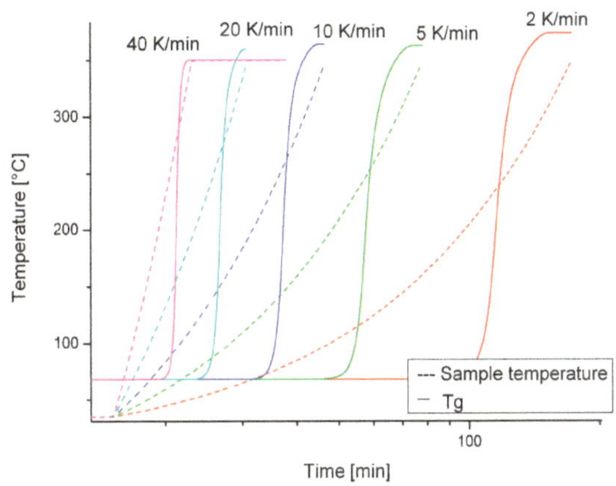

Figure 8. Prediction of the glass transition temperature for the non-isothermal experiments of photo BCB

Low heat rates or isothermal experiments allow to analyze better the influence of the virtification on the reaction rate. An isothermal measurement is based on a heat up phase, a soak temperature and a cool down phase. An isothermal experiment also gives a better prediction for the real cure process in a convection oven. But an isothermal measurement method has some disadvantages in contrast to an experiment with a continuous heat rate. The extent of cure is calculated by the integration of the reaction peak signal to get the reaction heat ΔH_{ds} over the soak temperature. The higher time resolution of the reaction signal is in relation to a much smaller reaction peak. The small peak and the possibility of not a fully polymerized layer at the soak temperature, caused by vitrification or short process time, are critical impacts in estimating the total heat reaction ΔH_{total}. A simple estimation of the total heat reaction can lead to an inaccurate interpretation of cure degree (see Eq.(3)). The reaction already starts during the heat up and cool down phases, which takes into account the integration of heat flow over time. The discontinuity of the temperature function produce parasitic signals. The DSC signal of a BCB soft cure at 210°C at 30 minutes is plotted in figure 9. The heat flow signal shows that there is a parasitic signal peak between isothermal and dynamic heat phases. The reaction start during the heat up phase, which can also be seen. The short process time leads only to a soft cured layer, which means that the measured reaction heat does not equates to the total reaction heat. After the soft cure, the sample was again heated up to 350°C with a constant heat rate. A leftover heat of reaction ΔH_{res} can be measured. The sum of ΔH_{ds} and ΔH_{res} leads to the total heat of reaction ΔH_{total} [12].

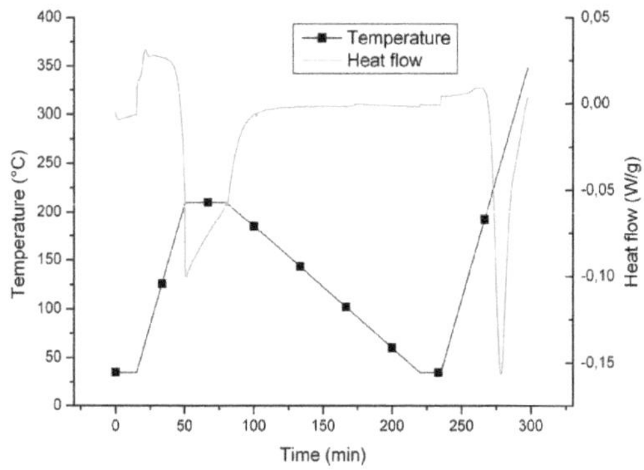

Figure 9. DSC measurement of a BCB soft cure process with additional non-isothermal heat up step to estimate the cure degree.

5. Development of a kinetic model of the polymerization reaction

The extent of cure for a dedicated cure program can be measured by DSC experiments. But the prediction of the extent of cure and the polymerization behavior based on any temperature – time program is necessary for any process optimization with respect to energy consumption, process time or material properties. On account of the DSC experiments, a kinetic model can be set up to characterize the polymerization reaction and can be used to design and optimize the cure process. The reaction is described by a mathematical relation between time, temperature and the extent of cure. In general, special software is required to develop the kinetic equation. The kinetic model in combination with a chemical-rheological model allows a simulation of the complete polymerization reaction [14] [8].

The kinetic models are generally sorted into two types:

A phenomenological model and
A chemical model.

The phenomenological model describes the reaction by a relative simple equation. This model type intentionally ignores details of the reaction with the benefit of simple application. Detailed relations and progression of the different chemical reaction steps and the relationship of the reaction groups will be ignored. The chemical reaction model takes into account the chemical steady state of the partial reactions. These types of models have a much better prediction and interpretation potential, but it is often not useable, especially for complex reactions. In contrast to a phenomenological model, the efforts involved for such a model are much higher, because many more reactions parameters are necessary, and the users need to do a chemical analysis of every reaction step. The phenomenological model can be used without chemical understanding, and can be created only by a comparison of reactants and products. The extent of cure behavior, which has been estimated by DSC experiments, was used to create a phenomenological model for BCB. The simple application makes the phenomenological model the most preferred for polymerization reactions [20] [21].

The two-step polymerization reaction of BCB can be described by a pseudo first-order equation. Based on that simplification, the BCB reaction could described as follows:

$$\frac{d\alpha}{dt} = k_e(T)f(\alpha) \tag{4}$$

Where α is the degree of cure, $k_e(T)$ is the reaction rate constant which depends on the temperature, and $f(\alpha)$ is the reaction model which describes the reaction order [22]. The kinetic equation such as Eq.(4) is a simplified assumption which describes the reaction rate as a product of a temperature-depending function and a fraction-transformed-depending function. There is a range of empirical mathematical forms for the reaction model [23]. Some particularly idealized reaction models are proposed in table 3.

Reaction model	$f(\alpha)$
Power law	$4\alpha^{3/4}$
Power law	$3\alpha^{2/3}$
One-dimensional diffusion	$1/2\alpha^{-1}$
Mampel (first order)	$1-\alpha$
Avrami-Erofeev	$4(1-\alpha)[-\ln(1-\alpha)]^{3/4}$
Three dimensional diffusion	$2(1-\alpha)^{2/3}[-\ln(1-\alpha)^{1/3}]^{-1}$
Contracting sphere	$3(1-\alpha)^{2/3}$

Table 3. A range of reaction models which are often used in kinetic analytics

The reaction rate constant could be realized for the simplest case by the Arrhenius relation:

$$k_e = A\exp\left(\frac{-E_A}{RT}\right) \tag{5}$$

Where R is the universal gas constant, E_A is the activation energy, T is the temperature and A is the reaction rate constant at an infinite temperature [24]. The Arrhenius equation describes a reaction in gas phase. In approximation, the relation can be used for liquid and solid materials. But influence by molecule mobility and vitrification are ignored and could be added by the adaptation of the reaction function.

The determination of the three important kinetic parameters, the two Arrhenius parameter E_A and A and the reaction model $f(\alpha)$ is an interlinking problem. The measurement of one of the parameters, and especially the accurate estimation, influence the other two parameters.

Based on this problem they are often called a kinetic triplet [25]. There are many different methods for the determination based on DSC measurements. The method of KISSINGER has been used for the investigation of the polymerization process of BCB. It allows for a simple application with a small inaccuracy, which is also described in detail by STARINK.

The KISSINGER method is focusing on the determination of activation energy as first parameter with the most accuracy possible. The reaction rate in Eq.(4) is calculated by the Eq.(1) from a DSC experiment. The Eq.(5) is inserted into the Eq.(4). The logarithm of Eq.(4) leads to the following equation:

$$\ln\frac{d\alpha}{dt} = -\frac{E}{RT} - \ln f(\alpha) \tag{6}$$

For non-isothermal experiments with constant heat rate (β=dT/dt) the Eq.(6) could be changed to:

$$\ln\frac{d\alpha}{dT}\beta = -\frac{E}{RT} - \ln f(\alpha)\,. \tag{7}$$

The activation energy could be obtained from the slope of plots of ln(dα/dT) versus 1/T. It is estimated without any knowledge of the reaction model f(α). In literature, the principle is also termed model free kinetic (MFK). The slope in the plot should be constant. If there is a change in the slope, it indicates a phase change in the reaction. The calculation of the reaction rate by Eq.(1) leads to an inaccuracy which could have a huge impact on the kinetic equation caused by the triplet relation. The assumption of the reaction rate can be replaced by correlation of a set of non-isothermal DSC experiments. Based on Kissinger, the Eq.(7) is integrated by separation of variables and the following approach was postulated:

$$\int_0^{\alpha_f}\frac{d\alpha}{f(\alpha_f)} = \frac{A}{\beta}\int_0^{T_{100\%}}\exp(-\frac{E_A}{RT})dT \approx \frac{AE_A\exp(\frac{-E_A}{RT})}{\beta R\cdot\exp(\frac{-E_A}{RT})^{\kappa}} \tag{8}$$

The Eq.(5) have been logarithm and lead to following equation:

$$\ln\frac{\beta}{T_f^{\kappa}} = -\frac{E}{RT} + C \tag{9}$$

C is a constant for different heating rates at a constant degree of polymerization and belongs to the reaction model. T_f means the temperature of experiments with different heat rates where the same extent of cure occurs. The variable κ is a correction factor in the range of 1,92 up to 2. With the Eq.(9) the E_A can determined as the slope of the plot $\ln(\beta/T_f^{\kappa})$ versus $1/T_f$ independent of constant C. Three or more experiments with different heating rate β are necessary to determine the activation energy with MFK. The correct estimation of T_f is the main challenge at this method. For the normal reaction, analyzing the maximum rate method or peak method is used, which is described also in the ASTM E698 norm [26]. The peak method avoids the defective calculation of cure rate by Eq.(1). It was found that the extent of cure is approximately fixed at the maximum reaction rate, but a small error will be introduced. The peak temperatures of the non-isothermal experiment are shown in table 2. The temperature of the peak is shifted with the heat rate. The results of Eq.(9) for the five non-isothermal experiments are plotted in figure 10.

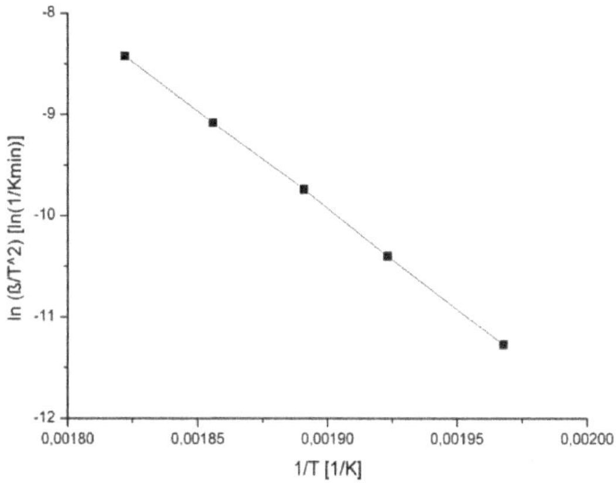

Figure 10. Non-isothermal measurement of BCB. The slope approximate the activation energy of the polymerization reaction cure degree of around 73%

A value of 149.5kJ/mol ±1.75 kJ/mol was measured, which acknowledges the declarations of DIBBS et. al.. Different literature sources describe an activation energy in a range from 146 kJ/mol up to 197,6 kJ/mol [27] [17] [28]. A disadvantage of this method is that the values are only estimated at one degree of cure (around 73 % conversion). Therefore, in relation to the used heat rates, the value describes only the chemically controlled kinetic. With a first-order reaction model, the A has a value of 1,35 h^{-1} for a reference temperature T_0 of 210°C:

$$\frac{d\alpha}{dt} = 1{,}35h^{-1}\exp\left[-\frac{149{,}5kJ/mol}{T}\left(\frac{1}{T}-\frac{1}{T_0}\right)\right](1-\alpha) \tag{10}$$

The plot for the non-isothermal experiments (Figure 11) shows that the model fits the chemically controlled regime very well. The estimated model is in good agreement with the already proposed kinetic model for BCB [2] [6]. There is a poor estimation for the diffusion controlled part, which is also described in previous publications [6] [28]. With the focus on lower cure temperatures, the influence of the diffusion controlled part increases.

The polymerization analysis aims for a better understanding of the influence of the vitrification. That demands a modulation of the normal kinetic model Eq.(10). There are many different methods for correcting the conventional reaction model by a conversion-dependent diffusion contribution. One of the most popular methods for thermosetting systems is to modify the kinetic equation by a "diffusion control function" which is described by SCHAWE [29].

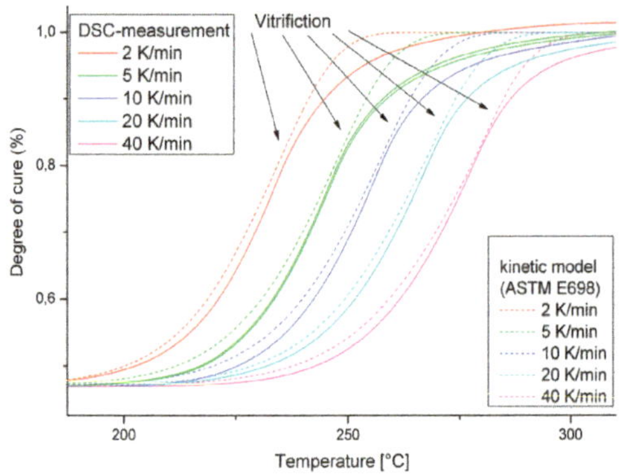

Figure 11. Extent of polymerization based on the non-isothermal experiments and calculation based on Eq.(10)

$$\frac{d\alpha}{dt} = k_e(T) f(\alpha) f_d(\alpha;T) \tag{11}$$

The diffusion control function works in an electro-technical analogy like a low-pass filter. During the polymerization up to the change in the solid state, the function should not influence the reaction which means $f_d(T<T_g)=1$. After the vitrification, the function decreases down to zero. A function which achieves this condition is proposed by SCHAWE:

$$f_d(\alpha;T) = f_d(T_g(\alpha)) = 1 - \left(1 + \left(\frac{T_0 - T_g(\alpha)}{D}\right)^3\right)^{-1} \tag{12}$$

where T_0 is the temperature at which the diffusion control function is zero and A is an empirical constant to smooth the function. The function is focused for a reaction which is nearly stopped in a glassy state. The difficulty of this method is to estimate the empirical diffusion control function which leads to the best fit. Another problem is the influence of the diffusion control function for the chemically controlled part.

MENCZEL describe an alternative model from RABINOWITCH for small-molecule reactions, where the rate constant k is based on a sum of a chemically controlled and diffusion controlled rate constant [8]:

$$\frac{1}{k_e(T,\alpha)} = \frac{1}{k_r(T)} + \frac{1}{k_d(T,\alpha)} \tag{13}$$

where k_e is the overall rate constant, k_r is the Arrhenius ratio constant for the chemical reaction and k_d is the diffusion rate constant. The overall reaction rate constant k_e is governed prior the vitrification by the Arrhenius rate constant, and after vitrification is dominated by the diffusion rate constant. The disadvantage is an extended interlink problem of the parameters.

A combination of both models from SCHAWE and RABINOWITCH has been used to create an advanced model for the polymerization reaction of CYCLOTENE. Two different rate constants (k_r;k_d) were calculated. The overall reaction rate constant is the sum of both, but the terms are combined with a step-function, which works like a switch and allows the change between gel and glassy state kinetic models.

$$\frac{d\alpha}{dt} = k_r(T) f(\alpha) f_{step}(\alpha;T) + k_d(T) f(\alpha)(1 - f_{step}(\alpha;T)) \tag{14}$$

The step-function belongs to the process temperature and the glass transition temperature and gets a value between one and zero. For both reaction regimes, a first-order reaction model was used based on a good fit, but the model allows for the declaration of two different reaction models for chemically controlled and diffusion controlled states. The following expression was used for the step-function:

$$f_{step}(\alpha;T) = \frac{1}{1 + \exp(-(T - T_g(\alpha))B)} \tag{15}$$

where B is a parameter to smooth the changeover between the two reaction functions. The value was set at 0.1 to get a change in temperature range of ±5°C around the glass transition temperature. The glass transition temperature depends on the extent of cure (see figure 5), and the following relation was estimated based on the measurements of TOEPPER:

$$T_g = \frac{35}{0.054}\alpha^2 - \frac{10865}{27}\alpha + \frac{3070}{27} \tag{16}$$

For the kinetic model, two Arrhenius terms need to be calculated, one for the chemically controlled part and one for the diffusion controlled part. The Eq.(9) allows the calculation of the activation energy for different extents of reaction. In contrast to the peak method, the extent of reaction need to be calculated by Eq.(3). The consideration of the vitrification-heat rate relation for the estimation of the activation energy of the two regimes is necessary to see any difference. For example, the highest heat rate of 40 K/min leads to a vitrification above 90% in contrast to the slowest heat rate of 2 K/min, which vitrificates around 80%. Therefore the heat rate of 2 K/min is used to calculate the activation energy for the chemically controlled state up to 80%, above this value it is used for the calculation of the activation energy of the diffusion controlled state. The calculated activation energies for different extents of reaction are shown in figure 12.

The change in the reaction kinetic is marked by the change in the activation energy. The activation energy for the chemically controlled region has an approximate value of 157.1 ± 3.6 kJ/mol. The activation energy is slightly higher than the value which was estimated by the peak method /ASTM E698 norm (pointed line). For the diffusion controlled kinetic, a higher activation energy of around 166.1 ± 8.1 kJ/mol was measured. The increase in the activation energy in a diffusion controlled regime shows a higher temperature dependency for the reaction, which could be explained by the more temperature-dependent molecule mobility.

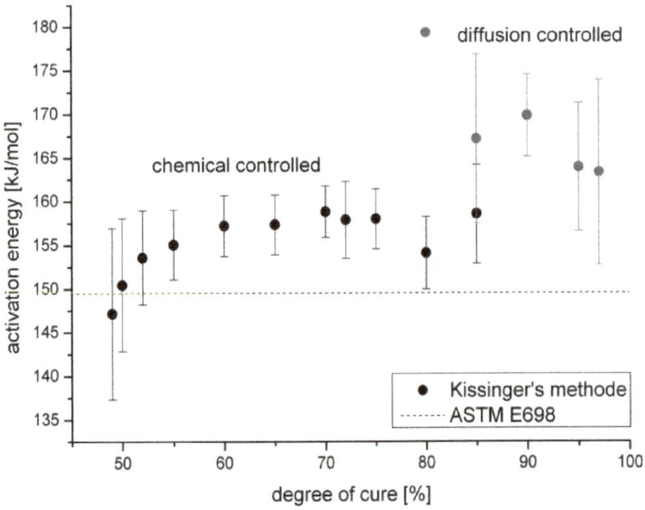

Figure 12. Activation energy estimated by the ASTM E698 norm and dependent of the degree of cure

The results for the reaction rate at the reference temperature are 1.41 hr^{-1} for the chemically controlled regime, and 0.27 h^{-1} for the diffusion controlled regime. In figure 13, the calculated degree of polymerization for the conventional (Eq.(10)) and modified kinetic model (Eq.14) is compared to a measured curve for a cure at 190°C. The modified model shows a much better fit in comparision with the real behavior above a degree of cure of approximately 80 %.

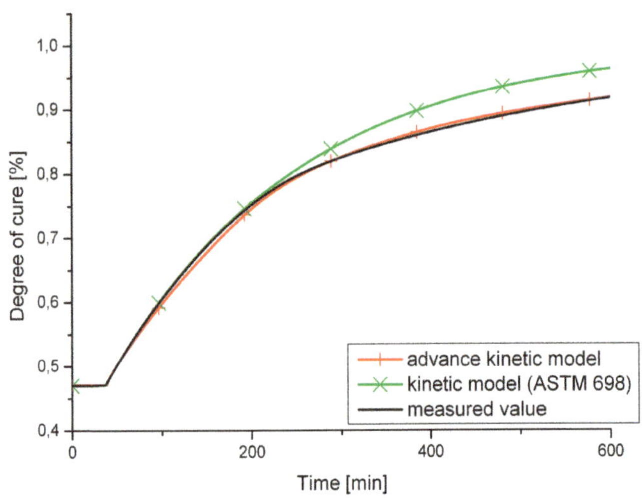

Figure 13. Degree of cure of a photo BCB sample at 190°C

6. Benefits of the kinetic model for the BCB polymerization and the relation for the polymer properties

The investigation of the cure behavior of BCB shows that an ongoing polymerization reaction in solid state is also possible with the requirement of a minimal process temperature above 150°C caused by the thermal activation. The enhanced model which is described in Eq.(14) allows a much better prediction for the polymerization of BCB in glassy state. The polymerization in vitrificated state is relative slow and should only use if the device is temperature critical. But the cure in glassy state could be interesting with the focus on the properties of the polymer. The stress development of BCB during the cure at 250°C is plotted in figure 14. The thermal stress of a polymer film on a substrate could be calculated as followed:

$$\sigma_{th} = \frac{E_f}{1-\upsilon_f}(\alpha_f - \alpha_s)\Delta T \tag{17}$$

Where E_f and υ_f are the Young's modulus and Poisson ratio of polymer. For silicon as substrate CTE α_s are approximately 2.6 μm/mK, and BCB has a CTE of 42 μm/mK. Delta T is the temperature difference between ambient temperature and the vitrificated temperature. The polymer is in gel phase during the heat up phase and the stress is nearly zero. When the polymer vitrificates, the stress state at that point is set to zero. During the cool down, the thermal tensile stress in the layer increases based on the coefficient of thermal expansion mismatch to the substrate [8] [30] [31]. A nearly fully cured BCB has a T_g near the decomposition temperature, which means the zero stress point could not reset by processing at a higher temperature. A reheating of the BCB after the cure at 250°C above the temperature where the polymer vitrificate leads to a compressive stress, which could be seen in figure 15. The stress measurements were done by Toho Technology with a Flexus-2320-S system. In comparing the stress values at room temperature between the measurement of figure 14 and figure 15 a slightly difference occurs. This could be explained by a visco elastic behavior of the BCB which is discussed elsewhere [32].

There are trends in microelectronics towards enlarging the wafer size in combination with thinner wafers to decrease costs. In contrast to this, the signal frequency gets higher and higher, which means packaging thick dielectric polymer layers to assure good signal integrity. It results in two major challenges. The first is in handling, because the stress leads to a bow of the wafer, which is a problem for automatic handling tools and also for the processes because of the topography. The second point is that the stress in the layer stack is a continual problem for reliability. It force cracks through the layers and delamination. The decrease of the stress in the layer stack becomes more and more a focus in the future. This could also be seen by the polymer supplier who presented a couple of new polymers with low temperature cure properties in the last few years.

The stress in the polymer layer can be minimized by the optimization of the cure process with the help of a kinetic model. There are already some works which try to decrease the stress by cure process optimization [33] [34], but they unaccounted the meaning of the vitrification for the stress formation. A set of BCB experiments was done to examine the

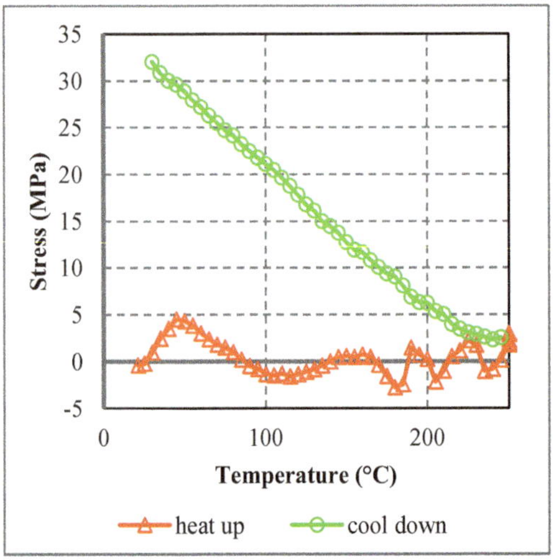

Figure 14. The stress in the material during a cure at 250°C

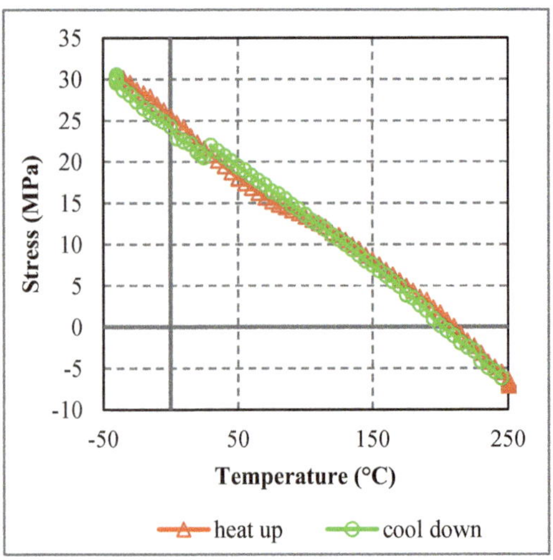

Figure 15. Stress curve of a cured BCB layer on a silicon substrate measured with Flexus-2320-S

potential of stress decreasing by process optimization. In relation to the lowest temperature where vitrification occurs, and a nearly fully cured layer is obtainable, a theoretical minimal stress of around 18 MPa should be achievable. There has been shown a possible decreasing of the stress in a BCB layer from 28 MPa based on a normal cure at 250°C down to a stress value of 19 MPa [32]. Nevertheless, the process time will increase exponentially with a decrease of the temperature, and for higher volumes a decrease of the temperature down to 150°C is not economically feasible. The kinetic model allows for the creation of curing processes with more than one soak temperature, with the benefit that the processing temperature before and after the vitrification is higher to decrease the processing time, and the temperature is set lower in the phase change to a solid state. The cure of the polymer is generally performed in convection ovens. The slow temperature change in a convection oven limits the difference between the high and low temperature levels. A programmable control of the oven makes a dynamic heat process possible, which reaches the shortest possible time in combination with low stress in the polymer layer. Such a theoretical temperature profile is plotted in figure 16.

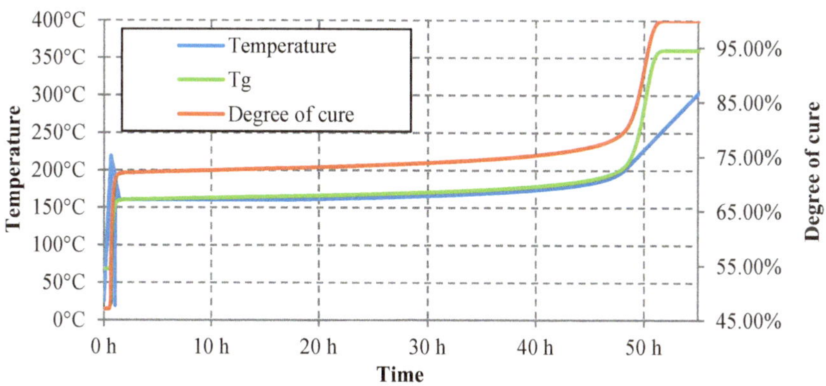

Figure 16. Calculated process profile with a focus on a vitrification at low temperature

The oven will heat up to a high peak temperature to achieve fast polymerization. After that the temperature is decreased to reach a vitrification at low temperature. The temperature will be ramped up after the vitrification, but the process temperature is controlled to be below the glass transition temperature, to decrease the process time and avoid devitrification. The described program aims for a vitrification at low temperature in relation to the stress in the polymer layer. DIBBS described a seven days process, which also aims to get a vitrification at a low temperature and slowly increase the temperature after that [2]. The process involves a long time processing at 150°C, and after that a slow ramp up to 250°C. The expected cure profile of DIBBS process is plotted in figure 17. The long heat up phase of 6 days results due the lack of a detailed kinetic model for the diffusion controlled part.

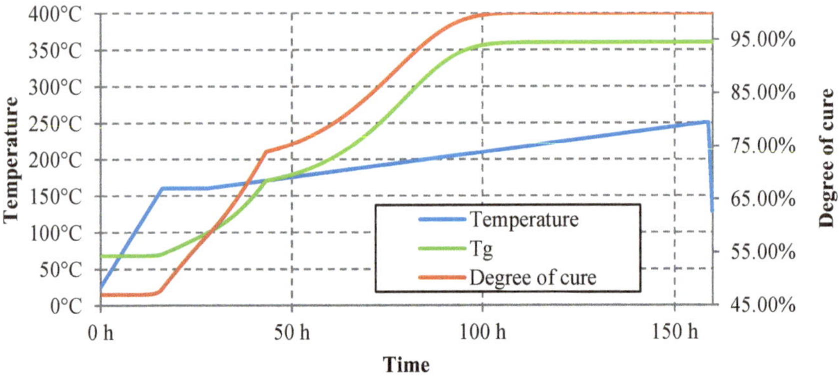

Figure 17. Calculated process profile of the low stress cure of DIBBS

He realized only a small decrease of the stress down to 24 MPa. The results of DIBBS and own experiments lead to the assumption that the ongoing of the polymerization in glassy state also influences the stress. A BCB layer on substrate was soft cured at 210°C up to 70%, which has a T_g of around 190°C. After soft cure, the layer was stored at 150°C for 100 hours and reached 93% polymerization. The stress in the layer decreased during the storage from 24 MPa down to 19 MPa [32]. A layer which was fully cured at 250°C shows no strong relaxation after the storage. This could be explained by two things. First, the partly cured film has a higher viscoplastic behavior and the material relaxed easier. Another possible reason is stress relaxation caused by the ongoing polymerization at low temperature. Further ongoing experiments are necessary to clarify in more detail the stress behavior in solid state BCB layers.

The polymer dielectric layer is an essential component in packaging technology. The temperature-activated polymerization processing of polymers has increasingly become a key parameter for the process workflow. The investigation of polymerization processes allows the optimization of the cure process down to lower temperatures. Based on the example of BCB, a simple model was developed, which also described the polymerization reaction of BCB in a solid state. The model allows for significantly decreasing the stress in the layer and also decreasing down to very low cure temperatures. The mechanical properties like Young's modulus or tensile strength of BCB will not be affected by the processing parameter changes [32]. In combination with the low-temperature curing possibilities of BCB, a decrease in the stress of about approximately 30% is possible.

Author details

Markus Woehrmann
Technische Universität Berlin (TUB), Berlin, Germany

Michael Toepper
Fraunhofer Institute for Reliability and Microintegration (IZM), Berlin, Germany

Acknowledgement

The authors want to emphasize the contributions from the colleagues at Fraunhofer IZM who are not mentioned here by name. Financial support from the Deutsche Forschungsgemeinschaft (International Research Training Group GRK 1215, "Materials and Concepts for Advanced Interconnects") is gratefully acknowledged.

7. References

[1] M. Toepper, T. Fischer, T. Baumgartner und H. Reichl, „A comparison of thin film polymers for Wafer Level Packaging," in *Electronic Components and Technology Conference (ECTC), Proceedings 60th*, Las Vegas, NV, USA, 2010.

[2] M. Dibbs, P. Townsend, T. Stokich, B. Huber, C. Mohler, R. Heistand, P. Garrou, G. Adema, M. Berry und I. Turlik, „Cure Management of Benzocyclobutene Dielectric for Electronic Applications," in *International SAMPE Electronic Materials and Processes Conference, Proceedings 6th* , Baltimore, MD, USA, 1992.

[3] N. Ghalichechian, A. Modafe, R. Ghodssi, P. Lazzeri, V. Micheli und M. Anderle, „Integration of benzocyclobutene polymers and silicon micromachined structures using anisotropic wet etching," *J. Vac. Sci. Technol. B*, Bd. 22, pp. 2439-2447, 2004.

[4] R. L. Hubbard und R. S. Garard, „Wafer Level Curing of Polymer Dielectrics with Variable Frequency Microwaves," in *International Conference on Advanced Thermal Processing of Semiconductors RTP, Proc. 15th* , Cannizzaro, Catania, Italy, 2007.

[5] G. Roelkens, J. Brouckaert, D. V. Thourhout, R. Baets und R. N. and, „Adhesive Bonding of InP/InGaAsP Dies to Processed Silicon-On-Insulator Wafers using DVS-bis-Benzocyclobutene," *Journal of The Electrochemical Society*, Bd. 153, pp. 1015-1019, 2006.

[6] M. Toepper, Entwicklung einer auf Photo-BCB basierenden Technologie für das Waferlevel Packaging, Phd thesis, Technische Universität Berlin, 2004.

[7] M. E. De, M. Moursb und H. H. Winter, „The Gel Point as Reference State: A Simple Kinetic Model for Crosslinking Polybutadiene via Hydrosilation," *Polymer Gels and Networks*, Bd. 5, pp. 69-94, 1997.

[8] J. D. Menczel und B. Prime, Thermal analysis of polymers: fundamentals and applications, John Wiley and Sons, 2009.

[9] W. R. Roth, M. Biermann, H. Dekker, R. Jochems, C. Mosselman und H. Hermann, *Das Energieprofil des ortho-Chinodimethan-Benzocyclobuten-Gleichgewichtes,* Chem. Bereichte 111, pp. 3892-3903, 1978.

[10] W. R. Roth, T. Ebbrecht und A. Beitat, Chinodimethan-Benzocyclobuten-Gleichgewicht; Sauerstoffabfangtechnik, Chem. Berichte 121, pp. 1357-1358, 1988.

[11] J. Yang, *Processing Procedures for CYCLOTENE 3000 Series Dry Etch Resins,* DOW Chemical Co., 2005.

[12] H. Schindlbauer, G. Henkel, J. Weiss und W. Eichberger, „Quantitative Untersuchungen zum Aushärtungsverhalten von Phenoplasten mittels differentialthermoanalytischer Messungen (DSC)," *Die Angewandte. Makromolekulare Chemie,* Bd. 49, Nr. 725, pp. 115-128, 1976.

[13] H. E. Kissinger, „Reaction Kinetics in Differential Thermal Analysis," *Analytical Chemistry,* Bd. 29, Nr. 11, pp. 1702-1706, Nov. 1957.

[14] G. Widmann und R. Riesen, Thermoanalyse, Hüthig, 1990.

[15] T. Stokich, W. M. Lee und R. A. Peters, „Real-Time FT-IR studies of the reaction kinetics for the polymerization of Divinyl Siloxane bis-Benzocyclobutene monomers," in *MRS Proceedings, 227,* 1991.

[16] J. Dupuy, E. Leroy und A. Maazouz, „Determination of Activation Energy and Preexponential Factor of Thermoset Reaction Kinetics Using Differential Scanning Calorimetry in Scanning Mode: Influence of Baseline Shape on Different Calculation Methods," *Journal of Applied Polymer Science,* Bd. 78, pp. 2262-2271, 2000.

[17] K. Chan, M. Teo und Z. Zhong, „Characterization of low-k benzocyclobutene dielectric thin film," *Microelectronics International,* Bd. 20, pp. 11-22, 2003.

[18] H. E. Bair und C. A. Pryde, „Curing and glass transition behavior of a benzocyclobutene polymer," in *Society Plastics Engineers Proceedings,* Montreal, Quebec, Canada, 1991.

[19] R. B. Prime, „An Introduction to thermosets," 2005.

[20] V. M. Gonzales-Romero und N. Casillas, „Isothermal and Temperature Programmed Kinetic Studies of Thermosets," *Polymer Engineering and Science,* Bd. 29, Nr. 5, pp. 295-301, 1989.

[21] A. Yousefi, P. G. Lafleur und R. Gauvin, „Kinetic studies of thermoset cure reactions," *Polymer Composites,* Bd. 18, Nr. 2, pp. 157-168, 1997.

[22] S. White und H. Hahn, „Process Modeling of Composite Materials: Residual Stress Development during Cure. Part I Model Formulation," *Journal of Composite Materials,* Bd. 26, pp. 2402-2422, 1992.

[23] S. Vyazovkin, „Modell-free kinetics," *Journal of Thermal Analysis and Calorimetry,* Bd. 83, pp. 41-51, 2006.

[24] K. J. Laidler, Chemical Kinetics, Harper & Row, 1987.

[25] M. J. Starink, „The determination of activation energy from linear heating rate experiments: a comparison of the accuracy of isoconversion methods," *Thermochimica Acta,* Bd. 404, pp. 163-176, 2003.

[26] B. Battenstein und H. Ritter, „Composite Membranen. Einfluß der Glastemperatur und der Zusammensetzung von Acrylsaure-Methylmethacrylat Copolymeren auf das Permeationsverhalten von Decan," *Makromol. Chem.,* Bd. 7, pp. 767-770, 1986.

[27] J. Kawahara, A. Nakano, K. Kinoshita, Y. Harada, M. Tagami, M. Tada und Y. Hayashi, „Highly thermal-stable, plasma-polymerized BCB polymer film," *Plasma Sources Sci. Technol.*, Bd. 12, pp. 80-88, 2003.

[28] P. E. Garrou, R. H. Heistand, M. G. Dibbs, T. A. Manial, C. E. Mohler, T. M. Stokich, P. H. Townsend, G. M. Adema, M. J. Berry und I. Turlik, „Rapid thermal curing of BCB dielectric," *IEEE Transactions on Components, Hybrids, and Manufacturing Technology*, Bd. 16, Nr. 1, pp. 46-52, 1993.

[29] J. E. K. Schawe, „A description of chemical and diffusion control in isothermal kinetics of cure kinetics," *Thermochimica Acta 388*, pp. 299-312, 2002.

[30] T. Hodge, N. Grove, B. Sinno, S. Bidstrup-Allen und P. Kohl, „Thermo-Mechanical Stresses of Spin-Cast Polymer Films on Silicon Substrates," *The International Journal of Microcircuits and Electronic Packaging*, Bd. 20, Nr. 1, pp. 12-20, 1997.

[31] A. M. Bowles, „Stress Evolution in Thin Film Polymer," Phd thesis, Harvard University, 2006.

[32] M. Woehrmann, M. Toepper, H. Walter und K.-D. Land, „Low temperature cure of BCB and the influence on the mechanical stress," in *Electronic Components and Technology Conference (ECTC), Proceedings 61th*, Orlando, FL, USA, 2011.

[33] A. K. Gopal, S. Adali und V. E. Verijenko, „Optimal temperature profiles for minimum residual stress in the cure process of polymer composites," *Composite Structures*, Bd. 48, pp. 99-106, 2000.

[34] S. White und H. Hahn, „Cure cycle optimization for the Reduction of Processing-Induced Residual Stresses in Composite Materials," *Journal of Composite Materials*, Bd. 27, pp. 1352-1378, 1993.

[35] S. Vyazovkin und W. Linert, „False isokinetic relationships found in the nonisothermal decomposition of solids," *Chemical Physics*, Bd. 193, pp. 109-118, 1995.

[36] S. Vyazovkin, „An approach to the solution of the inverse kinetic problem in the case of complex processes. Part 4. Chemical reaction complicated by diffusion," *Thermochimica Acta*, Bd. 223, pp. 201-206, 1993.

[37] S. Vyazovkin und A. Lesnikovich, „An approach to the solution of the inverse kinetic problem in the case of complex processes Part 1. Methods employing a series of thermalytical curves," *Thermochimica Acta*, Bd. 165, pp. 273-280, 1990.

[38] T. Ozawa, „Kinetic analysis of derivative curves in thermal analysis," *Journal of Thermal Analysis*, Bd. 2, pp. 301-324, 1970.

[39] K. Nawa, S. Ueda und H. Watanabe, „Advanced COPNA-resin as a low temperature curing resin for high-density electronic packages," *IEEE Transactions on Components, Packaging, and Manufacturing Technology, Part B: Advanced Packaging*, Bd. 19, Nr. 3, pp. 585-592, 1996.

[40] R. A. Kirchhoff, C. J. Carriere, K. J. Bruza, N. G. Rondan und R. L. Sammler, „Benzocyclobutenes: a new class of high performance polymers," *J. Macromol. Sci. - chem.*, Bd. A28, pp. 1079-1113, 1991.

[41] S. Hogmark, S. Jacobson und M. Larsson, „Design and evaluation of tribological coatings," *Wear,* Bd. 246, pp. 20-33, 2000.

[42] D. Dollimore, „The application of thermal analysis in studying the thermal decomposition of solids," *Therrnochimica Acta,* Bd. 203, pp. 7-23, 1992.

[43] M. Wagner, G. Czerlinskil und M. Pring, „The application of fitting methods to chemical kinetics," *Compur. Biol Med. Pergamon Press 1975. Vol 5. pp 105-117.,* Bd. 5, pp. 105-117, 1975.

[44] A. A. Skordos und I. K. Pazitridge, „Cure Kinetics Modeling of Epoxy Resins Using a Non-Pa rametric Nu me rica I Procedure," *Polymer Engineering and Science,* Bd. 41, Nr. 5, pp. 793-805, May 2001.

[45] S. Vyazovkin und C. A. Wight, „Model-free and model-fitting approaches to kinetic analysis of isothermal and nonisothermal data," *Thermochimica Acta,* Bd. 340, pp. 53-68, 1999.

Chapter 3

Photopolymers for Use as Holographic Media

Michael R. Gleeson, Jinxin Guo and John T. Sheridan

Additional information is available at the end of the chapter

1. Introduction

Photopolymers were first introduced as a holographic recording material by Close *et al.* in 1969, [1]. Since then numerous systems have been examined, but only a small number have become commercially available, [2]. Polymer materials have several advantages. Because thick layers can be fabricated they act as true volume materials giving high diffraction efficiency and good angular selectivity. Most of the materials are self–developing or require only some simple post–processing, such as an exposure to light or heat treatment. This eliminates the need for wet chemical development, which makes photopolymers suitable for applications such as holographic embedded photopolymer waveguides, [3-6] and holographic data storage [7-18].

Photopolymers generally consist of a monomer, a photosensitive dye and an initiator. They can either be liquid or dry layer systems. The dry photopolymers usually contain a polymeric binder in addition to the other components. As mentioned above the first photopolymer material used for holographic recording, was reported by Close *et al.* [1]. This liquid state material consisted of a mixture of acrylamide and metal acrylate monomers and a photocatalyst methylene blue. Sadlej and Smolinska [19] improved the original system proposed by Close by including a poly-vinylalcohol (PVA) binder which allowed the production of dry photopolymer layers.

In the eighties, Calixto [20] continued the work on acrylamide-based systems. The material contained acrylamide monomer, TEA as an electron donor, methylene blue photosensitizer and PVA as a binder. Blaya *et al.*, [21], improved the sensitivity of the acrylamide material for recording at 633 nm by changing the crosslinker, *N,N*-dihydroethylenebisacrylamide. A hybrid material containing acrylamide and acrylic acid as monomers was proposed by Zhao *et al.*, [22]. The material uses methylene blue as the photosensitizer, TEA and p-toluenesulfonic acid as sensitizers and gelatin as a binder.

While acrylamide based free-radical systems have received much attention in the literature other materials offer advantages when it comes to the development of practical storage media. Trentler *et al.*, [23] developed an epoxy resin photopolymer material with a solid matrix, which is formed in situ as the epoxy cures at room temperature. The unreacted vinyl monomers within the material are subsequently photo-polymerized during hologram recording. One of the key features of this type of material is the separation of the epoxy and vinyl polymerizations. This separation allows for a large index contrast to be developed in holograms when components are optimized. This material is functional in thick formats (several millimetres), which enables narrow angular bandwidth and high diffraction efficiency. A dynamic range (M/#) up to 13 has been measured in these materials.

More recently, works have been carried out on the development of a commercial grade acrylate photopolymer with the same two step polymerisation [24], offering very fast response of the refractive index modulation with respect to the recording dosage, particularly at lower power densities, very high storage resolution, and very large refractive index modulation [25,26].

Extensive work has been carried out, in both industry and academia, on the development of these photopolymer materials and more recently on the understanding of the photochemical kinetics associated with them. In order to maximise the potential of these materials for various applications, the necessity for a physically comprehensive theoretical model of the effects which occur during photo-polymerization is becoming ever more important. Providing such a model will enable potential trends in a material's performance to be recognized and optimised, [27]. Such models allow simulations of ratios of various key material components to be made, yielding indications of the most suitable material compositions in order to improve material performance.

In this chapter we examine some recently published results on the Non-local Photo-polymerization Driven Diffusion (NPDD) model, [26-32]. This model provides a comprehensive theoretical representation of the processes, which occur during free radical photo-polymerization. The physically realistic model enables predictions to be made about a number of very different photopolymer materials [26-32]. We present several extensions to the previous model in particular allowing for spatially and temporally varying primary radical generation, oxygen inhibition, dark reactions and chain transfer effects. We then apply this model to analyse a number of effects observed to take place during holographic grating formation in an acrylamide/polyvinylalcohol (AA/PVA) based photopolymer and compare experimental results and the predictions of the model with the aim of characterising these effects.

The chapter is structured as follows: In Section 2 we will briefly examine some of the methods used to measure and monitor the optical performance of holographic gratings in photopolymers. In Section 3 we briefly describe the fundamental photo-kinetic processes, which occur in photopolymers during holographic exposure. Then based on these photo-kinetic processes we construct a set of first order coupled differential equations which represent the temporal and spatial variation of the concentrations of constituents of the photopolymer, which form the basis of the Non-local Photo-polymerization Driven Diffusion (NPDD) model. We then analyse some of the simulated behaviour of this model under given physical conditions to examine its physicality and present some of the predictions made which offer potential methods to improve a photopolymers holographic performance. Following these predictions, in Section 4 we then attempt to improve an acrylamide/polyvinylalcohol (AA/PVA) based photopolymer's spatial frequency response which will increase its high density storage resolution. In Section 5 a brief conclusion is presented offering potential direction for future advancements in the area of photopolymer development.

2. Optical testing of photopolymers for holography

In the study of holographic recording materials it is common to record gratings in photosensitive materials, such as photopolymers, and to then optically examine the resulting gratings. The gratings produced are often electromagnetically modelled using Kogelnik's two-wave coupled wave theory. Kogelnik's two-wave coupled wave theory, [35], describes the efficiency with which thick holograms can diffract incident light. Analytic expressions for both the angular and wavelength dependence of the diffraction efficiency as the incident light deviates from the Bragg condition are derived. Thus the dependence of the diffraction efficiency, $\eta(t)$, on a number of grating parameters is known. For a lossless, unslanted transmission geometry grating, replayed on-Bragg with TE polarized probe light, it is shown, [35], that $\eta(t)$ is described by the following equation:

$$\eta(t) = \frac{I_D(t)}{I_{in}} = \sin^2\left[\frac{\pi n_1(t) d}{\lambda \cos\theta}\right], \tag{1}$$

where I_{in} and $I_D(t)$ are the incident and diffracted probe beam intensities respectively, d represents grating thickness, θ and λ are the Bragg angle and wavelength of incident probe beam inside the grating, and $n_1(t)$ is the refractive index modulation. In deriving Eq (1) all boundary reflections have been neglected.

Rearranging Eq (1) enables an expression for the temporally varying refractive index modulation, $n_1(t)$, to be obtained,

$$n_1(t) = \frac{\lambda \cos\theta}{\pi d} \sin^{-1}\left[\sqrt{\eta(t)}\right]. \tag{2}$$

By monitoring the amount of light diffracted from a weak probe beam during exposure, $I_D(t)$, growth curves of refractive index modulation against exposure time can be extracted. In the majority of the literature, such growth curves are used to monitor grating formation. A typical experimental set-up used to record and monitor these growth curves is presented in Figure 1.

During the grating recording process the evolution of the grating is monitored in real time. One of the main advantages of many photopolymer materials is that they are self-processing and thus, non-latent, [36], therefore the diffractive scattering properties are immediately available as the grating is being formed. This allows the evolution of the grating to be monitored by replaying the grating as it is being recorded using a probe laser with a wavelength, which lies outside the absorption spectrum of the photosensitizer used. This ensures that the probing does not affect the fabrication process. In the set-up presented in Figure 1 the probe laser is operating at λ = 475 nm, i.e. outside the absorption spectrum of the photosensitizer.

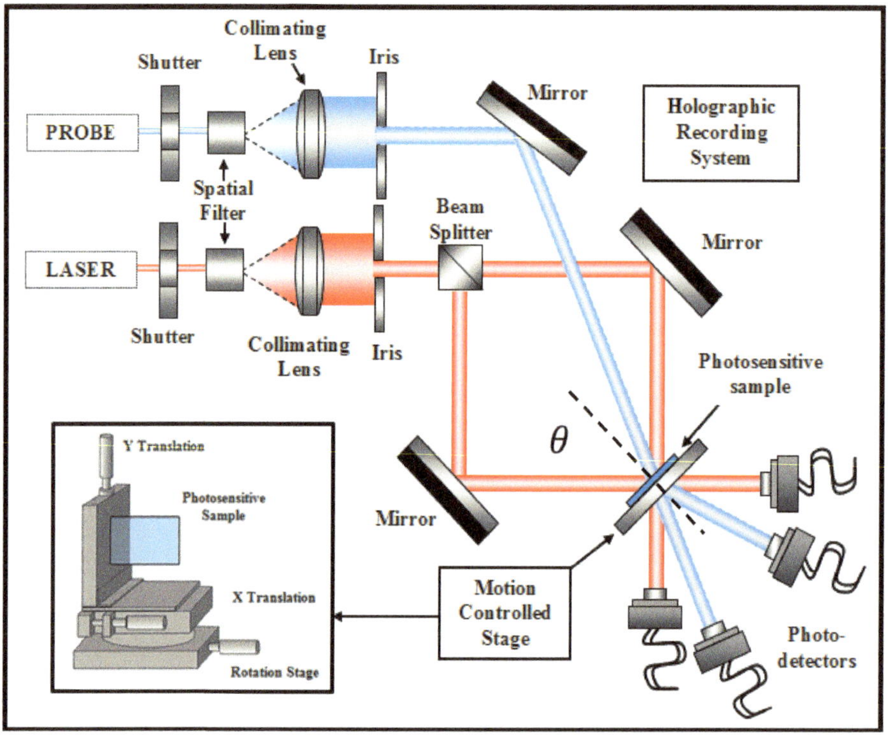

Figure 1. Typical experimental set-up used to record unlsanted volume transmission holographic gratings with a recording wavelength of λ = 633 nm.

Examining Eq (1) it is clear that the on-Bragg replay angle of the probe beam, θ_r, will be different from the recording angle, θ, due to the variation in the wavelength used, for example λ_{Record} = 633 nm, λ_{Replay} = 475 nm. Thus recreation of the object wave, as illustrated in Figure 1, occurs at a different angle due to the change in wavelength. The intensity of the resulting probe diffracted beam, $I_D(t)$, and therefore the diffraction efficiency, will depend on the strength of the grating, i.e. the refractive index modulation, $n_1(t)$. Thus, it is possible to monitor the grating formation (growth curve) by recording the intensity of the diffracted beam as suggested by Eq (1) and Eq (2).

In the above we have emphasised the recording of a single unslanted gratings and the capture of growth curves. However, the photosensitive sample can also be mounted on a rotation and translation stage (as shown in the inset of Figure 1), enabling the analysis of the angular response of the grating and the recording of grating arrays, [37]. In Holographic Data Storage (HDS) where the storage of multiple pages of information is required, these holograms (gratings) are angularly multiplexed on top of each other within the same volume of the holographic medium. It is the ability to achieve this which makes photopolymer materails an attractive media for optical storage.

It must also be noted that using this optical setup, the temporal variation in the absorbance of the photo-sensitiser can be examined. As the recording beams which are used to record the grating, transmit through the photopolymer sample, they can be collected in the photo-detectors shown in Figure 1. Thus enabling key material parameters related to the photo-absorption kinetics and sensitivity of the photopolymer, to be examined.

In the next section we will examine the photo-kinetic and photo-physical behaviour of these photopolymer materials. It is these reactions which are the basis of the theoretical models which are used as photopolymer material optimisation tools and are the fundamental building blocks of the Non-local Photo-polymerisation Driven Diffusion (NPDD) model [11,26-34].

3. Photo-kinetic behaviour

Let us begin this section with a review of the kinetic models of photopolymerisation which have been presented in the literature.

3.1. Review of kinetic models

The photochemical processes, which are present during photopolymerization, are complex [11,26-34,38-43], however an understanding of these processes is of utmost importance if a practical model is to be developed. In a recent review, [44] many of the assumptions made in developing photochemical models of free radical photo-polymerisation were discussed, [38-43]. A number of physical effects not included in the current models were listed, which indicated a lack of physicality under certain exposure conditions. Following the appearance of this review, a series of papers were published [30-32] which addressed many of these issues and provided a model containing a consistent set of chemical reaction equations to take into account many of these effects. These effects included;

i. Removal of the steady state approximation for macroradical concentration,
ii. inclusion of spatially and temporally non-local polymer chain growth,
iii. inclusion of time varying photon absorption,
iv. simultaneously including the effects of both primary, i.e. $R^\bullet$ - $M^\bullet$, and bimolecular, i.e. $M^\bullet$ - $M^\bullet$, termination,
v. inclusion of the changes in the polymerization kinetic constants caused by increased viscosity, and finally
vi. inclusion of polymerization inhibiting effects.

The resulting Non-local Photo-polymerisation Driven Diffusion (NPDD) model was then experimentally verified by applying it to study (a) normalised transmission curves, and (b) growth curves of refractive index modulation for both short and continuous holographic exposures, in two significantly different free radical photopolymer materials [23,30-32]. The quality of the fits obtained to both photopolymer materials, indicated the versatility and applicability of the NPDD model.

In the past number of years, extensive work has been presented in the literature to describe the time varying absorption effects, which occur in photopolymer materials during exposure, [21,28,29]. In all cases the aim has been to improve the understanding of the photo-kinetics occurring in these materials, and critically to enable accurate predictions of the generation of primary radicals. A model of photosensitiser behaviour proposed by Carretero *et al.* [45], has recently been extended to account for: (i) photon absorption, (ii) the regeneration or recovery of absorptive photosensitiser, and (iii) photosensitiser bleaching, [46]. Using this model an expression for the time varying absorbed intensity, $I_a(t)$ (Einstein/cm^{3}s), was derived and the values of key material parameters were estimated using non-linear fits of the dye model to experimentally obtained transmission curves, [32]. The processes of primary radical generation were then described using the simple expression,

$$R_i = 2\Phi I_a(t), \tag{3}$$

where R_i is the rate of generation of primary radicals, Φ is the number of primary radicals initiated per photon absorbed. The factor of two indicates that radicals are created in pairs, [45-48].

We will now examine the methods used to extend the NPDD model in by more accurately representing the temporal and spatial variation of the photosensitiser concentration and the associated temporal and spatial generation and removal of primary radicals. As a result the number of approximations made in modelling the photo-initiation kinetics are significantly reduced. Thus a more physically accurate representation of the photo-polymerization kinetics is produced. Crucially, the increased physicality of the proposed model enables a more accurate analysis of the process of inhibition.

3.2. Reaction mechanisms

The kinetic model presented in this analysis is based upon the following four reaction processes.

i. Initiation,

$$I \xrightarrow{h\nu} R^{\bullet}, \tag{4a}$$

$$R^{\bullet} + M \xrightarrow{k_i} M_1^{\bullet} = \text{Chain Initiator}. \tag{4b}$$

ii. Propagation,

$$M_n^{\bullet} + M \xrightarrow{k_p} M_{n+1}^{\bullet} = \text{Growing Polymer Chain} \, . \tag{5}$$

iii. Termination,

$$M_n^{\bullet} + M_m^{\bullet} \xrightarrow{k_{tc}} M_{n+m} = \text{Dead Polymer} \, , \tag{6a}$$

$$M_n^{\bullet} + M_m^{\bullet} \xrightarrow{k_{td}} M_n + M_m = \text{Dead Polymer} \, , \tag{6b}$$

$$M_n^{\bullet} + R^{\bullet} \xrightarrow{k_{tp}} M_n R = \text{Dead Polymer} \cdot \tag{6c}$$

iv. Inhibition,

$$D^* + Z \xrightarrow{k_{z,Dye^*}} LeucoDye + Z^* \, , \tag{7a}$$

$$R^{\bullet} + Z \xrightarrow{k_{z,R^{\bullet}}} \left(R + Z^{\bullet} \text{, and/or } RZ^{\bullet}\right) = \text{Scavenged Radical} \, , \tag{7b}$$

$$M_n^{\bullet} + Z \xrightarrow{k_{z,M^{\bullet}}} \left(M_n + Z^{\bullet} \text{, and/or } M_n Z^{\bullet}\right) = \text{Dead Polymer} \, . \tag{7c}$$

In the above set of chemical equations, I is the initiator concentration, $h\nu$ indicates the energy absorbed from a photon, M is the monomer concentration, Z is the inhibitor concentration, M_n, M_m, M_{n+m}, M_nR and $M_nZ^{\bullet}$ represent polymer species with no active propagating tip, i.e. *Dead Polymer*. D^* is the concentration of excited photosensitiser and Z^* is the concentration of singlet oxygen [38,40,48,49]. The term *Dead Polymer* signifies the cessation of the growth of a propagating macroradical of n monomer repeat units, [48], while the term *Scavenged Radical* signifies the removal of a primary radical, [38,48,49]. k_p, k_{tc}, k_{td}, $k_{z,M^{\bullet}}$, and $k_{z,R^{\bullet}}$ ($cm^3mol^{-1}s^{-1}$) are the rate constants of propagation, termination by combination, termination by disproportionation, inhibition of macroradicals and inhibition of primary radicals respectively.

3.3. Primary radical production

As can be seen in Eq (4), the initiation process involves two steps: The **first step** is the production of free radicals by homolytic dissociation of the initiator to yield an initiator (primary) radical, $R^{\bullet}$, i.e., Eq (4a). The **second step** is chain initiation, i.e., Eq (4b), in which the primary radicals produced due to the absorption of photons react with the monomer to produce the chain initiating species $M_1^{\bullet}$, [48]. The kinetic rate constant for this step is k_i ($cm^3mol^{-1}s^{-1}$), i.e. the chain initiation kinetic constant. As stated the main extensions to the previous model [30,31] involve improvements to the modelling of the temporal and spatial variations in primary radical production. Therefore, the main focus of this subsection will be the first step of the initiation mechanism, which is presented in Eq (4a).

In order to do this, we assume that the following photochemical reactions, take place upon illumination of a photopolymer layer sensitised with a xanthene or thiazine type photosensitiser [39], with an appropriate wavelength. These are as follows,

$$D + h\upsilon \xrightarrow{ka} D^*, \tag{8a}$$

$$D^* + Z \xrightarrow{kz,D} \text{Leuco Dye}, \tag{8b}$$

$$D^* \xrightarrow{kr} D, \tag{8c}$$

$$D^* + CI \xrightarrow{kd} R^\bullet + H^+ + D^{\bullet -} \longrightarrow R^\bullet + HD^\bullet, \tag{8d}$$

$$CI + HD^\bullet \xrightarrow{kb} H_2D + CI_{\text{int}}. \tag{8e}$$

In these equations D represents the concentration of photosensitiser (dye), $h\nu$ represents the photon energy incident on the material, D^* is the excited state of the dye, CI is the co-initiator, $R^\bullet$ represented the primary radical concentration, Z is the inhibitor, $HD^\bullet$ represents a radicalised dye, which has abstracted a hydrogen from the co-initiator, and H_2D is the di-hydro transparent form of the dye. CI_{int} is an intermediate form of the co-initiator, which is no longer available for reaction.

k_a (s^{-1}) is the rate of production of excited state photosensitiser, k_r (s^{-1}) is the rate of recovery or regeneration of photo-absorber, k_d ($cm^3mol^{-1}s^{-1}$) is the rate of dissociation of the initiator and $k_{z,D}$ ($cm^3mol^{-1}s^{-1}$) is the inhibition rate constant associated with the reaction with excited dye molecules. We note that previous models of the photo-initiation kinetics have not included all the reactions specified in Eq (8).

In order to use the proposed rate equations, it is first necessary to convert the exposure intensity I_0 (mW/cm^2) to the appropriate units (Einsteins/cm^3s). This can be done as follows, $I_0' = \frac{T_{sf}BI_0}{d}\left(\frac{\lambda}{N_a hc}\right)$, where λ (nm) is the wavelength of incident light, N_a (mol^{-1}) is Avogadro's constant, c (m/s) is the speed of light, and h (Js) is Plank's constant. $B = 1 - e^{-\varepsilon D_0 d}$, is the absorptive fraction which determines a material layer's initial absorptive capacity and is a function of the dye's initial concentration, T_{sf} is a fraction associated with the light lost by Fresnel and scattering losses, [29,32,45,46] D_0 (mol/cm^3), molar absorptivity, ε (cm^2/mol) and material layer thickness, d (cm).

The rate of production of the excited state photosensitiser, appearing in Eq (8a) can then be represented by $k_a = \phi\varepsilon d I_0'$ (s^{-1}), where ϕ (mol/Einstein) is the quantum efficiency of the reaction [48]. Therefore, if the photosensitiser's initial concentration, molar absorptivity, quantum efficiency, and layer thickness are known, the rate of generation of excited state photosensitiser, D^*, can be determined for a given exposure intensity.

3.4. Model development

In the case of holographic illumination, i.e. to record a holographic grating, there is a spatial distribution of irradiance, which in our case is typically cosinusoidal. In this case the incident intensity is represented as $I(x,t)=I_0^{'}\left[1+V\cos(Kx)\right]$, where V is the fringe visibility and $K = 2\pi/\Lambda$, where Λ is the grating period. The mechanisms, which are presented in Eq (8), can then be represented by a set of coupled differential equations. The combination of these equations is equivalent to the previous representation of primary radical production in time and space, which is presented in Eq (3). Combining these coupled differential equations with those previously presented in Ref [30-32] for the mechanisms of initiation, propagation, termination and inhibition, yields the following set of first-order coupled differential equations governing the photosensitiser:

$$\frac{dD(x,t)}{dt}=-k_a D(x,t)+k_r D^*(x,t), \tag{9}$$

$$\frac{dD^*(x,t)}{dt}=k_a D(x,t)-k_r D^*(x,t)-k_d D^*(x,t)CI(x,t)-k_{z,D}D^*(x,t)Z(x,t), \tag{10}$$

$$\frac{dCI(x,t)}{dt}=-k_d D^*(x,t)CI(x,t)-k_b HD^\bullet(x,t)CI(x,t), \tag{11}$$

$$\frac{dHD^\bullet(x,t)}{dt}=k_d D^*(x,t)CI(x,t)-k_b HD^\bullet(x,t)CI(x,t). \tag{12}$$

As in the previous analysis, [30-32] it is assumed that the effect of inhibition during exposure is due solely to the initially dissolved oxygen present within the photopolymer layer. The non-uniform recording irradiance causes concentration gradients of oxygen as it is consumed in inhibitory reactions. This then results in the diffusion of oxygen from the dark non-illuminated regions to the bright illuminated regions. As oxygen molecules are small compared to the other material components which constitute the photopolymer layer, it can be assumed that the oxygen is relatively free to diffuse rapidly, resulting in a one-dimensional standard diffusion equation for the concentration of inhibitor,

$$\frac{dZ(x,t)}{dt}=\frac{d}{dx}\left[D_z\frac{dZ(x,t)}{dx}\right]-k_{z,D}D^*(x,t)Z(x,t)$$

$$-k_{z,R^\bullet}Z(x,t)R^\bullet(x,t)-k_{z,M^\bullet}Z(x,t)M^\bullet(x,t), \tag{13}$$

where Z is the instantaneous inhibiting oxygen concentration and D_z is the diffusion constant of oxygen in the dry material layer, which in this analysis will be assumed to be time and space independent. This assumption is reasonable, as this fast rate of diffusion of the small oxygen molecule will not be significantly affected by any small changes in material viscosity. The inhibition rate constants, $k_{z,R^\bullet}$ and $k_{z,M^\bullet}$, will in general have different values (of reactivity) due to the differences in the relative molecular size, [48]. However in this analysis, for the sake of simplicity we assume $k_z = k_{z,R^\bullet} = k_{z,M^\bullet}$. Furthermore it is expected that the reactivity of oxygen with the excited state form of the photosensitiser will be much lower, i.e. $k_{z,D} << k_z$ and therefore we assume it is negligible in this analysis. As before [30-32], it is assumed that the inhibition rate constant can be expressed as,

$$k_z = k_{z,0}\exp\left(-E_z/RT\right), \tag{14}$$

where in this equation $k_{z,0}$ ($cm^3mol^{-1}s^{-1}$) is the Arrhenius pre-exponential factor, E_z = 18.23×10^3 ($Jmol^{-1}$) is the activation energy of oxygen, (i.e., the energy that must be overcome in order for oxygen to react with the given species), R = 8.31 ($JK^{-1}mol^{-1}$) is the universal gas constant, and T (K) is the local temperature [48].

The equation governing the concentration of primary radicals, including the new term for primary radical generation, is given by

$$\frac{dR^\bullet(x,t)}{dt} = k_d D^*(x,t)CI(x,t) - k_i R^\bullet(x,t)u(x,t) - k_{tp}R^\bullet(x,t)M^\bullet(x,t) - k_z R^\bullet(x,t)Z(x,t), \tag{15}$$

where $u(x, t)$ is the free-monomer concentration, (denoted earlier in the chemical reactions by M). This equation states that the rate of change of primary radical concentration is proportional to the concentration of primary radicals generated by photon absorption, minus the amounts removed by: (a) the initiation of macroradicals, (b) primary termination with growing polymer chains, and (c) inhibition by oxygen.

Including both types of termination mechanism (primary and bimolecular) and the effects of inhibition, the equation governing macroradical concentration is then

$$\frac{dM^\bullet(x,t)}{dt} = k_i R^\bullet(x,t)u(x,t) - 2k_t\left[M^\bullet(x,t)\right]^2 - k_{tp}R^\bullet(x,t)M^\bullet(x,t) - k_z Z(x,t)M^\bullet(x,t), \tag{16}$$

where the squared term represents the effects of bimolecular termination. The generation term in this equation previously appears as the removal term due to macroradical initiation in Eq (15).

The non-uniform irradiance creates monomer concentration gradients, and as a result monomer diffuses from the dark regions to the monomer depleted exposed regions. This results in a spatial polymer concentration distribution, which provides the modulation of refractive index in the material, i.e., the holographic grating. We represent the monomer concentration using the following 1D diffusion equation,

$$\frac{du(x,t)}{dt} = \frac{d}{dx}\left[D_m(x,t)\frac{du(x,t)}{dx}\right] - k_i R^\bullet(x,t)u(x,t) - \int_{-\infty}^{\infty} k_p M^\bullet(x',t)u(x',t)G(x,x')dx', \tag{17}$$

where $D_m(x, t)$ represents the monomer diffusion constant. $G(x,x')$ is the non-local material spatial response function given by [50]:

$$G(x,x') = \frac{1}{\sqrt{2\pi\sigma}} \exp\left[\frac{-(x-x')^2}{2\sigma}\right], \tag{18}$$

where σ is the constant non-local response parameter normalized with respect to the grating period, Λ. This non-local spatial response function represents the effect of initiation at location x' on the amount of monomer polymerized at location x.

The equation governing the polymer concentration is

$$\frac{dN(x,t)}{dt} = \int_{-\infty}^{\infty} k_p M^{\bullet}(x',t)u(x',t)G(x,x')dx' - \frac{d}{dx}\left[D_N(x,t)\frac{dN(x,t)}{dx}\right], \tag{19}$$

where $D_N(x, t)$ represents the polymer diffusion constant. As with the monomer above in Eq (17), the non-uniform irradiance creates a polymer concentration distribution. If the polymer chains are not cross-linked sufficiently, they will tend to diffuse out of the exposed regions in order to reduce the polymer gradient, [29]. If this takes place it will result in a decay of the grating strength with time. However here we assume there is sufficient cross-linking and that $D_N(x,t) = 0$, i.e., we record very stable gratings, as seen in the analysis presented in Ref [29], which uses the same material composition.

Since all the above equations presented in Eqs (9–13), (15-17) and (19), depend upon the spatial distribution of the exposing intensity, they will all be periodic even functions of x and can therefore be written as Fourier series, i.e., $X(x,t) = \sum_{j=0}^{\infty} X_j(t)\cos(jKx)$, where X represents the species concentrations, D, D^*, CI, $HD^{\bullet}$, $R^{\bullet}$, $M^{\bullet}$, u, N and Z. A set of first-order coupled differential equations can then be obtained in the same manner presented in Refs [30-32], by gathering the coefficients of the various co-sinusoidal spatial contributions and writing the equations in terms of these time varying spatial harmonic amplitudes. These coupled equations can then be solved using the following initial conditions,

$$Z_0(t=0) = Z_0,\ D_0(t=0) = D_0,\ CI_0(t=0) = CI_0,\ u_0(t=0) = U_0,$$

$$D_{n>0}(t=0) = D^*_{n\geq 0}(t=0) = HD^{\bullet}_{n\geq 0}(t=0) = CI_{n>0}(t=0) = 0, \text{ and}$$

$$Z_{n>0}(t=0) = R^{\bullet}_{n\geq 0}(t=0) = M^{\bullet}_{n\geq 0}(t=0) = N_{n\geq 0}(t=0) = 0. \tag{20}$$

As in previous analysis the Fourier series expansion of the monomer and polymer harmonics involves use of the non-local response parameter $G(x,x')$ which is represented in the coupled differential equations by $S_i = \exp(-i^2K^2\sigma/2)$.

3.5. Model simulations

In order to examine the general behaviour of this model, we now generate a number of theoretical simulations and analyse their predictions. In all theoretical simulations presented here, it is assumed that time varying viscosity effects are negligible and therefore, $D_m(x,t) = D_{m0} = 8.0\times10^{-11}$ cm²/s. All kinetic parameter values are assigned appropriate values, which are typical for the AA/PVA photopolymer material examined here, [30-32].

12 spatial concentration harmonics are retained in the simulations, solved using the initial conditions presented in Eq (20) with $U_0 = 2.83\times10^{-3}$ mol/cm³, $D_0 = 1.22\times10^{-6}$ mol/cm³, $CI_0 = 3.18\times10^{-3}$ mol/cm³ and $Z_0 = 1\times10^{-8}$ mol/cm³. Assuming typical recording conditions for an unslanted transmission type volume holographic grating, for $\Lambda = 700$ nm and fringe visibility $V = 1$, simulations of the temporal and spatial variation in the photosensitiser concentration, $D(x,t)$, are generated and presented in Figure 2a. The typical rate constants used were $k_p = k_i = 2.65\times10^7$ cm³/mol s, $k_t = 6\times10^9$ cm³/mol s, $k_{tp} = k_t \times 10$, $k_d = k_b = 1.6\times10^3$ cm³/mol s, $k_z = 3\times10^{12}$ cm³/mol s and $k_r = 1.22\times10^{-3}$ s⁻¹, [30-32]. For an exposure intensity of $I_0 =$ 1 mW/cm² and $\lambda = 532$ nm, the absorption parameters estimated from fits to normalised transmission curves for a material layer of thickness $d = 100$ μm were, $\varepsilon = 1.4\times10^8$ cm²/mol, $\phi = 0.066$ mol/Einstein and $T_{sf} = 0.76$, with $N_a = 6.02\times10^{23}$ mol⁻¹, $c = 3\times10^8$ ms⁻¹ and $h = 6.62\times10^{-34}$ Js, [32]. The oxygen diffusion coefficient was assumed to be $D_z = 1.0\times10^{-8}$ cm²/s, [51]. The parameter S_1, which quantifies the extent of the non-locality in the first harmonic coupled differential equation, was chosen to have a value of $S_1 = 0.94$, [29,50]. This corresponds to a non-local response length of $\sqrt{\sigma'} = 54$ nm, [29].

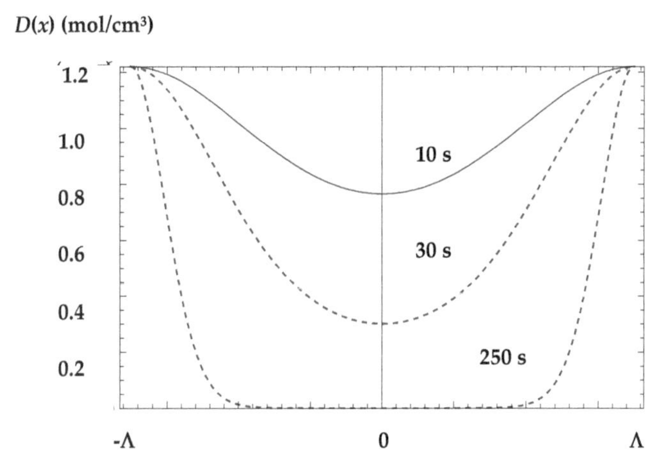

Figure 2. Simulation of the spatial variation of the ground state photosensitiser concentration for an exposure intensity $I_0 = 1$mW/cm², at $\Lambda = 700$ nm, for various exposure times, $t_{exp} = 10$s (joined line), $t_{exp} = 30$s (small dashed line), $t_{exp} = 250$s (dashed line).

As can be observed from Figure 2, the spatial sinusoidal variation in the exposing interference pattern causes a rapid consumption of the ground state dye in the bright illuminated regions. As the exposure time increases the sinusoidal variation of the dye concentration is distorted and the width of the non-illuminated dark bands narrows. This loss in sinusoidal fidelity results in a spatial production of primary radicals, as shown in Figure 3, which deviates significantly from the sinusoidal primary radical generation which would be generated using the term presented in Eq (3). Subsequently, this yields a non-linear material response, as the number of polymer chains initiated are not simply generated in direct proportion to the exposing interference pattern. This is an important prediction of the model, which agrees well with experimental observation.

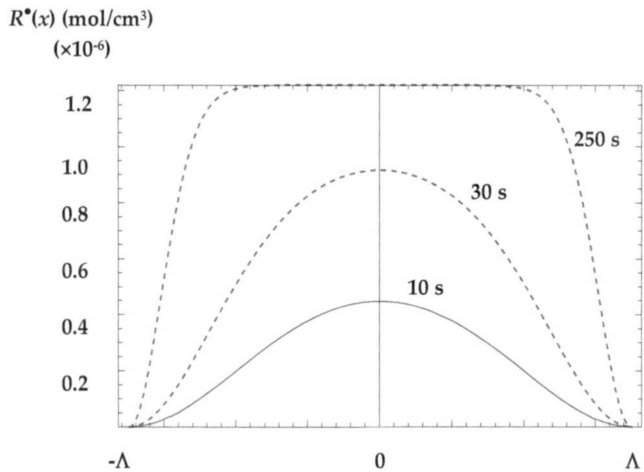

Figure 3. Simulation of the spatial variation of the generation of primary radicals for an exposure intensity $I_0 = 1\text{mW/cm}^2$, at $\Lambda = 700$ nm, for various exposure times, $t_{exp} = 10$s (joined line), $t_{exp} = 30$s (small dashed line), $t_{exp} = 250$s (dashed line).

Using the same parameter values used to generate Figures 2 and 3, Figure 4 shows a simulation of the amplitudes of the first two concentration harmonics of the monomer, u_0 and u_1, and the corresponding polymer variations, N_0 and N_1. The presence of a 'deadband' or inhibition period, t_i, can be observed at the early stages of exposure as a result of the inhibitory reactions. This behaviour is consistent with the reaction mechanisms discussed in earlier, where the primary- and macro-radicals are scavenged by oxygen, which is initially dissolved in the photopolymer layer.

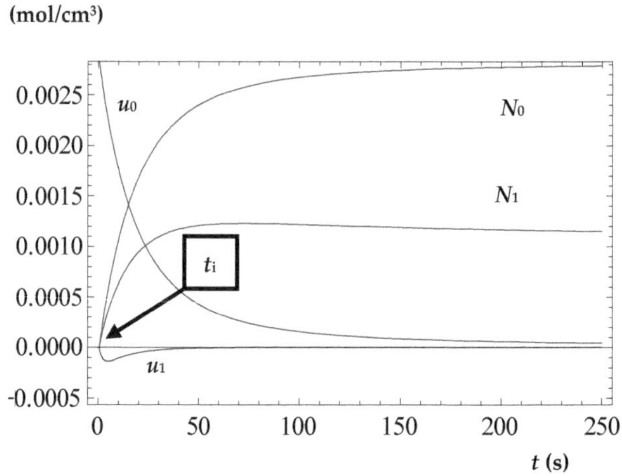

Figure 4. Simulations of the variation of the first two concentration harmonics of monomer and polymer using the theoretical model.

Figure 5, shows a simulation of growth curves of refractive index modulation with varying values of the concentration of initially dissolved oxygen, Z_0 (mol/cm^3), under the same conditions as the previous figures but with $Z_0 = 1\times10^{-8}$ mol/cm^3 (joined line), $Z_0 = 5\times10^{-8}$ mol/cm^3 (short dashed line), and $Z_0 = 1\times10^{-7}$ mol/cm^3 (long dashed line). As the concentration of inhibitor is increased, the inhibition time, t_i increases as expected, i.e., more inhibitor causes a greater scavenging of the primary- and macro-radicals.

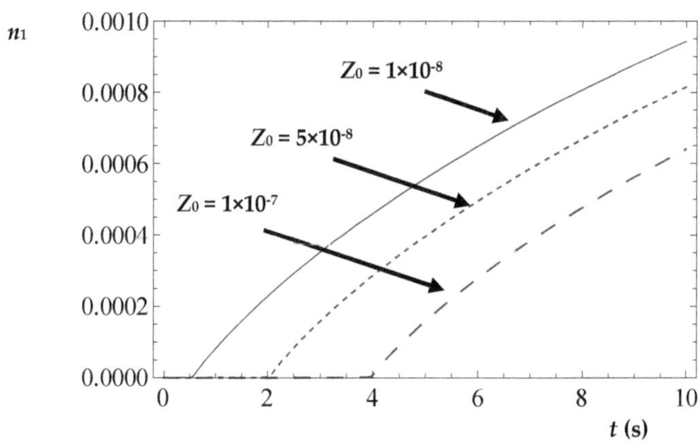

Figure 5. Simulations of the refractive index modulation with time, for various values of dissolved oxygen concentration. $Z_0 = 1\times10^{-7}$ mol/cm^3 (long dashed line), $Z_0 = 5\times10^{-8}$ mol/cm^3 (short dashed line) and $Z_0 = 1\times10^{-8}$ mol/cm^3 (joined line).

When comparing the experimental results obtained using the optical setup described above with the theoretical predictions generated by the extended model, it became clear that when using the model as presented, the trend of increased inhibition times, t_i, for reduced exposure intensities, did not satisfactorily replicate the experimental behavior observed. In order to achieve good fits to the experimental data, it was found necessary to increase the initial concentration of dissolved oxygen available in the photopolymer layer, Z_0, as the recording intensities were reduced. The variation between the experimental observation and theoretical prediction was as much as 8 s for the lowest recording intensity examined for unsealed photopolymer layers (not sealed from the environment). This divergence between experiment and prediction suggested that the model was incomplete and, that in order to mimic this physically observed behavior, amendments to the model were necessary.

In a previous paper published by the authors [49], it was found that by cover-plating or sealing the photopolymer layer with glass slides, the inhibition times observed during exposure compared with the uncoverplated or unsealed layers, were significantly reduced. These effects were attributed to the removal or reduction of oxygen diffusing in from the surrounding environment, which was replacing or replenishing the oxygen consumed during exposure. It must be noted at this point, that the experimental data examined here were uncoverplated photopolymer layers, which were subject to this potential external oxygen diffusion.

In order to represent this process in the model, an additive term representing the replenishing of inhibiting oxygen from the outside surrounding air, into the material layer, was included. Therefore, Eq (13) was revised and became,

$$\frac{dZ(x,t)}{dt} = \frac{d}{dx}\left[D_z \frac{dZ(x,t)}{dx}\right] - k_{z,D}D^*(x,t)Z(x,t) - k_{z,R^\bullet}Z(x,t)R^\bullet(x,t)$$

$$-k_{z,M^\bullet}Z(x,t)M^\bullet(x,t) + \tau_z\left[Z_0 - Z(x,t)\right], \quad (21)$$

where τ_z represents the rate of replenishing of oxygen into the material layer. We note that it is assumed that the oxygen concentration can never be larger than the original dissolved oxygen concentration, Z_0 (mol/cm^3) and that this additive term is assumed to be constant in space.

In order to illustrate these effects Figure 6 shows a simulation of the behavior of the oxygen concentration with varying values of the replenishing constant, τ_z, for an exposure intensity, I_0 = 0.04 mW/cm^2 and exposure time, t_{exp} = 30 s. As can be observed, an increase in τ_z results in: (i) an increase of the inhibition period, and (ii) an increase in the rate at which oxygen returns to its original dissolved oxygen concentration, post-exposure.

Implementing the appropriate Fourier series expansion to Eq (21) under the same initial conditions, the model is then applied to the experimental growth curves recorded in uncoverplated layers, yielding much more accurate fits to the data. Figure 7 shows a subset of this data with the corresponding fits obtained using the model. Some of the parameter

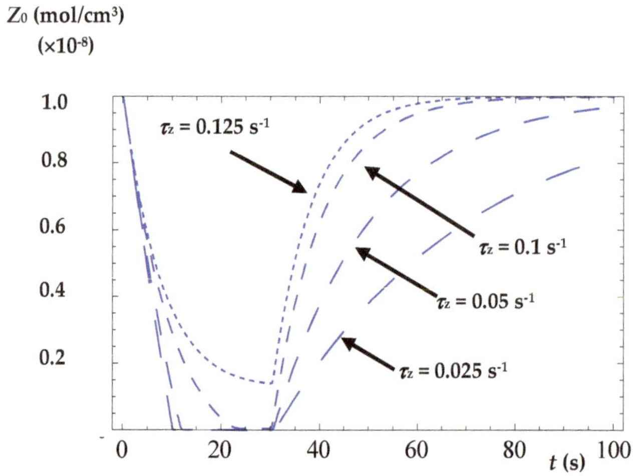

Figure 6. Simulation of the behaviour of the oxygen concentration with varying values of τ_z, for an exposure time, t_{exp} = 30 s and exposure intensity, I_0 = 0.04 mW/cm^2. τ_z = 0.125 s^{-1} (shortest dash), τ_z = 0.1 s^{-1} (short dash), τ_z = 0.05 s^{-1} (long dash), τ_z = 0.025 s^{-1} (longest dash).

values which were obtained from the fits to various intensities are, $k_d = 1.6\times10^3$ cm^3/mol s, $k_r = 1.2\times10^{-3}$ s^{-1}, $k_z = 3.0\times10^{12}$ cm^3/mol s, $D_z = 1\times10^{-8}$ cm^2/s, and it was assumed in all fits that $k_{tp} = 10\times k_t$ cm^3/mol s, and $k_i = k_p$ cm^3/mol s.

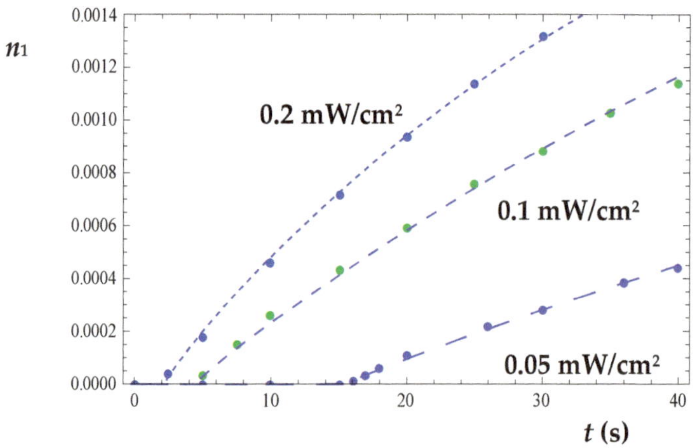

Figure 7. Experimentally obtained growth curves of refractive index modulation recorded in uncoverplated AA/PVA photopolymer material layers at a spatial frequency of 1428 lines/mm for 3 different exposing intensities, I_{01} = 0.2 mW/cm^2 (short dash), I_{02} = 0.1 mW/cm^2 (dashed) and I_{03} = 0.05 mW/cm^2 (long dash) with corresponding fits achieved with the theoretical model.

The most significant values extracted from the fits are presented in Table 1 along with the parameter search ranges, which were used to obtain a best fit between experimental and theoretical prediction. These search ranges are typical of the valued presented in the literature for similar photopolymer materials, [42,48,51]. The Mean Squared Error (MSE) between the fit and the data are also included to indicate the quality of the fits.

I_0 (mW/cm^2)	t_i (s)	k_p (cm^3/mols) (×10^7)	k_t (cm^3/mols) (×10^9)	D_{m0} (cm^2/s) (×10^{-11})	τ_z (s^{-1})	MSE (×10^{-11})
0.20	2.50	2.42	5.0	8.0	0.075	1.05
0.10	4.50	2.52	7.0	9.0	0.080	2.86
0.05	16.00	3.00	7.0	10.0	0.115	1.88
Search Range	-	0.1 - 5.0	0.1 - 9.0	1.0 – 12.0	-	-

Table 1. Parameters extracted from fits to experimentally obtained growth curves of refractive index modulation in uncoverplated photopolymer layers.

As can be observed from Figure 7, the fit quality is very good and the model predicts the observed trend, that a reduction in the exposure intensity causes an increase in the inhibition period due to (i) initially dissolved oxygen and (ii) oxygen diffusion into the material from the surrounding air. It can also be seen that there is a reduction in the propagation and termination rates with increasing exposure intensities. This is most likely due to the increased viscosity effects, which occur due to increased conversion of monomer to polymer [48]. This is consistent with the results obtained from the previous model, [30-32]. It must also be noted at this point that the estimates obtained for the rates of propagation and termination are slightly higher than those reported in the previous published work by the authors, [30-32]. This is as a result of a more physically accurate description of the primary radical generation introduced by the model development. However, the estimated values extracted from the fits still remain well within the accepted ranges presented in the literature for similar photopolymer materials.

In order to verify the necessity for the inclusion of the additive oxygen replenishing term in Eq (21), several growth curves of refractive index modulation were recorded in coverplated layers. These growth curves were recorded under the same conditions as the uncoverplated layers presented in Figure 7. Figure 8 shows experimental growth curves recorded at an exposure intensity of I_0 = 0.05 mW/cm^2, in both coverplated and uncoverplated layers. The subsequent fits to the experimental data, which were achieved using the revised model are represented as short dash line (coverplated) and long dash line (uncoverplated).

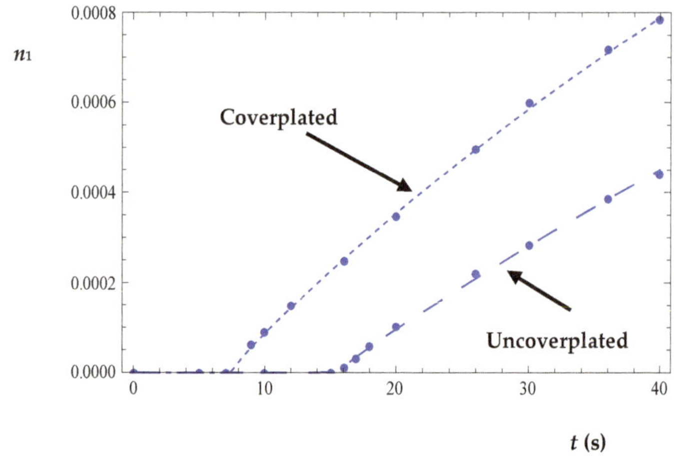

Figure 8. Experimentally obtained growth curves of refractive index modulation recorded in both coverplated (short dash) and uncoverplated (long dash) AA/PVA photopolymer material layers at a spatial frequency of 1428 lines/mm for a recording intensity $I_0 = 0.05$ mW/cm^2 with corresponding fits achieved with the model.

As can be observed from the figure there is a significant reduction in the inhibition period, from $t_i = 16$ s (uncoverplated) to $t_i = 9$s (coverplated). As stated above, this is attributed to a reduction in the amount of oxygen available to diffuse into the layer from the surrounding air. The estimated parameters extracted from these fits are presented in Table 2. The values determined for the replenishing rate τ_z, are consistent with what is experimental observed. In the case of the coverplated material layer, it is assumed that no oxygen can diffuse into the layer, i.e. $\tau_z = 0$.

	t_i (s)	k_p (cm^3/mols) (×10^7)	k_t (cm^3/mols) (×10^9)	D_{m0} (cm^2/s) (×10^{-11})	τ_z (s^{-1})	MSE (×10^{-11})
Coverplated	9.0	2.9	7.0	10.0	0.000	2.26
Uncoverplated	16.0	3.0	7.0	10.0	0.115	1.88
Search Range	-	0.1 - 5.0	0.1 - 9.0	1.0 - 12.0	-	-

Table 2. Parameters extracted from fits to experimentally obtained growth curves recorded at $I_0 =$ 0.05mW/cm^2 for a coverplated and uncoverplated polymer layer.

In this section, further developments of the Non-local Photo-polymerization Driven Diffusion (NPDD) model, were presented. For the first time, the spatial and temporal variations in primary radical generation were included. These extensions provided a more physically comprehensive theoretical representation of the processes, which occur during free radical photo-polymerization. A clearer more physical representation of the reactions, which take place during the photo-initiation stages, was also provided, including the spatial and temporal consumption and regeneration of the photosensitiser and the reactions between the excited dye molecules and the co-initiator. Simulations were presented, which highlight the loss of sinusoidal fidelity of the primary radical generation. This behaviour deviates from that which was previously predicted in the literature. Subsequently, this change in the spatial generation of primary radicals has a substantial effect on the distribution of the polymer chains formed and hence, on the resulting refractive index modulation recorded.

The model was then further extended to incorporate the effect of oxygen diffusion from outside the material layer by including a rate of oxygen replenishing. This allowed accurate modelling of the inhibition effects, which dominate the start of grating growth. The results obtained were consistent with previous studies where cover-plating techniques were used.

In the following section, we will examine the effects of adding chain transfer agents to an AA/PVA photopolymer material in order to reduce the average polymer chain length grown during grating fabrication. This will then cause a reduction in the extent of the non-local chain growth of these polymer chains and thus reduce the fall off in refractive index modulation at higher spatial frequencies.

4. Improving the spatial frequency response of photopolymers

The non-local spatial response function presented in Eq (18) (in Section 3.4) represents the effect of a chain initiation at location x' on the amount of monomer polymerized at location x, [50], where σ is the constant non-local response parameter which is normalized with respect to the period of the grating being recorded, Λ This is an important parameter when considering the data storage capacity or high density resolution of a photopolymer. The Non-local Photo-Polymerization Driven Diffusion (NPDD) model predicts that a reduction in this non-local response parameter within a photopolymer material will improve its high spatial frequency response. A point to note is that the non-local response of a given photopolymer is produced by a combination of several physical effects, which result in the smearing of the grating being recorded in the photopolymer. One such smearing effect is this growth of polymer chains away from the point of their initiation, into the dark less exposed regions of the material layer. This propagation out of the bright regions causes an increase in the average refractive index of the dark region and as a result, reduces the overall refractive index modulation achievable. This has been illustrated in discussed in detail in previous publications [29-31,50], and is more significant when recording high spatial frequency.

The introduction of a chain transfer agent acts to reduce the average molecular weight of polymer chains grown during free radical polymerization. Therefore a chain transfer agent (CTA) can provide a practical method to reduce the non-local response length. In this section an extended NPDD model is presented, which includes the chain transfer reactions and all major photochemical processes.

4.1. Chain transfer mechanism

In many polymerization systems, the average polymer weight is observed to be lower than predicted by the chain transfer reaction [29,52-55]. Generally, the chain transfer process causes the premature termination of a growing macro-radical chain and arises because of the presence of CTA [48]. Due to this reaction, a new radical is produced which is referred to in this work as a re-initiator. This re-initiator reacts with a monomer molecule to initiate a new growing macro-radical chain. Therefore we can write the chain transfer reactions as,

$$M_n^{\bullet} + (RI - X) \xrightarrow{k_{tr,S}} (M_n - X) + RI^{\bullet}, \tag{22a}$$

$$RI^{\bullet} + M \xrightarrow{k_{ri}} \left(RI - M^{\bullet}\right), \tag{22b}$$

where *RI-X* is the chain transfer agent, -*X* is the atom or species transferred and $RI^{\bullet}$ is the re-initiator which has a radical tip. $k_{tr,S}$ and k_{ri} are the transfer rate constant to chain-transfer agent and the re-initiation rate constant respectively. Due to the premature termination reaction with the chain transfer agent, *RI-X,* the propagating polymer chains will stop growing earlier than they would have if the CTA was not present. We assume that the free radical $RI\text{-}M^{\bullet}$ produced can be treated as acting chemically identical to a chain initiator $M^{\bullet}$. Therefore the re-initiator, $RI^{\bullet}$, simply initiates a new growing chain with a radical tip, $M^{\bullet}$. Thus, while the polymer chains are shortened, the amount of monomer polymerized and the rate of polymerization can remain high.

In the radical chain polymerization system [29,48], the polymerization rate can be expressed as:

$$R_P = k_p M^{\bullet}(x,t) u(x,t). \tag{23}$$

The polymerization rate, R_P, is also related to the number-average degree of polymerization, DP_n. DP_n is defined as the average number of structural units per polymer chain. It indicates the average length of polymer chain grown and therefore their molecular weight. According to the *Mayo Equation* [48,52],

$$\frac{1}{DP_n} = \frac{k_t R_P}{k_p^2 [u]^2} + C_u + C_{CTA}\frac{[CTA]}{[u]} + C_I\frac{[I]}{[u]}. \tag{24}$$

This quantifies the effect of the various chain transfer reactions on the number-average degree of polymerization. [*u*], [*CTA*], [*I*], represent the concentrations of monomer, chain transfer agent and initiator, respectively. The chain-transfer constants, C_u, C_{CTA} and C_I, for each particular substance are defined as the ratios of the rate constants for chain transfer of a propagating radical with that substance to the propagation rate constant, k_p. They can be expressed as:

$$C_u = \frac{k_{tr,u}}{k_p},\ C_{CTA} = \frac{k_{tr,CTA}}{k_p},\ \text{and}\ C_I = \frac{k_{tr,I}}{k_p}, \tag{25}$$

where $k_{tr,u}$, $k_{tr,CTA}$, $k_{tr,I}$ represent rate constants for chain transfer to monomer, to chain transfer agent and to initiator respectively. For the case examined here the chain transfer constants to monomer and initiator, can be omitted as they are typically very low for acrylamide [56,57], and therefore Eq. (24) can be simplified to:

$$\frac{1}{DP_n} = \frac{k_t R_p}{k_p^2 [u]^2} + C_{CTA} \frac{[CTA]}{[u]}, \tag{26}$$

which will be discussed in detail in next section.

4.2. Model development

In order to begin to examine the effects of the presence of CTA on the material non-local response length, $\sqrt{\sigma}$, (the actual physical length of the nonlocal response parameter, units of nanometres) we introduce a rate equation governing the CTA concentration:

$$\frac{d[CTA(x,t)]}{dt} = \frac{d}{dx}\left\{ D_{CTA}(x,t) \frac{d[CTA(x,t)]}{dx} \right\} - k_{tr,S}[CTA(x,t)]\,[M^\bullet(x,t)]. \tag{27}$$

It should be noted that, in the following analysis, we only consider chain transfer to chain-transfer agent, i.e., the chain transfer constants for monomer and initiator are assumed negligible. To further simplify the analysis in this work, we assume that $k_{tr} = k_{tr,CTA}$ and that the CTA diffusion rate, D_{CTA}, is similar to the diffusion rate of monomer, D_m, as their molecular weights are similar in the cases examined, i.e., $D_{CTA} \approx D_m$.

The equation governing the re-initiator concentration can be given by,

$$\frac{d[RI^\bullet(x,t)]}{dt} = k_{tr}[CTA(x,t)]\,[M^\bullet(x,t)] - k_{ri}[RI^\bullet(x,t)]\,[u(x,t)], \tag{28}$$

where $RI^\bullet$ denotes the re-initiator concentration. Since it is assumed that the initiator radical, $R^\bullet$, dominates the primary termination and inhibition processes, we only consider how the re-initiator, $RI^\bullet$, reacts with the monomer.

Furthermore the chain transfer and re-initiation reactions effect the variation of macro-radical, $[M^\bullet]$, and monomer, $[u]$, concentrations. Therefore we must change the coupled differential equations for monomer and macroradicals presented earlier in Section 3, giving,

$$\frac{d[M^\bullet(x,t)]}{dt} = k_i[R^\bullet(x,t)]\,[u(x,t)] + k_{ri}[RI^\bullet(x,t)]\,[u(x,t)] - 2k_t[M^\bullet(x,t)]^2$$

$$-k_{tp}[R^\bullet(x,t)]\,[M^\bullet(x,t)] - k_z[Z(x,t)]\,[M^\bullet(x,t)] - k_{tr}[CTA(x,t)]\,[M^\bullet(x,t)], \tag{29}$$

and,

$$\frac{d\left[u(x,t)\right]}{dt}=\frac{d}{dx}\left\{D_m(x,t)\frac{d\left[u(x,t)\right]}{dx}\right\}-k_i\left[R^{\bullet}(x,t)\right]\left[u(x,t)\right]$$

$$-\int_{-\infty}^{\infty}k_p\left[M^{\bullet}(x',t)\right]\left[u(x',t)\right]G(x,x')dx'-k_{ri}\left[RI^{\bullet}(x,t)\right]\left[u(x,t)\right]. \tag{30}$$

As before the concentrations of the components of the photopolymer and the amended equations appearing here in Eqs (27–30) will be periodic even functions of x and can therefore be written as Fourier series, i.e., $\left[X(x,t)\right]=\sum_{j=0}^{\infty}X_j(t)\cos(jKx)$, where X represents the particular species, i.e., CTA, $RI^{\bullet}$, $M^{\bullet}$, and u. A set of first-order coupled differential equations can then be obtained by gathering the coefficients of the various co-sinusoidal spatial components and writing the equations in terms of these time varying spatial harmonic amplitudes, $X_j(t)$.

For brevity we will assume here that harmonics of order greater than $j = 3$, are negligible and for illustration purposes we now present the first two harmonics of $[CTA]$, $[RI^{\bullet}]$, $[M^{\bullet}]$, and $[u]$, all of which are directly involved in reactions with the transfer agent.

Chain Transfer Agent Concentration: Retaining the first four concentration harmonic amplitudes in the analysis, the following first-order coupled differential equations govern the chain transfer agent concentration amplitudes, CTA_j:

$$\frac{dCTA_0(t)}{dt}=-k_{tr}CTA_0(t)M_0^{\bullet}(t)-\frac{1}{2}k_{tr}CTA_1(t)M_1^{\bullet}(t)-\frac{1}{2}k_{tr}CTA_2(t)M_2^{\bullet}(t)$$

$$-\frac{1}{2}k_{tr}CTA_3(t)M_3^{\bullet}(t), \tag{31a}$$

$$\frac{dCTA_1(t)}{dt}=-D_{CTA}K^2CTA_1(t)-k_{tr}CTA_1(t)M_0^{\bullet}(t)-k_{tr}CTA_0(t)M_1^{\bullet}(t)$$

$$-\frac{1}{2}k_{tr}CTA_2(t)M_1^{\bullet}(t)-\frac{1}{2}k_{tr}CTA_1(t)M_2^{\bullet}(t)$$

$$-\frac{1}{2}k_{tr}CTA_3(t)M_2^{\bullet}(t)-\frac{1}{2}k_{tr}CTA_2(t)M_3^{\bullet}(t), \tag{31b}$$

Re-initiator Concentration: The equations governing the re-initiator concentration amplitudes, $RI_j^{\bullet}$, are:

$$\frac{dRI_0^{\bullet}(t)}{dt}=k_{tr}CTA_0(t)M_0^{\bullet}(t)+\frac{1}{2}k_{tr}CTA_1(t)M_1^{\bullet}(t)+\frac{1}{2}k_{tr}CTA_2(t)M_2^{\bullet}(t)$$

$$+\frac{1}{2}k_{tr}CTA_3(t)M_3^{\bullet}(t)-k_{ri}RI_0^{\bullet}(t)u_0(t)-\frac{1}{2}k_{ri}RI_1^{\bullet}(t)u_1(t)$$

$$-\frac{1}{2}k_{ri}RI_2^{\bullet}(t)u_2(t)-\frac{1}{2}k_{ri}RI_3^{\bullet}(t)u_3(t), \tag{32a}$$

$$\frac{dRI_1^{\bullet}(t)}{dt}=k_{tr}CTA_1(t)M_0^{\bullet}(t)+k_{tr}CTA_0(t)M_1^{\bullet}(t)+\frac{1}{2}k_{tr}CTA_2(t)M_1^{\bullet}(t)$$

$$+\frac{1}{2}k_{tr}CTA_1(t)M_2^{\bullet}(t)+\frac{1}{2}k_{tr}CTA_3(t)M_2^{\bullet}(t)+\frac{1}{2}k_{tr}CTA_2(t)M_3^{\bullet}(t)$$

$$-k_{ri}RI_1^{\bullet}(t)u_0(t)-k_{ri}RI_0^{\bullet}(t)u_1(t)-\frac{1}{2}k_{ri}RI_2^{\bullet}(t)u_1(t)$$

$$-\frac{1}{2}k_{ri}RI_1^{\bullet}(t)u_2(t)-\frac{1}{2}k_{ri}RI_3^{\bullet}(t)u_2(t)-\frac{1}{2}k_{ri}RI_2^{\bullet}(t)u_3(t), \tag{32b}$$

Macro-radical Concentration: From Eq. (29), equations governing the $M_j^{\bullet}$, are:

$$\frac{dM_0^{\bullet}(t)}{dt}=-\frac{k_{tr}}{2}\Big[2CTA_0(t)M_0^{\bullet}(t)+CTA_1(t)M_1^{\bullet}(t)+CTA_2(t)M_2^{\bullet}(t)$$

$$+CTA_3(t)M_3^{\bullet}(t)\Big]-\frac{k_t}{2}\Big[2M_0^{\bullet}(t)^2+M_1^{\bullet}(t)^2+M_2^{\bullet}(t)^2+M_3^{\bullet}(t)^2\Big]$$

$$-\frac{k_{tp}}{2}\Big[2M_0^{\bullet}(t)R_0^{\bullet}(t)+M_1^{\bullet}(t)R_1^{\bullet}(t)+M_2^{\bullet}(t)R_2^{\bullet}(t)+M_3^{\bullet}(t)R_3^{\bullet}(t)\Big]$$

$$-\frac{k_z}{2}\Big[2M_0^{\bullet}(t)Z_0(t)+M_1^{\bullet}(t)Z_1(t)+M_2^{\bullet}(t)Z_2(t)+M_3^{\bullet}(t)Z_3(t)\Big]$$

$$+\frac{k_i}{2}\Big[2R_0^{\bullet}(t)u_0(t)+R_1^{\bullet}(t)u_1(t)+R_2^{\bullet}(t)u_2(t)+R_3^{\bullet}(t)u_3(t)\Big]$$

$$+\frac{k_{ri}}{2}\Big[2RI_0^{\bullet}(t)u_0(t)+RI_1^{\bullet}(t)u_1(t)+RI_2^{\bullet}(t)u_2(t)+RI_3^{\bullet}(t)u_3(t)\Big], \tag{33a}$$

$$\frac{dM_1^{\bullet}(t)}{dt}=-\frac{k_{tr}}{2}\Big[2CTA_1(t)M_0^{\bullet}(t)+2CTA_0(t)M_1^{\bullet}(t)+CTA_2(t)M_1^{\bullet}(t)$$

$$+CTA_1(t)M_2^{\bullet}(t)+CTA_3(t)M_2^{\bullet}(t)+CTA_2(t)M_3^{\bullet}(t)\Big]$$

$$-k_t\left[2M_0^\bullet(t)M_1^\bullet(t)+M_1^\bullet(t)M_2^\bullet(t)+M_2^\bullet(t)M_3^\bullet(t)\right]$$

$$-\frac{k_{tp}}{2}\Big[2M_1^\bullet(t)R_0^\bullet(t)+2M_0^\bullet(t)R_1^\bullet(t)+M_2^\bullet(t)R_1^\bullet(t)+M_1^\bullet(t)R_2^\bullet(t)$$

$$+M_3^\bullet(t)R_2^\bullet(t)+M_2^\bullet(t)R_3^\bullet(t)\Big]$$

$$-\frac{k_z}{2}\Big[2M_1^\bullet(t)Z_0(t)+2M_0^\bullet(t)Z_1(t)+M_2^\bullet(t)Z_1(t)+M_1^\bullet(t)Z_2(t)$$

$$+M_3^\bullet(t)Z_2(t)+M_2^\bullet(t)Z_3(t)\Big]$$

$$+\frac{k_i}{2}\Big[2R_1^\bullet(t)u_0(t)+2R_1^\bullet(t)u_0(t)+R_2^\bullet(t)u_1(t)+R_1^\bullet(t)u_2(t)$$

$$+R_3^\bullet(t)u_2(t)+R_2^\bullet(t)u_3(t)\Big]$$

$$+\frac{k_{ri}}{2}\Big[2RI_1^\bullet(t)u_0(t)+2RI_1^\bullet(t)u_0(t)+RI_2^\bullet(t)u_1(t)+RI_1^\bullet(t)u_2(t)$$

$$+RI_3^\bullet(t)u_2(t)+RI_2^\bullet(t)u_3(t)\Big]\ , \tag{33b}$$

Monomer Concentration: The coupled equations for the monomer concentration harmonics, u_j, are:

$$\frac{du_0(t)}{dt}=-\frac{k_p}{2}\left[2M_0^\bullet(t)u_0(t)+M_1^\bullet(t)u_1(t)+M_2^\bullet(t)u_2(t)+M_3^\bullet(t)u_3(t)\right]$$

$$-\frac{k_i}{2}\left[2R_0^\bullet(t)u_0(t)+R_1^\bullet(t)u_1(t)+R_2^\bullet(t)u_2(t)+R_3^\bullet(t)u_3(t)\right]$$

$$-\frac{k_{ri}}{2}\left[2RI_0^\bullet(t)u_0(t)+RI_1^\bullet(t)u_1(t)+RI_2^\bullet(t)u_2(t)+RI_3^\bullet(t)u_3(t)\right], \tag{34a}$$

$$\frac{du_1(t)}{dt}=-\frac{S_1k_p}{2}\Big[2M_1^\bullet(t)u_0(t)+2M_0^\bullet(t)u_1(t)+M_2^\bullet(t)u_1(t)+M_1^\bullet(t)u_2(t)$$

$$+M_3^\bullet(t)u_2(t)+M_2^\bullet(t)u_3(t)\Big]-\frac{k_i}{2}\Big[2R_1^\bullet(t)u_0(t)+2R_0^\bullet(t)u_1(t)$$

$$+R_2^\bullet(t)u_1(t)+R_1^\bullet(t)u_2(t)+R_3^\bullet(t)u_2(t)+R_2^\bullet(t)u_3(t)\Big]$$

$$-\frac{k_{ri}}{2}\Big[2RI_1^{\bullet}(t)u_0(t)+2RI_0^{\bullet}(t)u_1(t)+RI_2^{\bullet}(t)u_1(t)+RI_1^{\bullet}(t)u_2(t)$$

$$+RI_3^{\bullet}(t)u_2(t)+RI_2^{\bullet}(t)u_3(t)\Big]-K^2\left\{\left[D_{m,0}(t)-\frac{1}{2}D_{m,2}(t)\right]u_1(t)\right.$$

$$\left.+\left[D_{m,1}(t)-D_{m,3}(t)\right]u_2(t)-\frac{3}{2}D_{m,2}(t)u_3(t)\right\}, \tag{34b}$$

where $S_l=\exp\left(-l^2K^2\sigma/2\right)$ [30-32,50]. For generality, in Eq. (34b), we retain $D_{m,0}$, $D_{m,1}$, $D_{m,2}$ and $D_{m,3}$. However, in our simulations, we assume $D_{m,j>0}=0$ [30-32,50].

These coupled equations along with those presented in Section 3 are then solved under the initial conditions,

$$D_0(t=0)=D_0,\ CI_0(t=0)=CI_0,\ u_0(t=0)=U_0,\ CTA_0(t=0)=CTA_0,\ Z_0(t=0)=Z_0,$$

$$D_{n>0}(t=0)=CI_{n>0}(t=0)=u_{n>0}(t=0)=CTA_{n>0}(t=0)=Z_{n>0}(t=0)=0,\ \text{and}$$

$$D^*_{n\geq 0}(t=0)=HD^{\bullet}_{n\geq 0}(t=0)=R^{\bullet}_{n\geq 0}(t=0)=RI^{\bullet}_{n\geq 0}(t=0)=M^{\bullet}_{n\geq 0}(t=0)=N_{n\geq 0}(t=0)=0. \tag{35}$$

4.3. Numerical results

We will now examine some of the predictions of the extended model presented in this section. Unless otherwise stated all kinetic parameter values are assigned appropriate values, which are typical for the AA/PVA photopolymer material examined. The simulations are performed retaining four spatial concentration harmonics and the coupled differential equations are solved using the initial conditions given in Eq. (35). In all cases, $U_0 = 2.83\times10^{-3}$ mol/cm^3, $D_0 = 1.22\times10^{-6}$ mol/cm^3, $ED_0 = 3.18\times10^{-3}$ mol/cm^3, $CTA_0 = 1\times10^{-6}$ mol/cm^3, and $Z_0 = 1\times10^{-8}$ mol/cm^3 [30-34], where U_0, D_0, ED_0, CTA_0 and Z_0 represent the concentrations of monomer, photosensitizer, electron donor, transfer agent and inhibitor, respectively. We also assume that all time varying viscosity effects are negligible, i.e., $D_{m,j>0} = 0$, and that, $D_{m0} = 1.0\times10^{-10}$ cm^2/s. The exposure intensity chosen is $I_0 = 1$ mW/cm^2, the recording wavelength is $\lambda = 532$ nm and the layer thickness $d = 100$ μm. The absorption parameters are, $\varepsilon = 1.43\times10^8$ cm^2/mol, $\phi = 0.01$ mol/Einstein and $T_{sf} = 0.76$. The oxygen diffusion coefficient is $D_z(x,t) = D_z = 1.0\times10^{-8}$ cm^2/s and $\tau_z = 0$, i.e., sealed layers are used [32]. The typical rate constants used were: $k_p = k_i = 1\times10^7$ cm^3/mol s, $k_t = 3\times10^8$ cm^3/mol s, $k_{tp} = k_t \times 10$, $k_d = k_b = 1.6\times10^3$ cm^3/mol s, $k_{tr} = 1\times10^7$ cm^3/mol s, $k_{ri} = 1\times10^6$ cm^3/mol s, $k_z = 3\times10^{12}$ cm^3/mol s, and $k_r = 1.2\times10^{-3}$ s^{-1} [30-34,48]. Assuming typical recording conditions for an unslanted transmission type volume holographic grating, i.e., period $\Lambda = 400$ nm and fringe visibility $V = 1$, the resulting predictions of the temporal and spatial behaviour of the photochemical processes are now examined.

As different types of chain transfer agents will exhibit different kinetic behaviours, which result in variations in the polymerization rate and therefore changes to the number-average degree of polymerization, DP_n [48]. The results presented in Figure 9 demonstrate the effects of varying the initial CTA concentration on the DP_n when the re-initiation rate k_{ri} = 1×10^6 cm^3/mol s. For a particular type of CTA, i.e., when k_{tr} = 1×10^7 cm^3/mol s, it can be seen that increasing the initial CTA concentration leads to a rapid decrease in DP_n and that the value for DP_n decreases more slowly to lowest value shown. This result indicates that, for an appropriate concentration of CTA, i.e., $1\times10^{-6} < CTA_0 < 4\times10^{-6}$ mol/cm^3, DP_n is always reduced with the inclusion of CTA and that the reduction is larger for higher CTA concentrations. Furthermore, when $k_{tr} \geq 1\times10^7$ cm^3/mol s, the model predicts that above some specific CTA concentration a threshold exists and further increases do not result in any further significant reduction in DP_n, i.e., when $CTA_0 > 4\times10^{-6}$ mol/cm^3. Ideally, one wishes to identify the least amount of CTA required in order to achieve the largest reduction in the number-average of polymerization, DP_n. We also note that, for the same initial CTA concentration, a reduction in DP_n also takes place for an increase in chain transfer kinetic value, k_{tr}. Thus the addition of different types and concentrations of chain transfer agents are predicted to have different effects on the value of DP_n and therefore on the average polymer chain length in the otherwise identical photopolymer system. As discussed in earlier the non-local response length, $\sqrt{\sigma}$, and the number-average degree of polymerization, DP_n, are both related to the average polymer chain length. We would expect that any significant reduction in the number-average degree of polymerization, DP_n, should be accompanied by a reduction in the non-local response length, $\sqrt{\sigma}$, and therefore by an improvement in the refractive index modulation, n_1, which can be recorded at high spatial frequencies in the material.

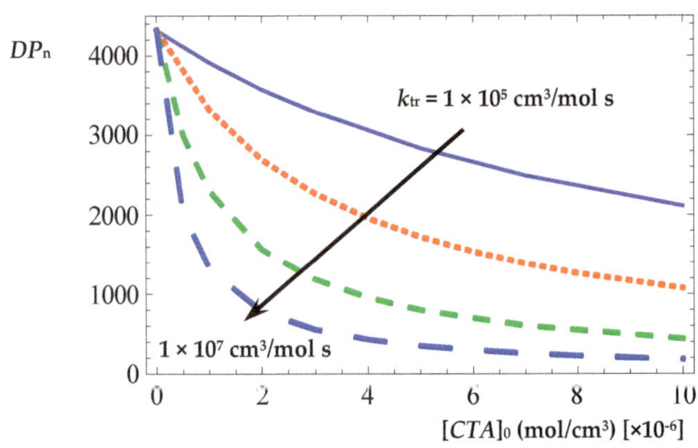

Figure 9. Effects of initial CTA concentration on number-average degree of polymerization, DP_n, and contributions of various rate constants of chain transfer, k_{tr}, [cm^3/mol s]: {1×10^5 (solid); 3×10^5 (short dashed); 1×10^6 (dashed) and 1×10^7 (long dashed)}.

In order to demonstrate the relationship between the non-local response length, $\sqrt{\sigma}$, and the refractive index modulation, n_1, Figure 10 shows four simulated growth curves of refractive index modulation, n_1, for four different values of $\sqrt{\sigma}$. In all cases the same rate constant values that were used above are employed. We see that larger values of $\sqrt{\sigma}$, lead to lower saturation (maximum) values of n_1 for the same grating period. In another words, the lower the value of $\sqrt{\sigma}$ the more localized the polymerization process and hence the shorter the polymer chains grown during holographic grating formation.

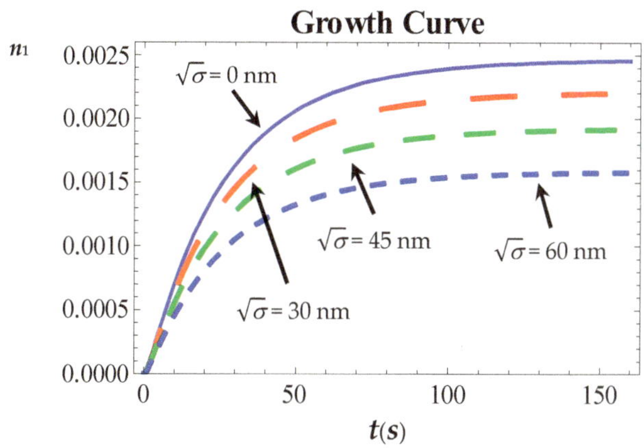

Figure 10. Simulations of the growth curves of refractive index modulation, n_1, at $\Lambda = 400$ nm for various values of non-local response length, $\sqrt{\sigma}$ nm: {0 nm (solid); 30 nm (long dashed); 45 nm (dashed) and 60 nm (short dashed)}.

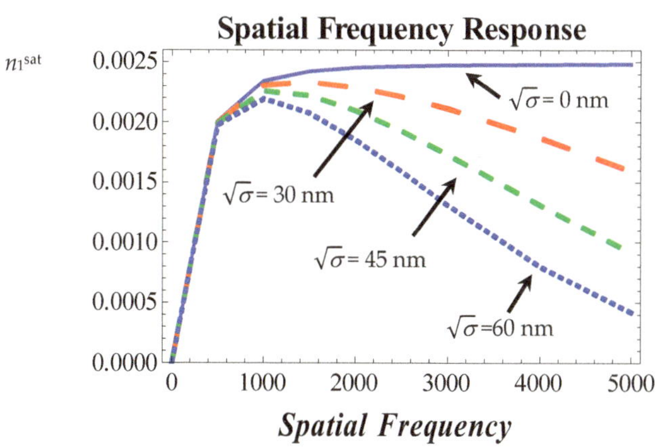

Figure 11. Simulations of the spatial frequency response of the refractive index modulation at saturation, n_1^{sat} for various values of the non-local response length, $\sqrt{\sigma}$ nm: {0 nm (solid); 30 nm (long dashed); 45 nm (dashed) and 60 nm (short dashed)}.

Figure 11 shows the saturation refractive index modulation, n_1^{sat} , plotted as a function of the grating spatial frequency, for the same grating parameter values, as those used in the generation of Figure 10. These results predict that lower $\sqrt{\sigma}$ values lead to a significant improvement in the high spatial frequency response of the material and therefore a reduction in the high spatial frequency roll-off observed experimentally. This is an important prediction of the NPDD model and motivates the study of the feasibility of applying chain transfer agents in free radical based photopolymer materials.

4.4. Experimental results

In this subsection, we aim to demonstrate and compare the effects of a number of CTAs on the spatial frequency response of the various photopolymer material compositions under examination. Following the analysis presented in Section 4.3, we aim to show that a reduction in average polymer chain length results in the predicted reduction in the non-local chain length $\sqrt{\sigma}$ and hence an increase in the material's response at high spatial frequencies. In order to achieve this, experimentally obtained growth curves and the saturation values of refractive index modulation, n_1^{sat}, for three photopolymer material compositions, over a range of spatial frequencies from 500 to 4799 lines/mm, are measured and compared [29,33,34].

The three material compositions examined are; (i) a standard acrylamide/polyvinylalcohol photopolymer [30-34], (ii) the standard acrylamide/polyvinylalcohol photopolymer with the addition of 0.96 g of sodium formate (CTA-1) [29,32,32] and (iii) the standard acrylamide/polyvinylalcohol photopolymer with 0.53 ml of 1-mercapto-2-propanol (CTA-2). .For each sample of these material composition examined, several growth curves were measured using a recording intensity of 1 mW/cm^2, at a recording wavelength of $\lambda = 532$ nm, in order to ensure experimental reproducibility. In all cases the diffraction efficiency of a probe beam, at $\lambda_p = 633$ nm was monitored during and post-exposure and angular scans of the gratings performed and analysed.

Applying the extended NPDD model presented in the previous subseciton, the experimental growth curve data was fit using a least squares algorithm, in which the Mean Square Error (MSE) between the prediction of the model and the experimental data was minimized to extract key material parameters. In order to carry out the fitting process, search ranges of typical parameter values based on data presented in the literature were used [30-34,48]. The model was then applied to analyse the temporal variation of the refractive index modulation n_1, for each of the materials at over the range of spatial frequencies examined. In this way the monomer diffusion constant D_{m0}, the propagation rate constant, k_p, the initial rate, k_i, bimolecular termination rate constant, k_t, the primary termination rate, k_{tp}, the chain transfer rate, k_{tr}, the re-initiation rate, k_{ri}, and the non-local response length, $\sqrt{\sigma}$, were all estimated.

The parameters estimated for the compositions studied are presented in Table 3 (standard AA/PVA), Table 4 (CTA-1) and Table 5 (CTA-2). For each spatial frequency the saturation refractive index modulation value, n_1^{sat} , is provided in the first column of each table.

Examining the estimated values for k_p, k_t and D_{m0} given in three tables, it can be seen that they are consistent and close to the values previously reported in the literature [30-34,48]. In particular we note that the mean values estimated for D_{m0}, given in all the tables, are very similar [30-34]. The values obtained for the parameters, k_{tr} and k_{ri}, in Table 4 and 5, lie within reasonable ranges based on the values reported in the literature [48]. They clearly indicate the different physical properties of the two CTAs when used in an AA/PVA based photopolymer material. For each CTA being examined, it should also be noted that the values found for k_{tr} and k_{ri}, are consistent for each of the different spatial frequencies.

SF (lines/mm) [n_1^{sat}×10^{-3}]	k_p (×10^7)	k_t (×10^8)	D_{m0} (×10^{-10})	$\sqrt{\sigma}$ (nm)	MSE (×10^{-10})
500 [2.07]	2.7	3.0	3.0	60	1.96
1000 [2.20]	2.3	3.6	1.0	68	1.64
1428 [2.36]	2.8	3.0	3.0	55	0.89
2000 [1.97]	2.2	3.8	2.0	60	1.09
2500 [1.56]	2.7	3.1	3.0	65	1.21
3000 [1.38]	2.6	3.2	3.0	60	2.72
Mean	**2.6±0.4**	**3.3±0.5**	**2.5±1.5**	**61.3±6.7**	**1.59±1.13**

Table 3. Spatial frequency parameter estimations for standard AA/PVA material.

SF (lines/mm) [n_1^{sat}×10^{-3}]	k_p (×10^7)	k_t (×10^8)	D_{m0} (×10^{-10})	k_{tr} (×10^6)	k_{ri} (×10^7)	$\sqrt{\sigma}$ (nm)	MSE (×10^{-10})
500 [1.90]	3.7	3.3	5.0	5.0	2.6	50	1.64
1000 [2.18]	3.5	3.4	3.0	2.0	1.0	51	2.72
1428 [2.39]	3.8	3.0	6.0	3.0	1.3	45	2.65
2000 [2.12]	3.2	4.0	6.0	2.0	1.7	45	2.36
2500 [1.87]	3.7	3.0	3.0	3.0	1.0	49	3.92
3000 [1.60]	3.8	3.6	1.0	2.0	1.0	47	1.17
Mean	**3.6±0.4**	**3.4±0.6**	**4.0±3.0**	**2.8±2.2**	**1.4±1.2**	**47.6±3.4**	**2.41±1.24**

Table 4. Parameter estimations for spatial frequencies in AA/PVA with sodium formate (CTA-1) material.

SF (lines/mm) [n_1^{sat}×10^{-3}]	k_p (×10^7)	k_t (×10^8)	D_{m0} (×10^{-10})	k_{tr} (×10^7)	k_{ri} (×10^6)	$\sqrt{\sigma}$ (nm)	MSE (×10^{-10})
500 [1.97]	2.0	3.5	5.0	1.0	8.0	40	1.98
1000 [2.09]	1.7	3.8	3.0	2.0	6.0	40	1.71
1428 [2.38]	1.6	3.9	2.0	2.0	3.0	42	1.53
2000 [2.23]	1.7	3.7	6.0	3.0	2.0	40	1.22
2500 [2.03]	2.1	3.0	2.0	1.0	3.0	38	5.53
3000 [1.68]	1.7	3.8	2.0	1.0	8.0	39	1.12
Mean	**1.8±0.3**	**3.6±0.6**	**3.3±2.7**	**1.7±1.3**	**5.0±3.0**	**39.8±2.2**	**2.18±3.35**

Table 5. Parameter estimations for spatial frequencies in AA/PVA with 1-mercapto-2-propanol (CTA-2) material.

In Table 3 the mean non-local response length in the standard AA/PVA material is estimated to be approximately $\sqrt{\sigma} \approx 61.3$ nm. This value agrees well with the previous estimates reported in the literature [29-34]. For the material containing sodium formate (CTA-1), the corresponding value is $\sqrt{\sigma} \approx 47.6$ nm, as given in Table 4, which corresponds to a reduction ~22.3% in the mean non-local response length, $\sqrt{\sigma}$. In Table 5, the $\sqrt{\sigma}$ value obtained for the material with 1-mercapto-2-propanol (CTA-2) is $\sqrt{\sigma} \approx 39.8$ nm. This corresponds to a ~35.0%, reduction in the mean $\sqrt{\sigma}$ value. Therefore significant improvements in the high spatial frequency responses with the addition of CTA can be clearly observed. CTA-2 is more effective in reducing the average polymer chain length. Furthermore, in very high spatial frequencies, the results consistently indicate an accompanying improvement in the magnitude of the saturation refractive index modulation. Based on the analysis above it is clear that shortening the polymer chains through the addition of chain transfer agents improves this photopolymer's spatial frequency response.

In order to more clearly illustrate the results presented in Tables 3 - 5 the spatial frequency response of each material are shown in Figure 12.

The improvements in the high spatial frequency response can clearly be observed. In order to demonstrate the improvements in AA/PVA material performance, numerical fits were carried out to the growth curves of all the three testing samples, for the 3000 lines/mm spatial frequency case. The results are presented in Figure 13 with the associated error bars, indicating the reproducibility of the experimental results.

In this section the NPDD model was extended to examine the effects of the addition of CTAs to an AA/PVA photopolymer. As the NPDD model predicts, a reduction in average polymer chain length grown during polymerisation will reduce the negative smearing effects which are caused by non-local polymer chain growth. As CTAs are known to reduce the average polymer chain length grown during polymerisation, a number of various types and concentrations of CTA were added to reduce these detrimental effects. Then using the NPDD model and fitting growth curves of refractive index modulation at various spatial frequencies for each of the compositions under examination, estimates of key material parameters were obtained. All results verified that the use of CTAs caused a substantial

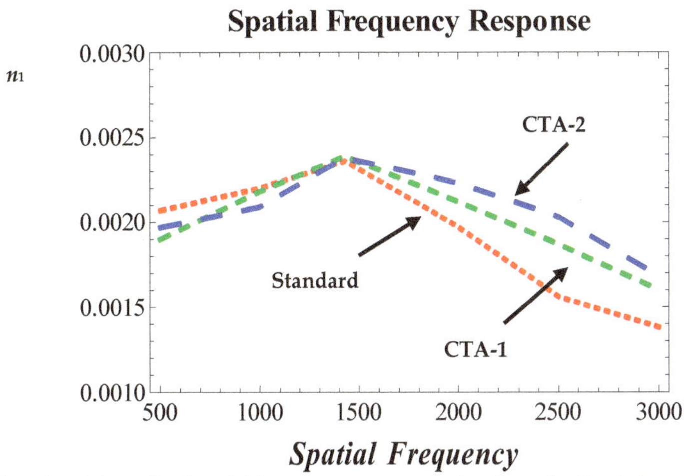

Figure 12. Experimental results of spatial frequency response of saturation refractive index modulation, n_1, variation, for standard material (small dashed line), standard material with 0.96g sodium formate (CTA-1) (medium dashed line) and standard material with 0.53ml 1-mercapto-2-propanol (CTA-2) (large dashed line).

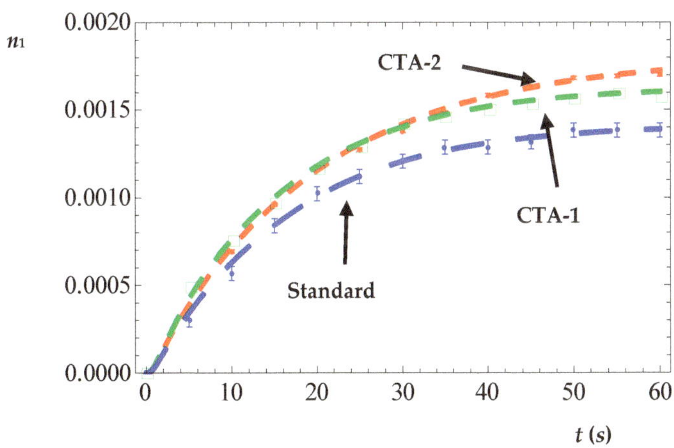

Figure 13. Experimental growth curve data with error bars and fits at 3000 lines/mm, standard material (large dashed line); with sodium formate, CTA-1, (medium dashed line and empty circle points) and with 1-mercapto-2-propanol, CTA-2, (small dashed line).

increase in the response of the AA/PVA photopolymer's spatial frequency response, particularly at higher spatial frequencies, which is in line with the NPDD model predictions.

5. Conclusion

In this chapter we briefly reviewed some of the developments made in the area of photopolymer material development. We also examined some of the extensions which have been made to the Non-local Photo-polymerisation Driven Diffusion (NPDD) model in order to increase its physicality, with the aim of producing a tool for photopolymer material optimization. A detailed understanding of the photochemical and photo-physical effects which take place during and after holographic exposure is of extreme importance in order to achieve such a tool. Some of the recent developments which have been made in order to achieve this were illustrated and their implications examined. Among these were the temporal and spatial primary radical, the effects of oxygen inhibition, non-local polymer chain growth and the addition of chain transfer agents to improve the spatial frequency response. As various photopolymer compositions have diverse chemical and structural characteristics, knowledge of the characteristics required when choosing these components will offer an informed choice to yield specific improvements in material performance. The implications of the predictions presented here suggest that there are many ways in which improvements can be made.

There still remain a number of effects which have not been included into the NPDD model which would increase its physicality and therefore make it a more powerful tool. One of these which has not been included here, is the addition of time varying viscosity effects. When polymerisation occurs, densification and crosslinking occurs, resulting in a reduction in the material's fractional free volume. This reduction causes an increase in the viscosity of the material which restricts the rate at which polymerisation proceeds. Accurate modelling of these effects would allow for more precise recording schedules for subsequent holographic exposures to be determined and would enable optimum concentration ratios of the main constituents of the material to be found.

Author details

Michael R. Gleeson
National University of Ireland Maynooth, Department of Computer Science, Faculty of Science and Engineering, Maynooth, Co. Kildare, Republic of Ireland

Jinxin Guo and John T. Sheridan
UCD School of Electrical, Electronic and Communications Engineering, College of Engineering, Mathematical and Physical Sciences, Communications and Optoelectronic Research Centre, SFI Strategic Research Centre in Solar Energy Conversion, University College Dublin, Belfield, Republic of Ireland

Acknowledgement

The authors acknowledge the support of the Irish Research Council for Science, Engineering and Technology through the Empower Postdoctoral research scholarship. The authors also acknowledge the support of Enterprise Ireland and Science Foundation Ireland through the national development plan.

6. References

[1] Close D H, Jacobson A D, Magerum R C, Brault R G, McClung F J (1969), Hologram recording on photopolymer materials. Appl. Phys. Lett. j. 14: 159-160.
[2] InPhase Technologies (2007), www.inphase-technologies.com Tapestry Media.
[3] Maruo S, Nakamura O, Kawata S (1997), Three-dimensional microfabrication with two-photon-absorbed photopolymerization. Opt. Lett. j. 22: 132-134.
[4] Saravanamuttu K, Blanford C F, Sharp D N, Dedman E R, Turberfield A J, Denning R G (2003) Sol-gel organic-inorganic composites for 3-D holographic lithography of photonic crystals with submicron periodicity. Chem. Mater. j. 15: 2301-2304.
[5] Gu M, Straub M, Nguyen L H, (2004), Complex-shaped 3-D microstructures and photonic crystals generated in a polysiloxane polymer by two-photon microstereolithography. Opt. Mater. j. 27: 359-364.
[6] Sullivan A C, Grabowski M W, McLeod R R, (2007), Three-dimensional direct-write lithography into photopolymer. Appl. Opt. j. 46: 295-301.
[7] Dhar L, Hale A, Katz H E, Schilling M L, Schnoes M G, Schilling F C (1999), Recording media that exhibit high dynamic range for digital holographic data storage. Opt. Lett. j. 24: 487-489.
[8] Dhar L, Hale A, Kurtis K, Schnoes M, Tackitt M, Wislon W, Hill A, Schilling M, Katz H, Olsen A (2000), Photopolymer recording media for high density data storage. IEEE Conf. Digest: Optical Data Storage 158-160.
[9] Schultz S, Glytsis E, Gaylord T (2000), Design, fabrication, and performance of preferential-order volume grating waveguide couplers. Appl. Opt. j. 39: 1223-1232.
[10] Sato A, Scepanovic M, Kostuk R (2003), Holographic edge-illuminated polymer bragg gratings for dense wavelength division optical filters at 1550 nm. Appl. Opt. j. 42: 778-784.
[11] Sheridan J T, Kelly J V, O'Brien G, Gleeson M R, O'Neill F T (2004), Generalized non-local responses and higher harmonic retention in non-local polymerization driven diffusion model based simulations. J. Opt. A: Pure Appl. Opt. j. 6: 1089-1096.
[12] McLeod R R, Daiber A J, McDonald M E, Robertson T L, Slagle T, Sochava S L, Hesselink L (2005), Microholographic multilayer optical disk data storage. Appl. Opt. j. 44: 3197-3207.
[13] Milster T D (2005), Horizons for optical storage. Opt. Photon. News. j. 16: 28-33.
[14] Barachevskii V A (2006), Photopolymerizable recording media for three-dimensional holographic optical memory. High Energy Chem. j. 40: 131-141.
[15] Sheridan J T, Gleeson M R, Kelly J V, Close C E, O'Neill F T (2006), Optimized holographic data storage: diffusion and randomization. J. Opt. A: Pure Appl. Opt. j. 8: 236-243.

[16] Toisihi M, Tanaka R, Watanabe K, Betsuyaku K (2007), Analysis of photopolymer media of holographic data storage using non-local polymerization driven diffusion model. Jap. J. Appl. Phys. j. 46: 3438-3447.

[17] Nagy Z, Koppa P, Ujhelyi F, Dietz E, Frohmann S, Orlic S (2007), Modelling material saturation effects in microholographic recording. Opt. Exp. j. 15: 1732-1737.

[18] Toishi M, Takeda T, Tanaka K, Tanaka T. Fukumoto A, Watanabe K (2008), Two-dimensional simulation of holographic data storage medium for multiplexed recording. Opt. Exp. j. 16: 2829-2839.

[19] Sadlej N, Smolinska B (1975), Stable photo-sensitive polymer layers for holography. Opt. Laser Technol. j. 7: 175-179.

[20] Calixto S (1987), Dry polymer for holographic recording. Appl. Opt. j. 26: 3904-3909.

[21] Blaya S, Mallavia R, Carretero L, Fimia A, Madrigal R F (1998), Highly sensitive photopolymerisable dry film for use in real time holography. Appl. Hys. Lett. j. 75: 1628-1630.

[22] Zhao F, Frietmann E E, Li X (1998), Novel type of red sensitive photopolymer system for optical storage. Proc. SPIE. j. 3486: 317-321.

[23] Trentler T, Boyd J, Colvin V (2000), Epoxy resin photopolymer composites for volume holography. Chem. Mater. j. 12: 1431-1438.

[24] Bayer MaterialScience AG, www.bayermaterialscience.com.

[25] Rolle T, Bruder F-K, Facke T, Weiser M-S, Honel D, Stoeckel N (2010), (2010), "Photopolymerzusammensetzungen für optische Elemente und visuelle Darstellungen. EP2 172 505 A1.

[26] Gleeson M R, Sheridan J T, Bruder F-K, Roelle T, Honnell D, Weiser M-S, Faecke T (2011), Comparison of a new self-developing photopolymer with AA/PVA based photopolymer utilizing the NPDD model. Opt. Exp. j. 19: 26325-26342.

[27] Gleeson M R, Guo J, Sheridan J T (2011), Optimisation of Photopolymers for Holographic Applications using the Non-Local Photo-Polymerization Driven Diffusion Model. Opt. Exp. j. 19: 22423-22436.

[28] Gleeson M R, Kelly J V, Sabol D, Close C E, Liu S, Sheridan J T (2007), Modelling the photochemical effects present during holographic grating formation in photopolymer materials. J. Appl. Phys. j. 102: 023108.

[29] Gleeson M R, Sabol D, Liu S, Close C E, Kelly JV, Sheridan J T (2008), Improvement of the spatial frequency response of photopolymer materials by modifying polymer chain length. J. Opt. Soc. Am. B. j. 25: 396-406.

[30] Gleeson M R, Sheridan J T (2009), Non-local photo-polymerization kinetics including multiple termination mechanisms and dark reactions: Part I. Modelling. J. Opt. Soc. Am. B. j. 26: 1736-1745.

[31] Gleeson M R, Liu S, McLeod R R, Sheridan T (2009), Non-local photo-polymerization kinetics including multiple termination mechanisms and dark reactions: Part II. Experimental Validation. J. Opt. Soc. Am. B. j. 26: 1746-1754.

[32] Gleeson M R, Liu S, Guo J, Sheridan J T (2010), Non-local photo-polymerization kinetics including multiple termination mechanisms and dark reactions: Part III. Primary radical generation and inhibition. J. Opt. Soc. Am. B. j. 27: 1804-1812.

[33] Guo J, Gleeson M R, Liu S, Sheridan J T (2011), Non-Local Spatial Frequency Response of Photopolymer materials containing Chain Transfer Agents: Part I. Theoretical Model. J. Opt. A: Pure Appl. Opt. j. 13: 095601.
[34] Guo J, Gleeson M R, Liu S, Sheridan J T (2011), Non-Local Spatial Frequency Response of Photopolymer materials containing Chain Transfer Agents: Part II. Experimental Results. J. Opt. A: Pure Appl. Opt. j. 13: 095602.
[35] Kogelnik H (1969), Coupled wave theory for thick hologram gratings. Bell Syst. Tech. J. j.48: 2909-2945.
[36] Syms R R A (1990), Practical Volume Holography, In Clarendon Press, Oxford.
[37] O'Neill F T, Lawrence J R, Sheridan J T (2000), Automated recording and testing of holographic optical element arrays. Optik. j. 111: 459-467.
[38] Fimia A, Lopez N, Mateos F, Sastre R, Pineda J, Amatguerri F (1993), Elimination of oxygen inhibition in photopolymer system used as holographic recording materials. J. Mod. Opt. j. 40: 699-706.
[39] Manivannan G, Lessard R A (1994), Trends in holographic recording materials. Trends Polym. Sci. j. 2: 282-290.
[39] Lawrence J R, O'Neill F T, Sheridan J T (2001), Photopolymer holographic recording material. Optik j. 112: 449-463.
[40] O'Brien A K, Bowman C N (2006), Modeling the effect of oxygen on photopolymerization kinetics. Macromol. Theor. Sim. j. 15: 176-182.
[41] Harbour S, Kelly J V, Galstian T, Sheridan J T (2007), Optical birefringence and anisotropic scattering in acrylate based holographic polymer dispersed liquid crystals. Opt. Comm. j. 278: 28-33.
[42] Bowman C N, Peppas N A (1991), Coupling of kinetics and volume relaxation during polymerizations of multiacrylates and multimethacrylates. Macromol. j. 28: 1914-1920.
[43]. Aubrecht I, Miller M, Koudela I (1998), Recording of holographic diffraction gratings in photopolymers: Theoretical modelling and real-time monitoring of grating growth. J. Mod. Opt. j. 45: 1465-1477.
[44] Gleeson M R, Sheridan J T (2009), A review of the modelling of free-radical photopolymerisation in the formation of holographic gratings. J. Opt. A: Pure Appl. Opt.: Special Issue, Optics of Nanocomposite Materials. j. 10: 024008.
[45] Carretero L, Blaya S, Mallavia R, Madrigal R F, Belendez A, Fimia A (1998), Theoretical and experimental study of the bleaching of a dye in a film-polymerization process. Appl. Opt. j. 37: 4496-4499.
[46] Gleeson M R, Liu S, O'Duil S, Sheridan J T (2008), Examination of the photoinitiation processes in photopolymer materials. J. Appl. Phys. j. 104: 064917.
[47] Gallego S, Ortuno M, Neipp C, Marquez A, Belendez A, Pascual I, Kelly J V, Sheridan T (2005), Physical and effective optical thickness of holographic diffraction gratings recorded in photopolymers. Opt. Exp. j. 13: 1939-1947.
[48] Odian G (1991), Principles of Polymerization. Wiley New York.
[49] Gleeson M R, Kelly J V, Close C E, O'Neill F T, Sheridan J T (2006), Effects of absorption and inhibition during grating formation in photopolymer materials. J. Opt Soc. Am. B. j. 23: 2079-2088.

[50] Sheridan J T, Lawrence J R (2000), Nonlocal response diffusion model of holographic recording in photopolymer. J. Opt. Soc. Am. A. j. 17: 1108-1114.
[51] Crank J (1956), The Mathematics of Diffusion. Oxford University Press.
[52] Mayo F R (1943), Chain Transfer in the Polymerization of Styrene: The Reaction of Solvents with Free Radicals. J. Am. Chem. Soc. j. 65: 2324-2329.
[53] Goretta L A, Otremba R R (1981), U.S. Patent No. 4,307,215.
[54] Gartner H A (1992), U.S. Patent No. 5,171,783.
[55] Fevola M, Hester R, McCormack C (2003), Molecular weight control of Polyacyrlamide with Sodium Formate as a Chain-Transfer Agent: Characterization via size Exclusion Chromatography/multi-angle laser light scattering and determination of chain-Transfer constant. J. Polym. Sci., Part A: Poly. Chem. j. 41: 560-568.
[56] Valdebenito A, Encinas M V (2005), Thiophenols as chain transfer agents in the polymerization of vinyl monomers. Polymer. j. 46: 10658-10662.
[57] Okaya T, Kikuchi K, Morii Y (1997), Polymerization of acrylamide in aqueous medium initiated with a redox system composed of cysteine and potassium bromate. Macromol. Chem. Phys. j. 198: 2027-2034.

Chapter 4

Conducting Polymers Application

Kareema Majeed Ziadan

Additional information is available at the end of the chapter

1. Introduction

Organic polymers are normally insulators, it can be presumed that conducting polymers must have an unusual structure. Polymers with conjugated π-electron (i. e. system have C=C conjugated bonds) backbones display unusual electronic properties such as low energy optical transition, low ionization potentials, and high electron affinities. The result is a class of polymers that can be oxidized or reduced more easily and more reversibly than conventional polymers. The effect of this oxidation or reduction on polymer is called doping, i. e. convert an insulating polymer to conducting one) Kroschwitze, 1988).

Conducting polymers (CPs) such as polypyrrole, polythiophene and polyanilines are complex dynamic structures that captivate the imagination of those involved in intelligent materials research. (Spinks, et al., 2000; Riley, & Wallace, 1991). The application of electrical stimuli can result in drastic changes in the chemical, electrical and mechanical properties of CPs. These complex properties can be controlled only if we understand, first, the nature of the processes that regulate them during the synthesis of the conducting polymers, and second, the extent to which these properties are changed by the application of an electrical stimulus. polyaniline and its derivative is one of important conducting polymer, it has many application such as organic light emitting diodes (OLEDs) (Burn et al, field-effect transistors (OFETs) (Nam et al 2011), corrosion (Solange, (2007) and solar cells (Alet, 2006. & McEvoy et al, 1994; Williams, 2005) In this chapter focusing on mechanism of conduction and mechanism of charge transport of organic conducting polymer. Also on application solar cell. chemical sensor and corrosion.

2. The mechanism of conduction

The polymer in their pure (undoped) state are describe as electronic insulators. When these polymer are doped the conductivity change from insulators to metals. The conductivity, б, is proportional to carrier concentration, n, and the carrier mobility, μ, i. e.

$$б = e\,n\,\mu \tag{1}$$

For intrinsic conductivity, n decreases exponentially with increasing band gap, since the conjugated polymers have relatively large band gap, consequently, n is very low in normal temperature, so that a low value of n leads to a low value of conductivity of undoped polymers even-though the polymers have high carrier mobility Kroschwitze J. I, 1988

In doped polymers, the doping of conjugated polymers generates high conductivities by increasing the carrier concentration n. This accomplished by oxidation or reduction with electron acceptors or donors respectively. The polymer is oxidized by the acceptors (removal of electron), thereby producing a radical cation (hole) on the chain.

The radical cation with lattice distortion around the charge is called polaron with positive charged hole site. This hole site moves through the polymer and contributes to the conductivity. This polymer is called p-type polymer. For donor doped polymer (n-type) that is obtained by reduction is adding electron to the chain. This process produces polaron with negative charge. The Hall effect measurement in polymer shows positively charged carriers for acceptor doped polymer (p-type) and negatively charged carrier for donor-doped polymer (n-type). The thermo power and junction measurement show the same result as that determined by Hall-effect (Krichelore, 1992).

The doping concentration in polymer is high compared with that in organic semiconductors (in parts per – million). In some case the doping reaches 50% of the final weight of conducting polymer. This can be determined by chemical or spectroscopic analysis or simply by weight up take. The conducting polymer doped can be return to insulating state by neutralization back to the uncharged stat. This return to neutrality is referred as compensation. Exposure of ox datively doped polymers to electron donors or conversely, of reductively doped polymer to electron acceptors effects compensation. This ability to cycle between charged and neutral states forms the base for the application of conducting polymer in rechargeable batteries (Kareema 1997).

The doping process produce number of carriers in polymer, but these carriers must be mobile in order to contribute to conductivity, eq. (1). The carriers transport in doped conjugated polymer are analogies to doped semiconductor. In both cases doping introduces new electronic states within the band gap of material. The difference is that in conductive polymer, the total oscillator strength dose not increased upon doping, and generated polaron density of state is created by shifted the band density of state to band gap. At high doping concentrations these states interact strongly with each other, and as a result, the overlap of their electronic wave functions yield a band of electronic state within the band gap instead of discrete levels. The mechanism of carrier transport in conducting polymer is probably more likely to that in amorphous semiconductors (hoppingtransport) than crystalline semiconductors (band transport). A conclusion may be drawn that the doping creates an active sites (polarons) which enable the carriers (electronic & holes) to move from one site to another. By hopping mechanism through these sites.

3. The mechanism of charge transport

The doping of conducting polymer induces charge transfer along the chains which leads to local relaxation. The equilibrium geometry in ionized states is different from that in ground state, and that the electronic structure is affected by the localized electron states in the gap which modify the π system. In order to understand the mechanism of conduction, we must have information about ground-state geometries and doped state. Polyacetylene has degenerate ground state (two geometric structures having the same total energy). The defect divides the chain of PA into two parts with the same energy. The movement of defect can be described by soliton scheme (1).

Scheme 1. Polyacetylene (degenerate ground stat)

Polythiophene, polypyrrole and heterocycle polymer processes a nondegenerate ground stats Kroschwitze J. I, 1988 scheme (2).

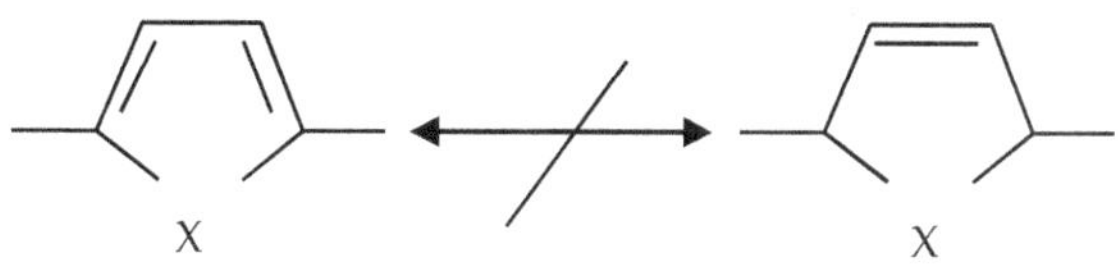

Scheme 2. Heterocyclic polymer x=S, O, NH (nondegenerate ground stats).

It has suggested that the stable defect state formed upon doping are polarons and bipolarons and optical data seem to support the evidence of the fommation of polarons (Single charge parameter state) and bipolarons. (Kareema 1997)

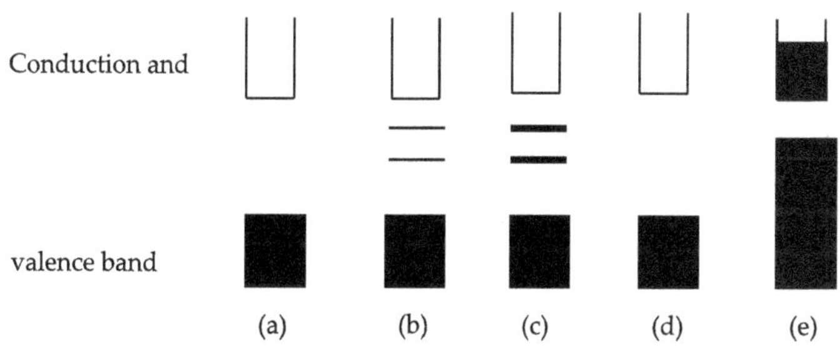

Figure 1. Schematic variation of band gap as a function of doping concentration \ (a) undoped (b) very small doping (polaron) (c) small doping (bipolaron) (d) high doping bipolaronic band (e) 100% doping (theoretically speaking.

Fig. (1-a) shows the neutral undoped heterocyclic polymer of band gap tabulated in table (1-1). Fig. (1-b) shows the doping levels of order (0. 1-1%) %. The appearance of an ESR signal and of optical bands in the gap is consistent with formation of polarons. Fig. (1-c) indicates the increasing of the doping give rise a few percent (1-5%), leading to drop in spin concentration and the presence, in the optical data, two peaks located respectively below the conduction band and above the valence band indicated that the polarons bind in pairs to form diamagnetic bipolaron state. Fig. (1-d) expresses when the doping levels reach (25 – 50) mol. % the bipolarons states overlap and form the bands in the gap. Fig. (1-e) shows when doping reach to 100% (theoretically speaking) the bipolaronic band may merge with conduction band and valence bands respectively leaving a reduced gap. This is consistent with optical data indicating that the band gap disappear (Skotheim, 1986).

4. Polymer solar cell

Polymer solar cells have attracted broad research interest because of their advantageous solution processing capability and formation of low-cost, flexible, and large area electronic devices. (Williams, 2005); Vignesh et al, 2006, Schiff, 2002). However, the efficiency of polymer solar cells is still low compared to that of inorganic solar cells. Therefore, it is a challenge to find a polymer that has all the required properties for high efficiency devices, such as strong and broad absorption, high carrier mobility, and appropriate energy levels. One possible solution to avoid the strict material requirements is to stack two or more devices with different spectral responses, which enables more efficient utilization of solar energy.

A typical example of the usability of organic semiconductors is the dye-sensitised solar cell which owes its first demonstration Gratzel and his co-workers (McEvoy et al., (1994); Michael, 2003).

Organic/inorganic hetero structure solar cells have also been studied recently and power conversion efficiency in the region of 2. 5% has been achieved by (Hussein, 2010). Solar cells based on conducting polymer/amorphous silicon (a-Si:H) structures have also been reported which demonstrate solar conversion efficiency in the range of 2% (Williams et al., 2005; Alet 2006). Blend heterojunctions consisting of a bulk mixture of poly (3-hexylthiophene) (P3HT) as donor and 6, 6-phenyl C61-butyric acid methyl ester (PCBM) as acceptor are very promising structures Brabec, C. J. (2004). Studies on blend nanoscale morphology Hoppe H. et al (2006) and stability of organic solar cells (Bettignies et al, 2006) are currently subjects of intense research.

In recent years, the development of thin film plastic solar cells, using polymer-fullerene (Gao et al., 1995); Shaheen et al., 2001) or polymer-polymer (Granström et al, 1998) bulk heterojunctions as an absorber (and transport layer at the same time), has made significant progress. Efficiencies between 1% and 2. 5% for laboratory cells under AM1. 5 illumination conditions have been reported (Gao et al., 1995;. Dyakonov. et al 2001). The typical structure of these devices consists of a composite of two materials with donor and acceptor properties, respectively, sandwiched between two electrodes. One advantage of this type of devices is their ease of processability. The active layer is solution processed by using spin-coating technique. Kareema et al, 2010). study photovoltaic properties of Polyaniline/Si solar cell in the dark and under illumination investigated hybrid and was found to deliver short circuit current density Jsc =45μA/cm2, open circuit voltage Voc = 400mV, and solar cell efficiency η=0. 3% under AM 1. 5 simulated solar light with the intensity of 100mW/cm2.

In this paragraph investigated some of the photovoltaic properties of organic polymer poly (o-toluidine) (POT) doped with para-toluene sulphonic acid (PTSA) deposited onto n-type silicon substrates (Kareema, 2012) (this study part of thesis done under my supervisor by Hussein, 2010).

Poly (o-toluidine) (POT) doped with para-toluene sulphonic acid (PTSA) were prepared following chemical methods described in the literature (Kulkarni, & Mulik, 2005). Powder of the doped polymer was dissolved in formic acid (HCOOH) in the concentration of 10 mg. ml-1. Single layer heterojunction solar cells were then prepared by spin casting solutions of POT-PTSA onto the silicon substrates using spin speeds in the range 1000-5000 rpm and spinning time of 60 sec. Thin films were then placed on a hotplate with temperature of 90°C for a period of 15 min for drying. Aluminum (Al) contact of about 90 nm in thickness was thermally evaporated onto the back (the unpolished side) of the Si substrate. This was carried out under vacuum of 10^{-5} -10^{-6} mbar and evaporation rate of 5 nm. sec^{-1}. Similar procedure was followed for the deposition of gold (Au) contacts onto the polymer film, with Au film thickness of about 20 nm evaporated through a suitable mask which provides device area of about $3x10^{-6}$ m^2. For DC electrical measurements of the solar cell devices a Keithley electrometer (Model 6517A) was used to measure current density (J) as a function of applied voltage in range ±1V and in steps of 0. 05V.

The photovoltaic properties of the solar cells were measured under illumination using a Bentham 605 solar simulator fitted with a xenon lamp. The photocurrent was measured for devices of different polymer film thickness and under four different light illuminating intensities between 10-100 mW/cm^2.

Figure (2) shows the J (V) characteristics of the fabricated POT-PTSA/n-Si solar cell structures, both in dark and under illumination. The polymer film thickness for this particular result is 35nm as determined by spectroscopic ellipsometry measurements and the illumination intensity is of 100 mW/cm^2. Both J (V) curves clearly possess good diode characteristics which clearly demonstrates the occurrence of a rectifying junction. This junction is expected to exist at the interface between the silicon substrate and the polymer film. This can be further justified by the fact that the silicon substrate used in this work is of n-type while the POT-PTSA films are considered as the hole transporting layer (Mangal et al., 2009). The POT-PTSA films are thin enough to allow the photon flux to reach the n-Si substrate and thus allowing the photocurrent current to saturate. Solar cell parameters, i. e., V_{oc}, J_{sc}, V_p. J_p. P_{max}, and FF have been determined and are summarised in Table I. The solar conversion efficiency η is given by the formula:

$$\eta = (FF\,V_{oc}\,J_{sc} / P_{in}) \qquad (2)$$

where P_{in} is the power of the incident light. A typical solar conversion efficiency value value of 2. 55% for the studied devices is found to be in line with data found in the literature and are thought to be limited by uncontrolled interface states at the POT-PTSA/n-Si junction

(Wang &. Schiff, (2007). A closely related structure of the type aluminium /polyaniline / GaAs metal-insulator semiconductor solar cell was found to give efficiencies in the region of 5%. The value of 0. 46 V for the short circuit voltage obtained in this work compares well with the value 0. 51V obtained for Pani /n-Si solar cell devices (Wang & Schiff 2007). A low values of FF of about 0. 43 is associated with a high series resistance and a low shunt resistance with typical values of 37Ω and 465Ω, respectively. High values for RS may originate from a poor absorber morphology limiting the electron hopping transport transport, (Levitsky et al., 2004; Riedel et al., 2004).

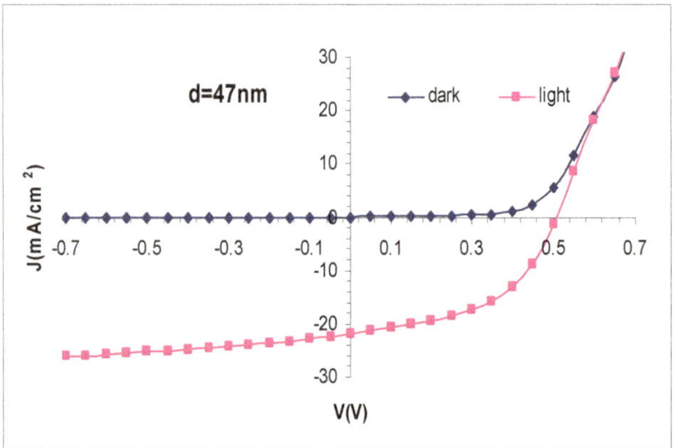

Figure 2. Current density as function of voltage for Au/POT-PTSA/n-Si/Al solar cell. The white light illumination intensity is 100mW/cm2.

Figure (3) shows the conversion efficiency as a function of film thickness of POT-PTSA spun films in the range 25-73nm. The efficiency is shown to increase sharply with increasing film thickness reaching a maximum value of 2. 55% for film thickness of 35nm, and then decreases for slightly larger thickness (43 nm), before it starts to increase again, but more gradually, for higher thicknesses. The organic thin film of POT-PTSA is typically characterized with a lower charge carrier density and charge mobility compared to polyaniline which is subjected to a similar doping treatment (Kulkarni. & Viswais, 2004).

It is expected that for thin films of relatively small thickness (in this case 35nm), the surface texture is more suitable for light trapping compared to smaller or larger film (25nm and 43nm) (Izabela et al., 2008; Haug, et al., 2009). Furthermore, the gradual increase in light conversion efficiency may be associated with the increase in grain size leading to reduced grain boundary charge scattering. Figure (4) shows the current density as a function of applied voltage with different illuminating light intensities. The parameters of solar cells are tabulated in Table (I). The short circuit current and open circuit voltage are decreasing with decreasing illumination intensity, however, the solar conversion efficiency is found to increase with decreasing light intensity with a maximum efficiency of about 5% observed under 100 mWcm^{-2}. Such effect has been associated with light dependent parallel resistance which affects the solar conversion efficiency of such devices.

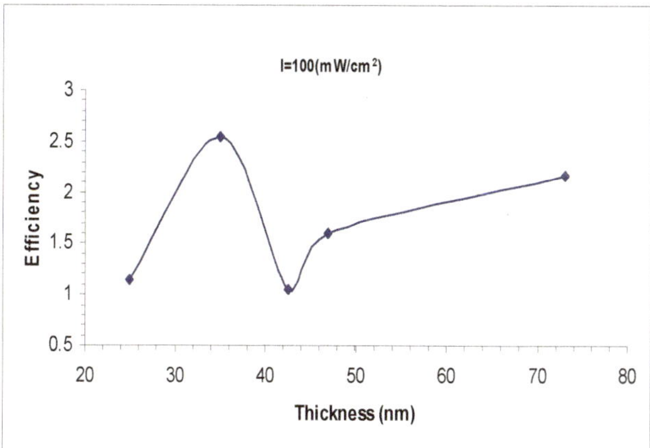

Figure 3. Efficiency versus film thicknesses of (POT-PTSA) /n-Si solar cell at room temperature.

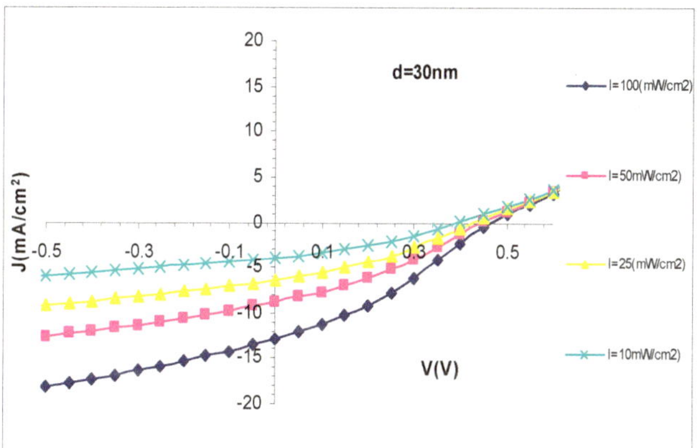

Figure 4. J (V) Characteristic of Al/n-Si/ (POT-PTSA) /Au solar cells measured at different light intensity with film thickness of 35nm.

Intensity	V_{0c} (V)	J_{sc} (mA /cm^2)	V_p (V)	J_p (mA /cm^2)	P_{max} (mW /cm^2)	FF	η	R_s Ω	R_{sh} Ω
100	0. 46	13	0. 34	7. 5	2550	0. 426	2. 5	37	465
50	0. 45	9	0. 3	6	1800	0. 444	3. 6	44	719
25	0. 44	6. 5	0. 24	4	960	0. 335	3. 84	46	917
10	0. 38	4	0. 22	2. 2	484	0. 318	4. 84	55	1666

Table 1. Parameter calculated form J (V) characteristics of (POT-PTSA) at different light intensity (film thickness 30nm

Photocurrent generation parameters of POT-PTSA/n-Si structures are expected to improve further by improving POT conductivity which can be achieved by increasing the ratio of the dopant (PTSA) concentration into POT solutions. Thin films of POT have been shown to have lower conductivity and charge mobility compared to Pani films due to the introduction of -CH3 groups into the Pani chains in order to produce POT (Gordon et al., 2003). It was demonstrated that Pani Conductivity in the range 10^{-5}-10^{1} S/cm can be achieved by increasing its Xylene concentration. In the current work poly (o-toluidine) doped by para-toluene sulphonic acid was prepared following the procedure of Kulkarni and his coworkers (. Kulkarni et al., 2004). The conductivity of 1. 93×10^{-3} S/cm was obtained for POT-PTSA samples in pellet form. Our preliminary study shows that solar cell properties based on hybrid structures incorporating POT-PTSA is quite promising. Therefore further work towards improving the electrical conduction properties of PTSA doped POT films is underway.

5. Chemical sensors

Conducting polymers, such as polypyrrole (PPy), polyaniline (Pani), polythiophene (PTh) and their derivatives, have been used as the active layers of gas sensors since early 1980. (McQuade et al, 2000].

There are two main types of applications for the Conducting organic polymers in electronics: first one is that a polymer can be used as a material for constructing different devices and as discriminating layers in electronic chemical sensors.

In both cases, interacting with surrounding gases is vital. It can possibly determine the performance of the devices that are depending on conducting polymers, while it is helpful and supportive in chemical sensors. Conductivity has been the main property of concern; the aim is of study and to identify the usability of the conducting polymers in the two kinds of electronic applications mentioned above.

Electronic Chemical Sensors for gases are thought of to be at the top of gaining the information related to the environment that we live in. The quality of the air that we breathe in our bodies is very important Issue and is a real concern of modern society.

The rich literature concerned about different applications of CPs could be classified within two groups: conducting polymers (CPs) in electronic (Angelopoulos, 2001), electromechanical (Gazotti, 2001), and optoelectronic (Otero, 2000) devices in the first group, and conducting polymers (CP's) in electronic chemical sensors that are based on the mechanisms of mechanical, electronic, or optical transduction (Bailey & Persaud, 2000). Sufficient operation of the first group depends on the chemical stability of the conducting polymers in the surrounding environment, where the applications of a sensor obtain benefit of the physical changes that occur to the CP as it is exposed to several chemical solvents. That property is due to the macroscopic and molecular structure of Conducting Polymers (CPs). They are quite open materials that allow gases to enter their inner structure (Fig5).

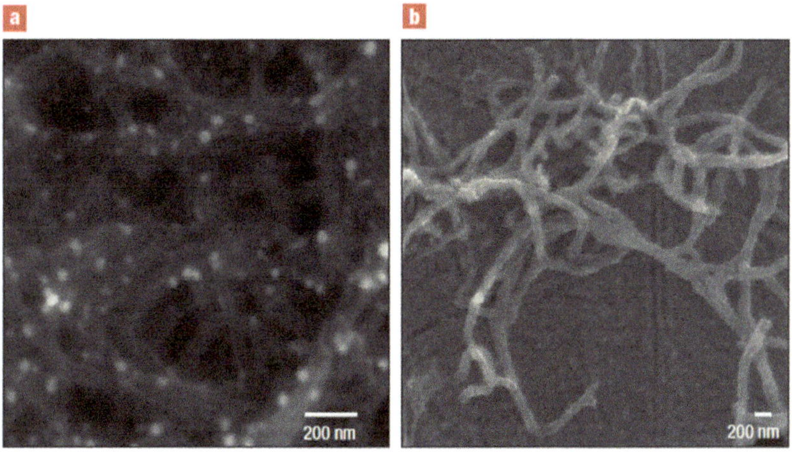

Figure 5. (a) polyaniline Open structure set by electropolymerization. The white dots are gold clusters created by immersion of the film (b), Filament growth onto compact and smooth 300-nm-thicklayer of electropolymerized polyaniline

On the contrary, the resistance of usual "silicon solid-state electronics" to changes in the surrounding chemical environments is up to the exceptional passivating properties of the layers of the fine inorganic surface (Huber, 1985).

This statement will only discuss sensors that are based on the chemical modulation of the CPs electronic properties which result from interacting with gases. Recently, reviews have been about changes of other properties that are used as principles of transduction, e. g., mass (Milella, & Penza, M. SAW (1998) (36) or optical properties Leclerc, M (1999) (37), and in liquid applications (Michalska, & Lewenstam, 2000)

Carbon nanotube and single polyaniline nanofiber gas sensors were investigated (Da-Jing CHEN et al., 2012) in this study. Carbon nanotube gas sensor was constructed by means of dielectrophoresis. Single nanofiber was deposited as nanofiber sensor across two gold electrodes by means of near-field electrospinning without conventional lithography process. The nanofiber sensor showed a 2. 7% reversible resistance change to 1 ppm NH3 with a response time of 60 s. Carbon nanotube sensor showed good linearity in the concentration range above 20 ppm with response times between 100 and 200 s. The fibers with smaller diameter showed quicker response to NH3 on the basis of gas diffusion mechanism. As such, CNT sensor and nanofiber-based sensors could be promising for gas sensing array and multi-chemical sensing applications.

Arenas (Arenas et al., 2012) studied conducting Polyaniline films (Pani) on Corning glass substrates, produced using either an in-situ doping process or a co-doping process, were prepared by the oxidative polymerization of aniline in N, N, dimethylformamide. Bicyclic aliphatic camphorsulfonic acid (CSA), aromatic toluenesulfonic acid (TSA) and carboxylic trifluoroacetic acid. The codoped process reduces the roughness of the CSA-doped films by 50%, but the conductivity depends on the acid type used for this process (TSA or TFA). The optical gas sensor response of the films is related to both the morphology and the degree of protonation. In this study, Pani with a microfiber morphology obtained from TSA-doping is the most sensitive to ammonia gas sensing, and Pani with flower-like morphology is the least sensitive.

The performance at room temperature of nanostructured polyaniline (Pani) –titanium dioxide (TiO2) ammonia gas sensors was investigated by (. Pawar1 et al., 2012) The PANi–TiO2 thin-film sensors were fabricated with a spin-coating method on glass substrates. PANi–TiO2 (0–50%) sensor films were characterized for their structural, morphological, optical, and various gas-sensing properties. The gas-sensing properties showed that the sensors exhibited selectivity to ammonia (NH3) at room temperature

Patil (Patil S. V et al., 2012) studied room temperature (300 K) liquefied petroleum gas (LPG) sensor based on n-polypyrrole/p-polyaniline (n-PPy/p-Pani) heterojunction has been fabricated using simple inexpensive electrodeposition method. The n-PPy/p-Pani heterojunction was fabricated by depositing polyaniline over predeposited polypyrrole thin film. The n-PPy/p-PANI heterojunction showed selectivity towards LPG as compared to N2 and CO2. The room temperature maximum gas response of 33 % (± 3 %) was achieved upon exposure of LPG at 1040 ppm

Conductive polyaniline (Pani) and single-walled carbon nanotube (SWNT) composite materials and its sensing property when NH3 and CO gases co-existed investigated by (Hyang Hee Choi et al., 2012). To improve the gas sensor properties, we deposited PANI using a drop-casting method to warp the Pani surrounding the SWNT. The Pani/SWNT composite material sensors showed a faster response to NH3 gas than CO gas. The CO gas increased the composite conductance, while the NH3 gas had the opposite effect.

Conducting polymers (CP's) could be utilized as discriminating layer in a sensor or to be the transducer itself. Therefore, for instance, change in conductivity of a Conducting Polymer as it is exposed to a gas is the mechanism of sensing in a chemiresistor or field effect transistor characteristic.

5.1. Chemiresistors

The most widespread group of sensors is those that use Conducting Polymers. They are cheaply and quite easily made-up. In addition they utilize of the main property of Conducting Polymers (their conductivity). See figure. 6 shows a schematic diagram of a chemiresistor.

At its basic, a Chemiresistor is simply formed by two electrodes as contact points with the conducting polymer (CP) put onto an insulate substrate. When applying a constant current, the probable difference occurs on the electrodes represents the response output signal.

Although. The easiness of the sensing concept and its recognition does not come without a price. Since it is rarely to know what is occurring between the two electrodes. There will be a number of spots whereby the chemical change of a signal might initiate (Fig7) (Janata, 2002). For the measurements that use a constant current the capacitors used with equivalent resistors could be ignored. However, these capacitors have very significant role if an alternating current is used to excite those chemiresistors as well as if transient signals are involved.

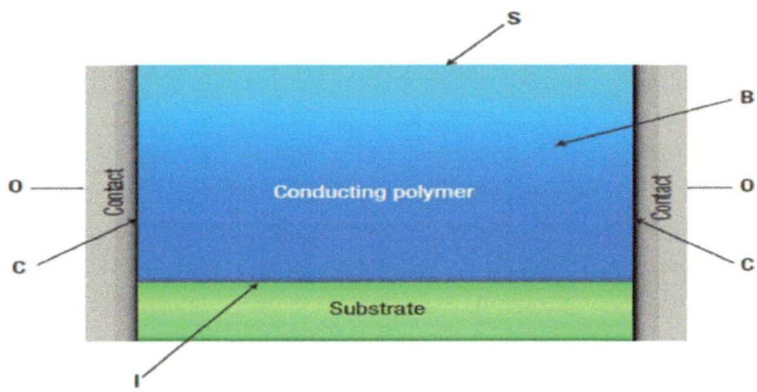

Figure 6. "Chemiresistor; B: bulk of the CP, S: surface, I: interface with the insulating substrate. C: interface with the contacts".

Polyaniline by chemical method then used as ammonim gas sensor (Kareema et al., 2010; Rawnq, 2009)

The conducting polymer CP can be acting as electrons donor or electrons acceptor when it is interacting with gaseous forms. The Hole Conductivity of a conducting polymer increases when it acts as an electron donator to the gas, on the other hand, the conductivity of the same conducting polymer decreases when it accepts electronics from the gas. Beside the changes of the carriers' number, changes could occur to their mass mobility. This is normally because of the adaption changes of the CP backbone (Zheng, 1997). Drawback of initiating the response in the mass of a Conducting Polymer is relatively long time (parts of second till a few minutes), usually coupled with a delay. Those consequences happen because of the slow diffusion of the gas into the Conducting Polymer. Due to the secondary doping results of modulation of the carrier number and the mobility, their related role in the total conductivity change differs. Change in the conductivity on the CP/electrode contacts could attribute to modulation of the Schottky barrier height. The contact barrier value is defined by the difference into work functions of both the organic semi-conductor and the metal. At this contact, there is a space-charge area generated into the interface of the semi-conductor and the metal. This means that the effecting resistance depends on the applied bias voltage as the measurement is being carried out. It was calculated (Leisinget al., 1998) that for a Conducting Polymer with about 3Venergy gap and carrier concentration of $n>10^{17}$ cm^{-3} on contact with a small work function metal (e. g., 2. 7 eV to calcium) the "depletion width" was <30 nm. The conductivity changes on the surface are much more complicated to be interpreted due to the complexity of the surface morphology.

Even though it is possible to describe the basic interaction with the analyte by one of adsorption isotherm forms, its outcome onto the surface layer conductivity is much more complicated compared with the case of semi-conducting oxide sensors (Barsan& Weimar, 2001). It is an acknowledged fact that the conductivity of matching samples of a Conducting Polymer differs from preparation to another. Such variability is referred to the sort preparation method and the film thickness (Stussi et al., 1997), which has a great effect on the surface morphology (Fig. 1b). Fine films (<300) of PANI electro-chemically developed are soft and compacted, in contrast the thick films have dendritic structure of the surface (Josowicz,. & Janata, 2002). The interface amid the Conducting Polymer and the insulating substrate is an additional place that could add to the total conductivity. Substrates such as (quartz, glass, sapphire... etc) are normally oxides. It is acknowledged (Domansky, et al. 1993), that the conductivity of the surface for oxides changes with the hydration degree. Therefore, the well known interferant to a chemi- resistor run at room temperature is the water vapour. To avoid this problem, it is recommended to make the substrate surface hydrophobic when performing the deposition of the conducting polymer.

All Those things and their unexpected behavior do not allow the precise analysis of the chemiresistor results to be easy, and several supportive techniques are often required to achieve additional detailed characterization. Because of all of that, the following techniques have been in use; Spectroscopy for Impedance (Ogura et al., 2001; Musio, et al., 1995), methodical changes of the chemiresistor geometry, changes of the electrodes geometry (Ingleby et al., 1999) and their materials. However, the relationship between the analyte gas concentration and the measured bulk conductivity (the response of chemiresistor) is constantly experimental.

5.2. Field-Effect Transistor (FET) sensors

The interaction among the neutral gases and organic semi- conductors has been utilized as the principle of transduction in Field-Effect Transistor (FET) sensors from the late 1980s (Josowicz & Janata, 1986), while it has been almost uncared for within non-sensing applications.

The category of sensors that are based on work function modulation contains three kinds of Micro Fabricated Devices which are; "Chemically sensitive Diodes, chemically sensitive capacitors, and chemically sensitive FET's (CHEMFETs) "

Several systems of CHEMFETs are found for both applications of liquid and gas species (Josowicz. & Janata, 1988). It is really key point to differentiate whether the current runs throughout the silicon or throughout the conducting polymer CP. In this case they could be divided in more detailed categories as follows; (a) thin film transistors (TFT) (Covington et al., 2001) as seen in figure 7 and (b) insulated gate Field-Effect Transistors (IGFET) (Janata, 1989) as shown in figure 8

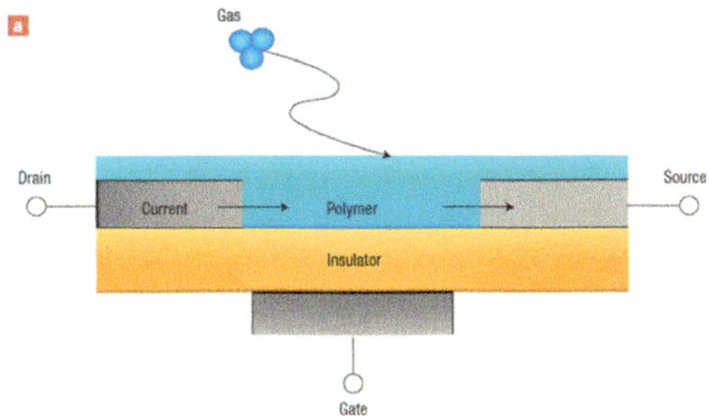

Figure 7. Shows C P'S in field-effect transistors, thin film transistor

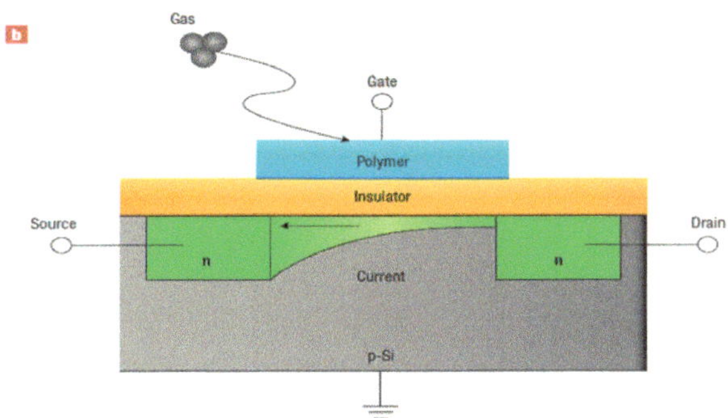

Figure 8. Shows CP's in field-effect transistors, insulated gate

For the TFT the current flows throughout a conducting polymer (CP) that its conductivity produced from the reaction with the analytes and/or by the electric fields. So, it can be said that the response signal depends on two things; the work function and the conductivity of the Conducting Polymer. Therefore, the analysis of the TFTs chemical response will be complex due to the same reasons for the chemiresistors, specifically, the division of different types of conductivity modulation, and of various forms of work function. The work function values can be affected by the interpreted energy states but the conductivity of the Conducting Polymer (CP) cannot be (Polk et al., 2002).

Conversely, in other usual silicon based transistors the situation will be very simple when the Conducting Polymer is employed only as a gate conductor. Then, the current will flow from side to side using the silicon channel, and the CP conductivity will not make any difference.

The Output response of these sensors relies just on the chemical modulation of a conducting polymer's work function. The possible outcome of the tuning of work function is the probability of building multi-sensing micro fabricated arrays. Look at figure (9) which shows a general platform for a similar array has been constructed.

Figure 9. A sensor array consisting of eight CHEMFETs, for more details look at (Polk, et al., 2002).

6. Corrosion

Conducting polymers of various forms will be electrodeposited onto oxidisable metals and electrochemical and environmental means will be used to access their applicably for corrosion protection.

Polyaniline (Pani) (Zhong et al., 2006;Wejood. 2005) and its derivatives (Bernard etal 2006) are among the most frequently studied CPs used for corrosion protection. In addition, the use of PANI for corrosion protection of metals has been of wide interest since the works by (DeBerry 1988; Mengoli et al. 1981) reporting that electro active coatings of the Pani could provide adequate protection against corrosion of stainless steels and iron sheets, respectively. Ever since, numerous studies have been published in which various CPs in different configurations were evaluated for corrosion protection of different metals and alloys: mild steel (Pritee, etal., 2006;Wejood, 2005, Solange de Souza, 2007).

(Pritee et al. 2006), study electrochemically synthesized of polyaniline on mild steel from an aqueous salicylate medium. The extent of the corrosion protection offered by polyaniline coatings to mild steel was investigated in aqueous 3% NaCl solution, 0. 01M Na2SO4 solution and in aqueous solutions of $NaCl+Na_2SO_4$ with different concentrations by potentiodynamic polarization technique and electrochemical impedance spectroscopy (EIS). The results of these studies reveal that the corrosion resistance of the polyaniline-coated mild steel is significantly higher and the corrosion rate is considerably lower than that of uncoated steel.

Polyaniline (Pani) nanofibers were were successfully fully synthesized by a modified rapid mixing method,) Sude Ma et al. 2012), that is, by the rapid mixing of solutions of aniline and ammonium peroxydisulfate in either hydrochloric acid or filtrates of oxidative polymerization of aniline. Composite coatings were fabricated with the dispersions of nanofibrous PANI and solutions of epoxy. Greatly enhanced corrosion protection performances were demonstrated by the coatings loaded with a small quantity of nanofibrous Pani.

Kamaraj et al. (2012) have been syntheses polyaniline (Pani) and polypyrrole (Ppy) and studied the effect of electropolymerised Pani films on corrosion protection performance of epoxy coating on AA 2024 and AA 7075 aluminium alloys. Polyaniline was electropolymerised on both the alloys by galvanostatic method. A post treatment of cecrium was given to seal the pinholes of PANI film. Epoxy coating was applied over these films and their corrosion protection performance was found out by EIS studies in 3% NaCl and salt spray test. EIS studies have shown that the coating resistance (*R*c) of PANI with the epoxy coated aluminium alloys has remained above 106 Ω cm2 whereas the alloys coated with epoxy alone have shown the *R*c values less than 104 Ω cm2. Besides, the salt spray tests showed a better corrosion protection of PANI with epoxy coated aluminium alloys.

Good research investigates the possibility of improving the corrosion resistance of buried steel by coating it with polyaniline (Pani) layer, (A. H. El-Shazly& H. A. Al-Turaif, 2012). The formed Pani layer was examined for its corrosion resistance while coupled with stainless steel cathode and buried in sand containing different known amounts of moisture, salt (NaCl) and sulphuric acid (H2SO4) using the potentiodynamic examination test. The results show that coating steel with Pani layer can improve its corrosion resistance against NaCl, H2SO4 and water by factors up to 1. 88, 1. 89 and 1. 54 respectively

Solange de Souza (2007) work vary good research, he studies the electrochemical behavior of a blend formed by camphor sulphonate-doped polyaniline and poly (methyl methacrylate) used for iron, copper and silver corrosion protection in acidic environments with or without chloride ions. The results obtained showed the good efficiency of these polymeric coatings against metal corrosion, proved by open circuit potential, linear voltammetry, electrochemical impedance spectroscopy, and scanning electron microscopy. It was observed that the protection depends on the formation of a passive film between the polymeric coating and he metallic substrate. SEM technique. Show at Fig. 10 (A) and (B), the micrographs reveal polymeric films with quite mogeneous morphologic appearance. As a consequence of the high miscibility of both acrylic polymers (PMMA) and the conducting [Pani (CSA)] amidst organic solvent, the presence of polymeric agglomerates is not observed in the films of Pani (CSA) –PMMA.]. After OCP and LV measurements, scanning electron microscopic micrograph of the Fe/Pani (CSA) -PMMA electrode reveals that surface is covered by homogeneous film [Fig. 10 (C)]. On the other hand, after OCP and LV measurements, visual observation of the Fe/PMMA electrode reveals a damaged film with many cracks [Fig. 10 (D)]. This damage may explain the lower OCP (around -0. 56 V (SCE), value corresponding to iron dissolution) and the higher current density value of approximately 2. 0×10^{-2} A cm^{-2}. More information about the ability of this Pani (CSA) -PMMA blend to provide stabilization and enhancement of the iron passivity against pitting corrosion was obtained by peeling off the Pani (CSA) –PMMA layer and observing the metal substrate. After removing the organic film and the powder precipitated between this film and the iron surface, the formation of a second physical barrier was observed. Fig. 11. (A) shows an appearance of little iron dissolution, which is attributed to Pani (CSA) polymer and to The coating based on polyaniline acrylic blend stabilized the potential of the metals Cu and Ag in a passive regime, maintaining a protective layer on the metal. The SEM micrographs of Pani (CSA) –PMMA-coated Cu and Ag electrodes Fig. 12 (A) and (B), respectively indicated that, after 85 days of immersion of these electrodes in acid solution, the morphology of Pani (CSA) –PMMA films did not show significant inference in comparison with the Fe/Pani (CSA) –PMMA electrode[Fig. 10 (C)]. After removing the polymeric coatings, the forming of a second physical barrier between the copper and polyaniline film was observed [Fig. 12 (C)]; whereas, for Ag/Pani (CSA) –PMMA electrode, there was no precipitate at the interface between polymeric coating and silver substrate [Fig. 12 (D)]. second layer that avoided penetration of aggressive ions.

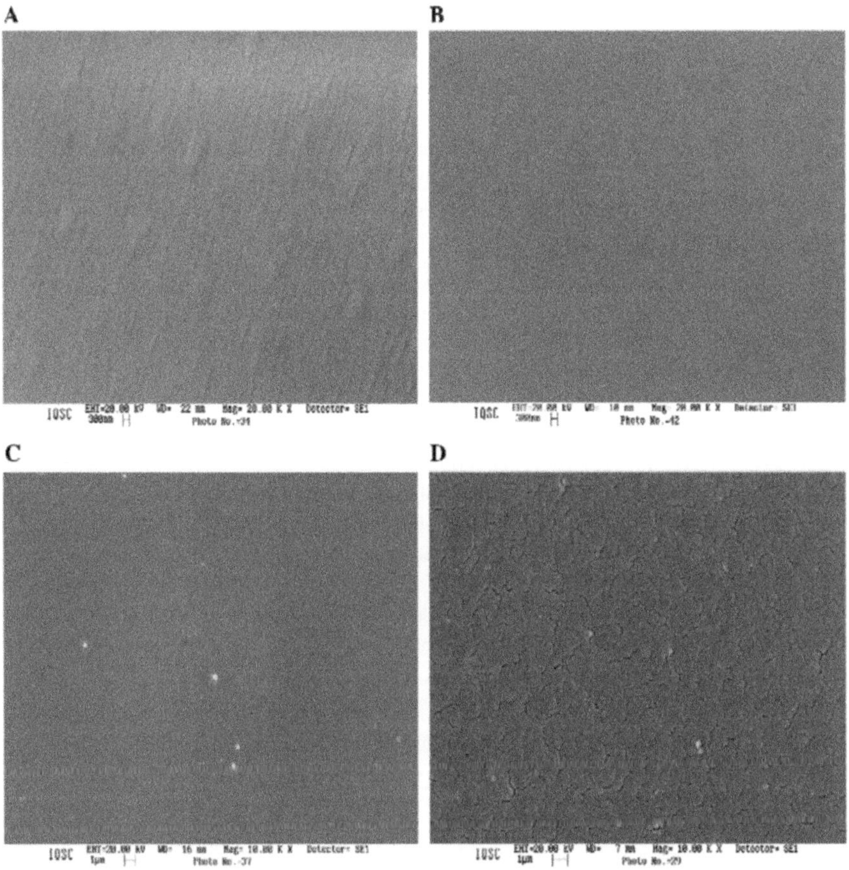

Figure 10. SEM micrographs of the (A) Pani (CSA) –PMMA film on Fe and (B) PMMA film on Fe, before OCP and LV measurements, and (C) Pani (CSA) –PMMA film on Fe and (D) PMMA film on (Solange de Souza 2007)

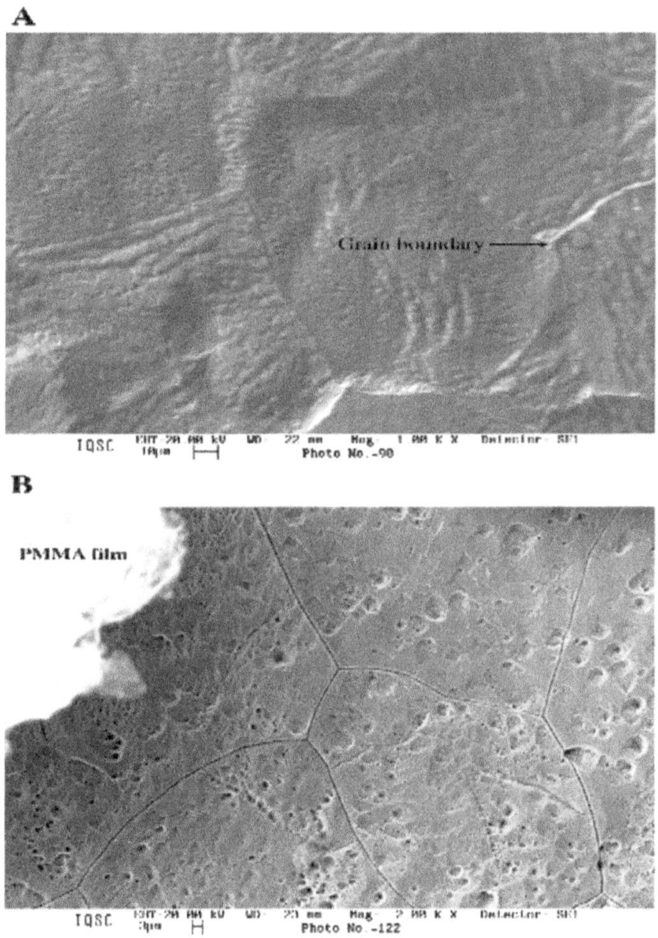

Figure 11. SEM micrographs of the (A) Fe/Pani (CSA) –PMMA interface, after removing of the polyaniline film and powder precipitated and (B) Fe/PMMA interface, after removing of the acrylic film, after OCP and LV measurements (Solange de Souza 2007)

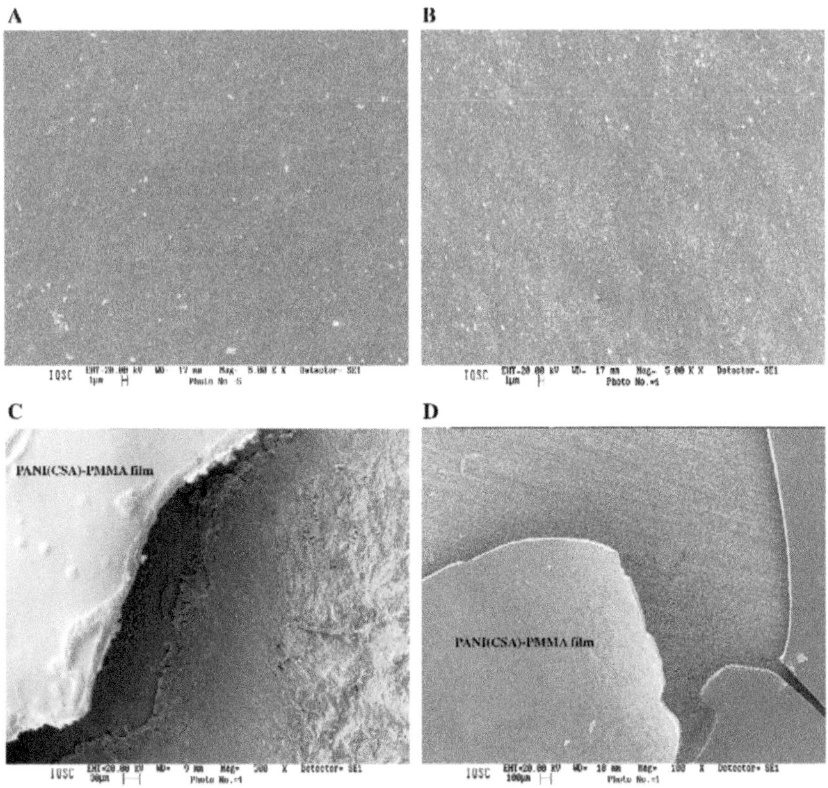

Figure 12. SEM micrographs of the (A) PANI (CSA) –PMMA film on Cu, (B) Pani (CSA) –PMMA film on Ag, (C) Cu/Pani (CSA) –PMMA interface, after removing of the polyaniline film and (D) Ag/Pani (CSA) –PMMA interface, after removing of the polyaniline film, after OCP and LV measurements. (Solange de Souza 2007)

7. Conclusion

Conducting polymers have an immense advantage of being simple to synthesis, with their chemical structure tailored to alter their physical properties, such as their band gap. They exhibit an extensive rangeof electrical conductivity and can exhibit metallic to insulator property (10^5- 10^{-9} S/cm). Further to theirease of synthesis and with lower cost, they are known to have low poisoning effects. So that they have many application such as solar cell,, sensor and corrosion Solar cells devices incorporating thin films of poly (o-toluidine) doped in p-toluene sulfonic acid (PTSA) spin-coated onto n-Si substrates have been produced and their photovoltaic characteristics have been studied and discussed. Solar conversion efficiency of about 2. 55% is Acknowledgments obtained for films with thickness of 35 nm. This efficiency is found to depend on film morphology as determined by the effect of film thickness and that lower incident light intensity is shown to cause an increase in these devices efficiency.

Author details

Kareema Majeed Ziadan
University of Basrah, College of Science, Department of Physics, Basrah, Iraq

Acknowledgement

I would like to thanks MyPh. student Dr Hussein F. Hussein. AlsoI grateful to Dr. Hussein. Assel Hassan and Dr. Abass Hashem at Sheffield helm university for used their Lab. in measurement and desiccation.

8. References

Alet P. -J, S. Palacin, P. I. C. Roca, B. Kalache, M. Firon, R. de Bettignies (2006). *Hybrid solar cells basedon thin-film silicon and P3HT-A first step towards nano-structured devices.* Eur. Phys. J. Appl. Phys;36, 231234

Angelopoulos, M., (2001) *Conducting polymers in microelectronics.* IBM J. Res. Dev;45.

Arenas, M. C. ; Sánchez, Gabriela; Nicho, M. E. ; Elizalde-Torres, Josefina; Castaño, V. M. (2012). Engineered doped and codoped polyaniline gas sensors synthesized in N, N, dimethylformamide media. Applied Physics A; 106: 901-908

Bailey, R. A. & Persaud, K. C. in Polymer Sensors and Actuators, (2000) (eds Osada, Y. & De Rossi, D. E.) 149–181.

Barsan, N. & Weimar, U. (2001), *Conduction model of metal oxide gas sensors. J. Electroceram;.* 7 (1) 143–167

Bettignies R. De, J. Leroy, M. Firon, C. Sentein. (2006) *Accelerated lifetime measurements of P3HT:PCBM solar cells.* Synth. Metals; 156: 510-513

Burn, Paul L., Lo, Shih-Chun and Samuel, Ifor D. W. (2007) *The development of light-emitting dendrimersfor displays. Advanced Materials*; 19: 16751688.

Covington, J. A., Gardner, J. W., Briand, D. & de Rooij, N. F. A (2001), polymer gate FET sensor array for detecting organic vapors. Sens. Actuat. B 77,.

Da-Jing CHEN, Sheng LEI, Ren-Hui WANG, Min PAN, Yu-Quan, CHEN (2012), Dielectrophoresis Carbon Nanotube and Conductive Polyaniline Nanofiber NH3 Gas Sensor, Chinese Journal of Analytical Chemistry; 40 :145-147.

Dimitrakopoulos, Ch. D. & Malenfant, P. R. L. (2002) Organic thin film transistors for large area electronics. Adv. Mater. 14:111-1222,

Domansky, K., Leng, Y., Williams, C. C., Janata, J. & Petelenz, D. (1993). Appl. Phys. Lett. 63,

Dyakonov V, I. Riedel, C. Deibel, J. Parisi, C. J. Brabec1, N. S. Sariciftci1, J. C. (2001), Hummelen, Electronic Properties of Polymer-Fullerene Solar Cells. Mat. Res. Soc. Symp. Proc.; 665: C7. 1. 1-12.

El-Shazly A. H., H. A. Al-Turaif (2012), Improving the Corrosion Resistance of Buried Steel by Using Polyaniline Coating, Int. J. Electrochem. Sci; 7 : 211 – 221

Gao G. Yu, J, J. C. Hummelen, F. Wudl, A. J. Heeger (1995). Polymer photovoltaic cells: enhancedefficiencies via a network of internal donor-acceptor heterojunctions. Science 270: 1789-1791.

Gazotti, W. A. et al. Handbook of Advanced Electronic and Photonic Materialsand DevicesVol. 10 (ed. Nalva, H. S.) (Academic, New York, (2001),

Gordon G. Wallace, Geoffrey M. Spinks, Leon A. P. Kane-MaguirePeter R. Teasdale (2003) Conductive electroactive polymers : intelligent materials systems. --2nd, CRC Press U. S. A

Granström M., K. Petritsch, A. C. Arias, A. Lux, M. R. Andersson, R. H. Friend. Laminatedfabrication of polymeric photovoltaic diodes. Nature 1998; 395: 257-260.

Haug F. J., T. Söderström, D. Dominé, C. Ballif., (2009.) " Light trapping effects in thin film silicon solar cells ", Journal: Manuscript ID: 576929. R1 MRS Spring Meeting, 2-13,

Huber, R. J. i (1985). n Solid State Chemical Sensors (eds Janata, J. & Huber, R. J.) 119–162 (Academic, New York,

Hussein F. H. preparation of conducting poly (o-toluidine) and study of its electrical and optical properties and application as solar cell. Ph. D. thesis university of Basrah, Basrh, Iraq (2010)

Hyang Hee Choi1, Junmin Lee1, Ki-Young Dong2, Byeong-Kwon Ju2, and Wooyoung Lee (2012), Gas Sensing Performance of Composite Materials Using Conducting Polymer /Single-Walled Carbon Nanotubes Macromolecular Research;. 20 : 143-146.

Ingleby, P., Gardner, J. W. & Bartlett, P. N. Effect of micro-electrode geometry on response of thin-film poly (pyrrole) and poly (aniline) chemoresistive sensors. Sens. Actuat. B 57, (1999)

Izabela Kuzma-Filipek, Filip Duerinckx, Kris Van Nieuwenhuysen, Guy Beaucarne, and Jef. Poortmans, (2008) " Plasma Texturing and Porous Silicon Mirrors for Epitaxial Thin Film Crystalline Silicon Solar Cells", Mater. Res. Soc. Symp. Proc., Materials Research Society, Vol. 1101 C.

Janata, J. (1989). Principles of Chemical Sensors 81–168 (Plenum, New York,

Janata. J. (2002). Electrochemical sensors and their impedances: A tutorial. Crit. Rev. Anal. Chem. 32,

Josowicz, M. & Janata, J. (1986). Suspended gate field-effect transistor modified with polypyrrole as alcohol sensor., Anal. Chem. 58,

Josowicz, M. & Janata, J. (1988). in Chemical Sensor Technology (ed. Seiyama, T.) 153–177, New York,

Josowicz Li, G.,, M. & Janata, (2002) J. Electrochemical assembly of conducting polymer films on an insulating surface. Electrochem. Solid State 5,

K. Kamaraj, V. Karpakam, S. Syed Azim, S. Sathiyanarayanan (2012). Electropolymerised polyaniline films as effective replacement of carcinogenic chromate treatments for corrosion protection of aluminium alloys, Synthetic Metals, 126,, 536–542

Kareema M. Ziadan. 'The electrical and optical properties conducting polymers (PT, PT/PVCand PT/PTFE) of and Their application in Rechargeable Batteries; Ph. D thesis university of Basrah, Basrah Iraq (1997).

Kareema M. Ziadan, H. F. Hussein, R. A. Tali, A. K. Hassan (2011) Synthesis and characterization of (Pani/n-si) solar cell Energy Procedia (85–91)

Kareema M. Ziadan Krabit A. Haykaz and Ali Q. Abudalla (2012) Some electrical properties of soluble conducting polymer polyHexylthiophene (PHT) prepared byElectrochemical polymerization, Energy Procedia 6 (2012) in print.

Kareema M. Ziadan Hussein. F. H, and K. I. Ajeel. (. 2012), Study of the electrical characteristics of poly (o-toluidine) and application in solar cell, Energy Procedia 6 (2012) in print.

Kareema M. Ziadan Hussein. F. H, and K. I. Ajeel (. 2012) The electrical characteristics of poly (o-toluidine) doped with HCl (POT-HCl) and application in solar cell The 2nd International Conference on Renewable Energy: Generation and Applications March 4-7, 2012, United Arab Emirates University, Al Ain, UAE

Kareema M. Ziadan, R. A. Talib and Sattar (2010), Using polyaniline doped with formic acid as a gas seneor) J. kuha-physices A special ISSUE for Kufa first conference for physics6-7 oct. 2010.

Kareema. M. Ziadan a, Wjood T. Saadon, (2012), Study of the electrical characteristics of polyaniline prepeared by electrochemical polymerization, MEDGREEN-SY 2011:2nd Conference & ExhibitionSeptember 19th - 21th 2011, Aleppo – SyriaEnergy Procedia 6 (2012) in print.

Krichelore H. R., (1992) Hand book of polymer synthesis part B, Mercel Dekker N. Y, p1883

Kroschwitze J. I, (1988), Electronical and Electronic properties of polymer, John Wiley and Sons INC P57.

Skotheim T.. (1986). A, Hand book of polymer synthesis part B, Mercel Dekker N. Y

Kulkarni M. V., A. K. Viswais Nath (2004), Comparative studied of chemically synthesized polyaniline and poly (o-tolidine) doped with p-toluene sulphonic acid, European polymer journal, 40, 379-384

Leclerc, M. (1999). Optical and electrochemical transducers based on functionalized conjugated polymers. Adv. Mater. 11,

Leising, G., Tasch, S. & Graupner, W. in1998). Handbook of Conducting Polymers 2nd edn (eds Skotheim, T. A. Elsenbaumer, R. L. & Reynolds, J. R.) 854 (Marcel Dekker, New York,

Levitsky I. A., W. B. Euler, N. Tokanova, B. Xu, J. Castracane (2004). Hybrid solar cells based on porous Si and copper phthalocyanine derivatives. Appl. Phys. Letters; 85: 6245-6247.

McQuade, D. T., Pullen, A. E. & Swager, T. M (2000), Conjugated polymer-basedchemical sensors. Chem. Rev. 100,

McEvo A. J y, M. Gratzel, H. Wittkopf, D. Jestel, J. Benemann. (1994) Nanocrystalline electrochemical solar cells, IEEE First World Conference on Photovoltiac Energy Conversion, 5-9 December, USA 1779;178

Michalska, A. & Lewenstam, A. (2000). Potentiometric selectivity of p-doped polymer films. Anal. Chim. Acta 406,

Milella, E. & Penza, M. SAW (1998). gas detection using Langmuir–Blodgett polypyrrole films. Thin Solid Films 329,

Musio, F., Amrani, M. E. H. & Persaud, K. C. D. (1995). High-frequency AC investigation of conducting polymer gas sensors. Sens. Actuat. B 23,

Nam s, J. Jang, K. Kim, W. M. Yun, D. S. Chung, J. Hwang, O. K. Kwon, T. Chang and C. E. Park (2011). Solvent-free solution processed passivation layer for improved long-term stability of organic field-effect transistors. Journal of Materials Chemistry; 21: 775-780.

Ogura, K., Tonosaki, T. & Shiigi, H. AC (2001) Impedance spectroscopy of a humidity sensor using poly (o-phenylenediamine) /poly (vinyl alcohol) composite film. J. Electrochem. Soc. 148,.

Otero, T. F. (2000). in Polymer Sensors and Actuators (eds Osada, Y. & De Rossi, D. E.) 295–324 (Springer, Berlin,

Pritee Pawara, A. B. Gaikawadb, P. P. Patil (2006), Electrochemical synthesis of corrosion protective polyaniline coatings on mild steel from aqueous salicylate medium, Science and Technology of Advanced Materials 7 732–744

Patil S. V& P. R. Deshmukh, C. D., (2012) Lokhande Room Temperature Liquefied Petroleum Gas Sensing Polymer (n-polypyrrole/p-polyaniline) Based Heterojunction, Sensors & Transducers Journal;173: 104-114

Pawarl S. G, M. A. Chougule, Shashwati Sen, V. B. Patil, 2012, Development of nanostructured polyaniline–titanium dioxide gas sensors for ammonia recognition, Journal of Applied Polymer Science 1022: 1418–1424,

Polk, B. J., Potje-Kamloth, K., Josowicz, M. & Janata, J. (2002). Role of protonic and charge transfer doping in solid-state polyaniline. J. Phys. Chem. 106,

Rawnq A. Talib.. preparation of conducing polymr (PAni) by chemical method, Study some of physicalproprtis and its application as ammonia gas sensor. MSc Thesis, Basrah University, Basrah, Iraq (2009).

Riedel I., N. Martin, F. Giacalone, J. L. Ssegura, D. Chirvase, J. Parisi, and V. Dyakonov, (2004), Polymer solar cells with novel fullerene-based acceptor, Science direct, Thin solid films, 451-452, p. 43-47

Sariciftci N. S.. (2006) Efficiency limiting morphological factors of MDMO-PPV:PCBM plastic solar cells. ThinSolid Films; 511-512: 587–592.

Schif E. A f, A. R. Middya, J. Lyou, N. Kopidakis, S. Rane, P. Rao, Q. Yuan, and K. Zhu, A. R. Middya, (2002) Electroabsorption and Transport Measurements and Modeling Research in Amorphous Silicon Solar Cells. Final Technical Report 24 March 1998—15 August (National TechnicalInformation Service, Document DE2003;p1500360.

Shaheen S. E., C. J. Brabec, F. Padinger, T. Fromherz, J. C. Hummelen, Sariciftci.. (2001) 2. 5%efficient organic plastic solar cells. Appl. Phys. Lett; 78: 841843

Solange de Souz, (2007) Smart coating based on polyaniline acrylic blend for corrosion protection of different metals a Surface & Coatings Technology 201 7574–7581

Sude Ma, Guolin Song, Ningbo Feng, Pei Zhao, 2012, Journal of Applied Polymer Science; 125 : 1601–1605

Stussi, E., Stella, R. & De Rossi, D. (1997) Chemoresistive conducting polymer-based odour sensors: influence of thickness changes on their sensing properties. Sens. Actuat. B 43,

Vignesh Gowrishankar, Shawn R. Scully, Michael D. McGehee, QiWang, Howard M. Branz. (2006) Excitonsplitting and carrier transport across the amorphous-silicon/ polymer solar cell interface. Appl. Phys. Lett.; 89: 252102

Wang W., E. A. Schiff. (2007) Polyaniline on crystalline silicon heterojunction solar cells. Appl. Phys. Lett; 91p 133504.

Williams E. L., G. E. Jabbour, Q. Wang, S. E. Shaheen, D. S. Ginley, E. A. Schiff. (2005) Conducting polymerand hydrogenated amorphous silicon hybrid solar cell. Appl. Phys. Lett.;87: 1-3.

Zheng, W. (1997) Effect of organic vapors on the molecular conformation of nondoped polyaniline. Synth. Met. 84,

Chapter 5

Hydrogel Films on Optical Fiber Core: Properties, Challenges, and Prospects for Future Applications

Sergey V. Kazakov

Additional information is available at the end of the chapter

1. Introduction

1.1. Properties of hydrogels

Gels are most generally understood to be 3D-polymeric networks with cross-linked long chain molecules, which absorb a large amount of solvent to cause macroscopic changes in dimensions. Polymer network immersed in an aqueous medium is referred to as a hydrogel. The three-dimensional network of a hydrogel can be stabilized as a gigantic single molecule by chemical (covalent bonds) or/and physical (ionic bonds, entanglements, crystallites, charge complexes, hydrogen bonding, van der Waals or hydrophobic interactions) cross-links. A hydrogel is also considered as an open container with semipermeable boundaries, across which water and solute molecules can move whereas charged (ionizable) groups fixed on the network chains cannot move (Figure 1).

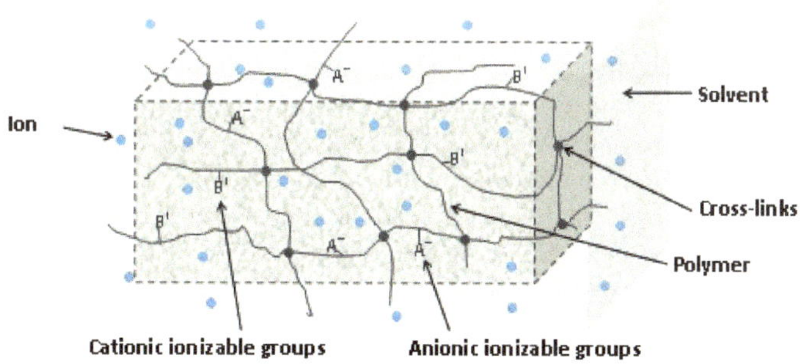

Figure 1. A schematic presentation of a gel immersed in a solvent.

Networks of cross-linked polymers (hydrogels) exhibit both liquid-like and solid-like behavior [1-4]. Polymer networks are able to absorb water up to a thousand-fold of the dry weight of a polymer [5]. The unique property of hydrogels is the abrupt volume changes under transition from their collapsed to swollen state in response to external stimuli [6-14], see Figure 2. Due to this property, hydrogel is an "intelligent" material suitable for design of regulated systems. Hydrogels are especially desirable for bioanalytical applications because of their high water content and elastic nature similar to natural tissue.

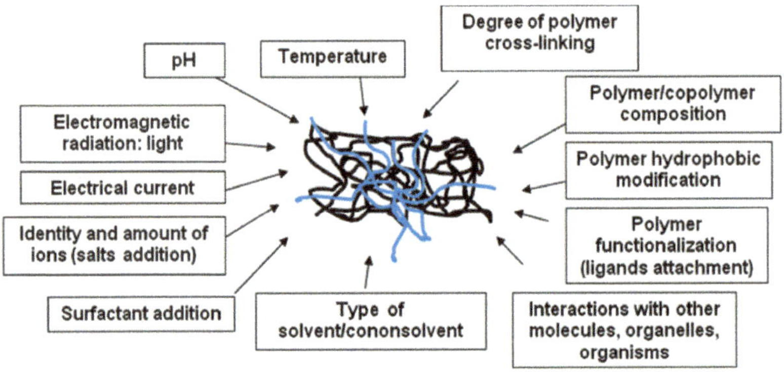

Figure 2. Different effectors of the swelling capacity of polymer networks (hydrogels).

Thermodynamically, the ability of hydrogels to swell in the course of water absorption is the matter of solubility of macromolecules comprising their 3D networks. We have already described elsewhere [15] possible types of solubility phase diagrams available within a window of the experimentally accessible thermodynamic parameters (e.g. temperature, pressure, concentration).

Various types of hydrogels based on both natural (e.g., hyaluronic acid, collagen, chondroitin sulfate, alginates, fibrin, and chitosan) and synthetic polymers made of neutral (e.g., 2-hydroxyethyl methacrylate, N-alkylmethacrylamides, N-alkylacylamides, N,N-dialkylacrylamides), acidic (e.g., acrylic acid, metacrylic acid, 2-acrylamido-2-methyl propane sulfonic acid), and basic (e.g., N,N-dialkylaminoethyl methacrylate, 1-vinylimidasole, methacryloyoloxyethyltrialkylammonium bromide) monomers have been prepared, studied, and used in numerous applications (bioseparation, tissue engineering, sensing and molecular recognition, drug and gene delivery, controlled release, artificial muscles, flow control). It is not our intention to present a comprehensive review on hydrogel properties and applications. In this section, we would like to let the readers get a feeling of how significant the volume of hydrogel matrix can change in response to different stimuli.

1.2. Effectors of volume change in hydrogels

Temperature. The most pronounced volume changes of hydrogels are observed at the volume phase transition corresponding to the solubility phase diagram of a polymer in the hydrogel network. Many polymers exhibit coexistence curves with LCSP, so that the range of transition temperatures (T_c) and concentrations are dependent of the chemical nature of the polymer. As a consequence, a temperature range where the hydrogel shrinks expelling water upon heating intrinsically depends on the type of polymer constituting its network. Figure 3 shows the examples of temperature dependencies of the swelling ratio for different polymeric hydrogels.

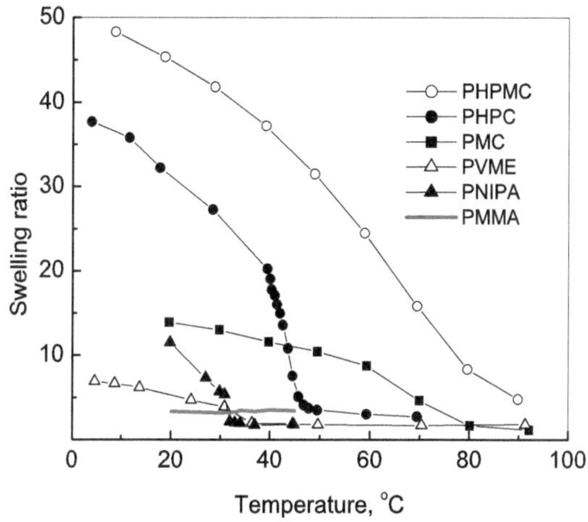

Figure 3. Relative volume changes as a function of temperature for hydrogels of different types: MC – methylcellulose, HPMC – hydroxypropylmethylcellulose, HPC – hydroxypropylcellulose, NIPA – *N*-isopropylacrylamide, MMA – methacrylic acid, VME – vinyl methyl ether. The hydrogel transition temperature is close to the polymer LCST. PMAA hydrogel is not temperature sensitive because PMAA polymer is supposed to not exhibit a solubility phase transition at the experimentally achieved thermodynamic conditions. Data are adapted from [16-18].

The studies [16,19,20] show that increased cross-linking may significantly decrease swelling ability of hydrogel, especially, below T_c, but has a little effect on the value of T_c. These conclusions agree with our results (Figure 4).

It has been shown [21] that incorporation of a small amount of ionizable groups into the nonionic gel network drives the volume phase transition from continuous volume changes toward discontinuous one. Figure 5 explicitly demonstrates that an increase in the portion of sodium acrylate with carboxylic groups on the PNIPA network increases T_c and extent of swelling ability below the transition temperature.

It is worthy to highlight, that the volume changes in a water-swollen hydrogel are expected to be continuous as a function of environmental stimuli, if the system remains totally miscible at given thermodynamic conditions. On the contrary, if changes in chemical nature of the polymer network, solvent quality, or environmental stimulus "push" the system into a two-phase (unstable) region of the solubility phase diagram, a swollen hydrogel collapses to a shrunken state due to reduction in solubility. During such volume phase transition, one can expect that properties of the hydrogel, most notably its volume, change discontinuously.

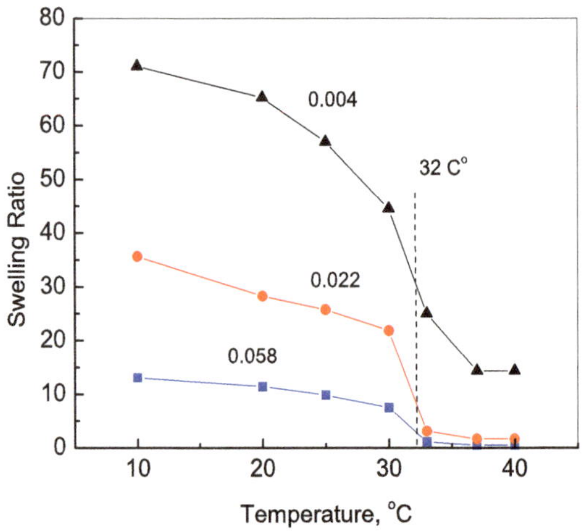

Figure 4. Swelling ratio [Q = (W-W₀)/W₀, W₀ is the weight of dry gel] of PNIPA hydrogel cylinders (Ø10 mm × 10 mm) as function of temperature for different relative amount of the cross-linker (methylenebisacylamide, MBA) in the hydrogel-forming solution. The numbers near the curves are the values of ratio: $n_{MBA}/(n_{MBA}+n_{NIPA})$, where n_{MBA} and n_{NIPA} are the number of moles of MBA and *N*-isopropylacrylamide (NIPA), respectively, at constant $n_{MBA}+n_{NIPA} \approx 0.66$.

Ions and ionic strength. Figure 5 also suggests that incorporation of charged (anionic, cationic, or both) groups on the polymer network makes the volume transition temperature and degree of swelling dependent on pH and ionic strength. Indeed, the poly(*N*-isopropylacrylamide-*co*-methacrylic acid) (PNIPA-MAA) microgel particles [22] at pH 3.4 exhibited a decrease in T_c from 33.5 to 28 C° with an increase in MAA content, whereas at pH 7.5, the higher MAA content resulted in the higher T_c. In weakly charged PNIPA hydrogels, addition of ionizable groups on the polymer network pronounced the volume changes when temperature crossed T_c [21,23,24]. The experimental studies [25] revealed that even distribution of ionic groups in the network affects the temperature of volume change transition.

The type of ionizable groups on the polymer networks makes the maximum swelling ability of a gel strongly dependent on pH. For example, the anioic PNIPA-MAA microgels exhibited the maximum swelling in the range of pHs from 6.5 to 10 [22], whereas for the cationic PNIPA-VI nanogels (VI stands for 1-vinylimidazole) the maximum swelling ratio was observed in the range of pHs from 6 to 3.5 [26]. It becomes even more intriguing if the so-called polyampholyte hydrogels are designed [27] by addition of both cationic (VI) and anionic (AA) groups on the network. The polyampholyte gel was in a shrunken state in the vicinity of the isoelectric point (pH ~ pI), and it swelled at both higher and lower pHs. It is interesting that such designed polyampholyte gels can work like biochemo-mechanical systems in which the enzymatically induced pH changes control the volume of polyampholyte network or, in opposite direction, the pH sensitive volume changes control the activity of enzymes immobilized into the gel [28].

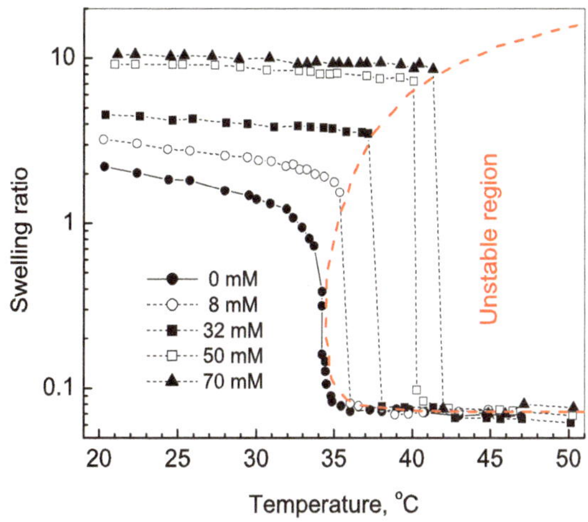

Figure 5. The degree of swelling of the poly(N-isopropylacrylamade-co-sodium acrylate) gel in water as a function of temperature. Numbers are the molar concentrations of sodium acrylate in the preparations. Data are adapted from [21].

There are experimental evidences [29-31] of de-swelling effects of different monovalent (Li^+, Na^+, K^+, Cs^+) and divalent (Ca^{2+}, Mg^{2+}, Sr^{2+} , Ba^2) ions in hydrogels of different chemical nature, including the biologically relevant gels [32] and cross-linked DNA [33]. Interestingly, at the same molar ratios of divalent to monovalent cations, ~ 1 mM to 30 mM, respectively, similar volume changes were observed in biological polyelectrolyte systems during physiological processes like nerve excitation, musle contraction, and cell locomotion [34-40].

Surfactants. The extensive theoretical [41-43] and experimental [44-48] studies have shown that addition of anionic, cationic, and nonionic surfactants to the solution containing a gel can also influence the volume phase transition temperature and swelling degree of hydrogels depending on their hydrophobicity and charge of the polymer network. In general, addition of anionic or cationic surfactant to the solution of nonionic hydrogel rises the transition temperature as well as the swelling range, whereas the nonionic surfactant does not affect the transition temperature or the volume change. The surfactants with ionic head groups convert the neutral hydrogels to a polyelectrolyte gels when bind to the nonionic polymer networks, so that the transition temperatures elevate due to introduction of additional osmotic pressure by ionization. The changes in the volume phase transition are also dependent on the length of hydrophobic tail of ionic surfactants and the critical concentration of micelle formation. Indeed, it was found [45] that the transition temperature of nonionic poly(acryloyl-L-proline methyl ester) hydrogels increases more drastically in solutions of anionic sodium alkane sulfonate surfactants with the higher number of methylene units in the tail and at lower concentration than those with the shorter tails. The other interesting findings are [47]: (i) the amount of an ionic surfactant bound onto the swollen network of the nonionic PNIPA hydrogel is much greater than that to the collapsed one, (ii) on the contrary, the amount of nonionic surfactant bound onto the collapsed network of the PNIPA gel is greater than that on the swollen one, (iii) the transition temperature changes with the amount of the surfactant bound much stronger for anionic surfactants (e.g., sodium dodecyl sulfonate) than for cationic ones (e.g., dodecylamine hydrochloride), (iv) the distribution of anionic surfactant (e.g. sodium dodecylbenzene sulfate) bound within the PNIPA hydrogel is appeared to be heterogeneous [48] - the surfactant concentration is higher in the vicinity of the gel surface, whereas a central region of the gel may not contain any bound surfactant molecules. Therefore, there can be a gradient of swelling ability along the depth of surfactant penetration into the gel, namely: the peripheral layers are in the more swollen state with a higher transition temperature in comparison with the hydrogel core.

The uptake of a surfactant into the ionic and neutral polymer networks has been found to be different [49,50]. Particularly, when the cationic surfactant (e.g., N-alkylpyridinium chloride, C_nPyCl, n = 4, 8, 10, 12, 16) was absorbed from the solution into the anionic hydrogel [e.g., poly(sodium 2-acylamido-2-methylpropane sulfonate) (PNaAMS) gel], the polymer network was collapsed not swelled, herein the degree of shrinking increased with the length of the surfactant alkyl tail. Surprisingly, the surfactant influx (mol/s/cm^2) was enhanced by the increase in the network density. It was also observed that ionic strength significantly affects the uptake of surfactants into PNaAMS hydrogel, namely: (i) in pure water, the hydrogel began absorbing surfactants at concentrations below 10 μM and the level of fluxes was high for alkyl tails of different lengths, (ii) in the 10 mM NaCl solution, the uptake of surfactants was practically absent at the concentrations lower than 100 μM of $C_{12}PyCl$ and 400 μM of $C_{10}PyCl$, but abruptly increased at higher concentrations, especially for surfactants with longer tail, e.g., the surfactant flux jumped ~ 3-fold for the surfactants with the number of carbon atoms in the alkyl tail greater than 10. The kinetics of surfactant uptake clearly indicated [50 and references therein] that binding process significantly increases the amount of surfactant absorbed by the gel and slows down the diffusion of surfactant inside the gel. Theoretical and experimental analyses [43,48-50] also showed that the closer to the surface of the gel, the higher is the concentration of the bound surfactant. Herein, this non-uniformity of surfactant binding increases with the cooperativity of binding in the presence of a salt. Thus, it was predicted that the progress of surfactant binding within the gel can be observed as the movement of the front of network collapse from the surface toward the center of gel.

Light. Hydrogels sensitive to light have been reported as well. For example, photosensitive gels with incorporated photosensitive molecules, such as leucocyanide and leucohydroxide, into the gel network have been reported [51]. These gels underwent volume changes upon irradiation and removal of ultraviolet light as a result of ionization reaction and internal osmotic pressure initiated by UV light. Significant volume changes in hydrogels can be also induced by visible light [52]. The mechanism of volume transition is different; it is due to direct heating of the polymer network by light. By the way, the fact that this process is extremely fast is of great importance from the practical point of view.

More recent reports [53] show that a focused laser beam was able to generate the radiation force which induces reversible shrinkage in polymer gels. Control experiments confirmed that the laser-induced volume phase transitions are due to radiation forces, rather than local heating, modifying the weak interactions in the gels. Due to shear-relaxation processes, gel shrinkage occurs up to several tens of micrometers away from the irradiation spot. The compaction was also found in acrylamide gels, which are not temperature sensitive. In hydrogels with temperature sensitive volume phase transitions, such as PNIPA gel, it was found that the radiation force of the laser beam not only induces the volume phase transition, but also lowers the transition temperature T_{tr} by about 10 °C at an irradiation power of 1.2W (λ = 1064 nm).

The light induced volume changes in photosensitive hydrogels have received increasing attention because of their scientific and technological significance for the so-called shape-memory polymeric systems [54]. Being exposed to the light with lower wavelengths, the shape-memory materials become deformed and their temporary shape is fixed due to cross-linking. When irradiated with higher wavelengths, they recover their initial shape because of cleavage the cross-links.

Electrical field. Back in 1950s, it was found that contraction, oscillation, and bending of polyelectrolyte gels can be activated electrically [55-57]. In particular, gels prepared from polymers and copolymers that contain ionizable groups exhibited remarkable contraction when placed between a pair of electrodes connected to a direct current source. Polymer gels containing no ionizable groups showed no volume change under electrical field applied. The extent and rate of volume change of the polyelectrolyte gels were shown to increase with increasing electrical field [58]. An increase in the ionic strength (e.g., an addition of NaCl) also increases the rate of gel shrinkage, whereas an addition of organic solvent (e.g., acetone or ethanol) decreases both the extent and the rate of shrinking. In different types of hydrogels, an electrically activated volume changes were associated with the induction of the medium pH change by the electric field [59], the electrically initiated volume phase transition [60], and the so-called electrokinetic phenomena – ion transport of counter-ions in the electric field [61].

The reversible shrinking and swelling of the polyelectrolyte gels stimulated by an electric field have become a basis for development of the so-called mechanochemical (chemomechanical) devices (engines) [62] functioning as actuators and artificial muscles [63-65] and drug delivery systems [66].

1.3. Supported hydrogel films

Polymer networks being capable of stimuli responsive volume changing and controllable mass transferring have been increasingly recognized as building blocks in microfluidic devices (sensors, actuators, fluid pumps, and valves) [67], scaffolds in tissue engineering and regenerative medicine [68-70], actuators in artificial muscle development [63-65], and targeted and controlled release elements in drug delivery systems [17], including iontophoresis [71]. Hydrogels are flexible and fragile and require mechanical support. The idea of a supported hydrogel has been employed for decades in the planar configurations. Deposition of planar polymer networks onto supports allows for

i. biofunctionalization of surfaces, providing a natural environment for immobilization of highly specialized proteins, lipids, carbohydrates, and nucleotides
ii. preparation of flat homogeneous surfaces with nanoscale smoothness and heterogeneous substrates with more complex structures
iii. development of "cushions" that effectively separate from the solid surface more fragile layers, such as lipid bilayer membranes with reconstituted proteins
iv. detection of the changes in electrical or optical signals induced by the events of binding/dissociation or variations in the integral characteristics (hydrophobicity, charge, etc.) in relation to varied degrees of swelling
v. miniaturization of the abovementioned devices and systems.

1.4. Methods for planar hydrogel micropatterning

In the development of the intelligent biomaterials and biochips, based on the supported polymer networks, one goal is to synthesize regions with different physico-chemical properties: e.g., hydrophobic/hydrophilic, a wide range of shapes, thicknesses and dimensions, sensitive to different stimuli (pH, temperature, electrical field, radiation, etc.). Coupled with the need to synthesize regions of different properties, there is also a need in techniques to form high aspect ratio structures. Optical lithography or photolithography is the most suitable technique for micropatterning cross-linked polymer networks as high aspect ratio structures with features of controlled surfaces. The change in the solubility of polymer-coated substrate after exposure to radiation allows the selective removal of one part of the polymer coating, if irradiated part becomes soluble in the developer, it is called a positive-tone resist. When it becomes insoluble, it is called a negative-tone resist. The latter one provides a tool for preparation of patterned hydrogels, when water-soluble polymer subjected to the cross-linking step [72]. State-of-the-art photolithographic techniques are capable of mass-producing patterned structures in thin films with feature sizes as small as ~250 nm with possible extension to features as small as ~100 nm by use of a combination of deep UV light and improved photoresists [73]. However, the so-called 100 nm barrier cannot be easily surmounted. Advanced lithographic techniques, such as extreme UV (EUV) lithography, soft X-ray lithography, e-beam writing, focused ion beam (FIB) writing, and proximal-probe lithography (74, 75), have the capability of generating extremely small features (as small as a few nm), but their development for manufacturing of nanostructures requires substantial effort and they, as the conventional photolithography, are poorly suited for patterning non-planar surfaces.

1.5. Evanescent wave as a tool for nanoscaling

To improve the resolution, a procedure for the fabrication of very thin nanoparts has been proposed – photopolymerization by evanescent waves (PEW) [76].

The total internal reflection (TIR) is the phenomenon guiding light in a higher refractive index (RI) medium [77]. Under conditions of TIR, light can totally reflect from a dielectric interface, remaining inside the medium of a higher RI (n_1). The critical angle (θ_c), at which the reflected ray will travel along the boundary surface, is determined by the ratio n_2/n_1, where n_2 is the lower RI of the second medium:

$$\theta_c = \arcsin(n_2 / n_1), \tag{1}$$

During the propagation of electromagnetic radiation inside high RI medium, a portion of the electric field can penetrate outside, to the medium of the lower RI. The penetrated electric field is called the **evanescent wave**. The intensity of the evanescent wave decays exponentially [77] as follows:

$$E(z) = E_0 \exp(-z / d), \tag{2}$$

where E_0 is the amplitude at the interface and z is the distance outside the high RI medium. The depth of penetration of the evanescent wave is the distance at which the intensity of the electric field has decayed to 1/e of that at the interface:

$$d = \frac{\lambda}{2\pi n_1 \sqrt{\sin^2\theta - (n_2 / n_1)^2}}, \tag{3}$$

where θ is the incident angle and λ is the wavelength of the radiation. The penetration depth of the evanescent field can be interpreted as the thickness of the layer in which acting light is confined. Typically *d* is about few tenth of λ. Within the depth of penetration, the evanescent wave can interact with molecules located nearby the interface. Thus far, the evanescent wave can be absorbed within the polymerizable formulation to initiate a polymerization reaction around the area of the interface exposed to the light (Figure 6).

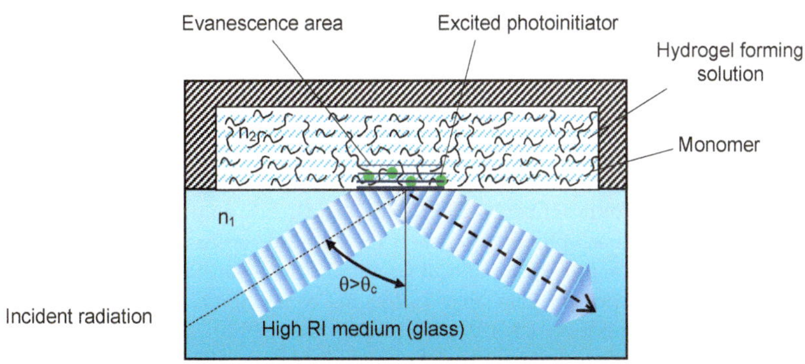

Figure 6. Schematic representation of photopolymerization by evanescent waves in the planar configuration. The laser beam is totally reflected at the interface between the high refractive index glass and the hydrogel forming solution, as the angle of incidence is greater than a critical angle θ_c. Close to the interface, the electromagnetic field is non-zero in the polymerizable medium but decays exponentially to zero.

It has been shown [76] that the thickness and the shape of the hydrogel planar films obtained by this lithographic technique correlated with the properties of the incident radiation. In particular, a linear dependence of the film thickness on the logarithm of the exposure time was observed. The film thickness varied with the angle of incidence of the laser beam at the surface of the photopolymerizable mixture. When the wavelength of laser used was 514 nm, the films of 30 nm thickness were obtained. To fabricate homogeneous planar films, the light intensity had to be constant over the whole irradiated area. On the other hand, it was demonstrated that 2D and 3D photolithography were also possible, i.e. when the intensity of the incident radiation was not uniform at the interface, the polymerized parts were non-planar. The light intensity at the interface was modulated using photomasks or causing interfering pattern in the evanescence plane. It was also possible to control the thickness of the hydrogel layer locally by varying the exposure time. A diaphragm of variable diameter was imaged at the glass/solution interface, so that the central part was illuminated for a longer time than the peripheral area – a stepwise structure with controllable height of every step was obtained. These experiments present the indications that PEW may provide a simple method of stereophotolithography well suited for patterning non-planar surfaces.

1.6. Hydrogel/optical fiber assembly

Cylindrical surface of optical fiber core seems to be the next step in the development of methods for micropatterning non-planar supporting surfaces. Cylindrical surfaces of optical fiber core offer several advantages over conventional planar configuration, e.g.:

- Optical fibers provide the method for monitoring optical signal of different spectral composition
- They have potential for a higher density of extremely small features (as small as a few nm) deposited on the cylindrical surface
- Regions with different physico-chemical properties can be synthesized and multiple reagents can be immobilized on their surfaces
- Simultaneous response to more than one analyte can be recorded, stored, and interpreted
- They allow devising the miniature devices which can be used for monitoring parameters *in vivo*

The idea of a hydrogel/optical fiber assembly has been already employed for decades mostly in the context of the fiber optic chemical and biochemical sensors [78-81]. The features of hydrogels relevant to the purpose of optical fiber coating are: satisfactory light transmission, low refractive index (lower than that of glass), compatibility with glass substrates, permeability for water-born analytes, and mechanical stability due to cross-linking. The work on sensors development has also shown that cross-linked polymer matrix on the optical fiber is of great potential due to its functionality:

1. hydrogel capable of swelling and shrinking can produce microbending in fiber [82],
2. it can absorb the evanescent tail of the light propagating within the fiber core [83,84], and

3. hydrogel matrix can serve as a container for ligands, which are normally adsorbed onto [85], entrapped into [86-88] or covalently bound to [89-91] the supporting matrices, so that biological recognition processes through antigen-antibody, enzyme-substrate, or ligand-receptor interactions can be incorporated to identify and quantify biochemical substances.

There exist numerous sensing schemes [78-81], however, a simple analysis shows that only two methods of deposition of hydrogel matrix on an optical fiber have been employed: (i) hydrogel spots were polymerized on fiber distal ends (not on the side wall of a fiber core) by direct light [78,92,93] and (ii) hydrogel was deployed on side-surface of a fiber core by spreading a preliminary prepared polymer followed by its further cross-linking [94].

The disadvantage of the polymer matrices deposited by those "macroscopic" methods is in that they are relatively thick (microns to cm). As a result, the polymer matrix exhibits a prolonged response time (minutes to days), hampered expansion and contraction, and hindered mass transfer. Thus, it is advantageous to have a cross-linked polymer matrix of nanometer thickness (hydrogel nanofilm) as a prerequisite for future applications. The bottom up question is how to prepare the nanometer sized hydrogel layers with controllable thickness. Polymerization by evanescent waves (PEW) has been shown [76] to be a versatile and convenient method for fabrication of nano- and microparts with controllable submicrometer heights on the planar surface of glass slide. We have proposed the employment of evanescent waves for preparation of ultrathin hydrogel films in the proximity of the side surface of a fiber core, *i.e.* in cylindrical configuration [95]. We believe that cylindrical surface of an optical fiber core is more practical compared to a flat surface to support ultrathin cross-linked polymer fabrics. First, the *cylindrical architecture* of the "nanofilm/fiber structure" will not allow the polymer "cylindrical sock" to come out of the fiber core. Second, optical fiber connected to a spectrometer is a polymerization setup and an analytical device in one. Considered below are the distinct prospects of using the combination of a spectroscopic tool (optical fiber) and environmentally responsive element (cross-linked polymer matrix on side wall of optical fiber core) in various bioanalytical applications. The cylindrical configuration of the hydrogel/fiber structure is a promising approach to miniaturization of lab-on-a-chip devices.

2. Basic model of polymerization by evanescent waves

2.1. Evanescent waves produced by TIR

The evanescent waves have been widely used for the spectroscopic detection of chemical and biological species by fiber-optic sensors [78-81]. In our previous work [95], evanescent waves were used to initiate a polymerization reaction around a fiber core. This procedure is really new: (i) for the first time, EW were used to fabricate very thin polymer films of cylindrical configuration, (ii) for the first time, the PEW procedure is developed for a photopolymerizable formulation sensitive to ultraviolet light which is preferred for a higher spatial resolution, (iii) it requires no preforming of the polymer layer, (v) the thickness is just governed by the penetration of acting light into the polymerizable formulation, which in turn is a function of the radiation properties (intensity, exposure time, spectral composition), optical properties of fiber core and hydrogel forming solution (refractive indexes), and geometry of the fiber core (diameter). In other words, because of the propagation mode of EW produced by TIR, the PEW procedure is a unique way to reduce the interaction volume of light with the reactive formulation to the utmost, i.e. down to the nanometer scale – it looks possible to obtain cylindrical cross-linked polymer films with the record thickness of 1 to 5 nm.

Consider a light beam that propagates in a higher index medium (refractive index n_1) of the fiber core and reflects at the interface with a lower index medium (refractive index n_2) of the hydrogel forming solution (Figure 7). The beam propagates by a series of total internal reflections at the core/solution interface. The interference between the incident and the reflected waves gives rise to standing waves inside the fiber core, perpendicular to the interface, and to the generation of an evanescent field inside the solution. The intensity of the electromagnetic field transmitted by EW decreases exponentially as it propagates farther away into the sample and is expressed as follows:

$$I(z) = I_0 \exp(-2z / d) \tag{4}$$

with d expressed by Eq (3) has the dimension of a length and is called the characteristic penetration depth.

The evanescent wave can interact with molecules located within the depth of penetration. Thus far, the evanescent wave can be absorbed, scattered or can excite molecules around the area of a fiber core exposed for polymerization (Figure 7).

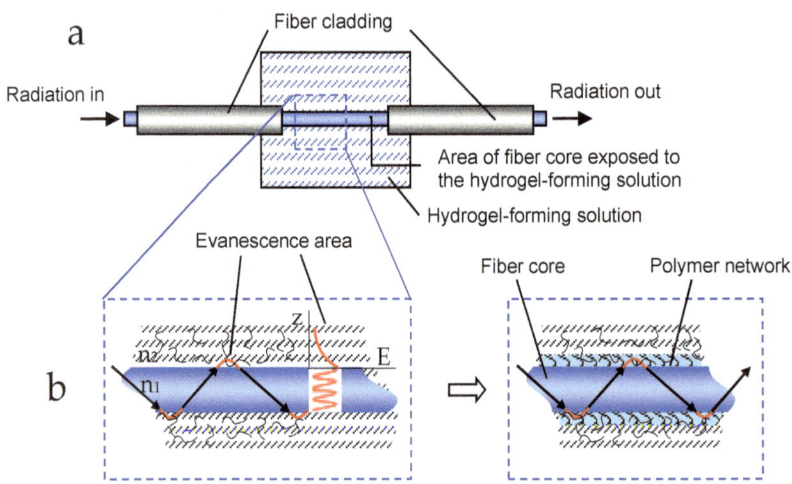

Figure 7. Optical fiber arrangement for polymerization by evanescent waves: (a) free-cladding area of fiber core exposed to the hydrogel-forming solution (cylindrical configuration); (b) in the course of the propagation of electromagnetic radiation inside the optical fiber, a portion of the electric field, which penetrates out of the fiber core (evanescent wave) initiates the polymerization reaction in nanometer layer around the fiber core.

2.2. Polymerization process

Radical initiation of the photopolymerization generally provides high polymerization yield, and photosensitivity for radical polymerization can be extended in the visible range by adding a suitable sensitizer [96]. The radical photopolymerization requires three essential elements: electromagnetic radiation (hν), photoinitiators (PI), and mono/macromers (M) with unsaturated moieties such as a vinyl group and its derivatives. Ultraviolet (190 – 400 nm) and visible light (400 – 800 nm) can be harnessed for photopolymerization. Accordingly, the maximal absorption wavelength (λ_{max}) of the photoinitiator is in the relatively high energetic UV or near UV visible range.

2.2.1. *Photopolymerization reaction scheme*

The photopolymerization of a monomer (M) in the presence of a photoinitiator (PI) can be represented schematically by the following equations:

Step 1. Radical formation and initiation of the polymerization

$$PI \xrightarrow{hv} R\bullet + R\bullet \xrightarrow{[O_2]} deactivation$$
$$\downarrow$$
$$R\bullet + M \xrightarrow{hv} \bullet M_1$$

It is known [76] that in the early stages of irradiation, the absorbed light is used mainly to consume oxygen, which acts as a strong deactivating agent by trapping the formed radicals through peroxidation. This phenomenon is one of the main reasons accounting for a delay between the beginning of the irradiation and the start of polymerization. After a while, as the concentration of oxygen decreases, it becomes low enough and the reaction of initiating radicals with monomers competes successfully with the deactivation processes. The dose of energy absorbed before polymerization occurs, the so-called the threshold dose E_{th}, is the dose of energy necessary to consume a sufficient amount of oxygen. By the way, polymerization by evanescent waves allows one to circumvent the problem: the total internal reflection provides the thinnest diffusion layer with the highest rate of oxygen consumption. Moreover, polymerization in N_2 atmosphere (nitrogen-blanketing) is a precaution that allows substantial overcoming O_2 inhibition of photopolymerization [98]. Therefore, assuming that the formation of monomer radicals $\bullet M$ is much faster process than the formation of radicals $R\bullet$, the initiation of polymerization in the absence of O_2 is the rate-limiting process determined by the rate of $R\bullet$ formation:

$$r_I = k_I[PI], \quad (5)$$

where PI is the concentracion of a photoinitiator, k_I is the rate constant of the initial step. As monomer radicals $\bullet M$ are produced, the absorbed energy is used to form the 3D polymer network.

Step 2. Propagation

$$M + \bullet M_1 \xrightarrow{h\nu} \bullet M_2$$
$$M + \bullet M_2 \xrightarrow{h\nu} \bullet M_3$$
$$------------$$
$$M + \bullet M_{n-1} \xrightarrow{h\nu} \bullet M_n$$

Polymer chains start growing and the cross-linking reaction proceeds rapidly to reach the gel point and finally vitrify. Letting $[\bullet M]$ represent the sum of concentrations of all radicals including $\bullet M_i$ and $R\bullet$, the rate of propagation determines the rate of polymerization:

$$r_P = k_P[M][\bullet M], \tag{6}$$

where k_P is the rate constant of the propagation step. Mechanical resistance appears only when the dose of energy received by the system is greater than E_{gel}.

Step 3. Termination

$$\bullet M_n + \bullet M_m \rightarrow \begin{cases} M_{n+m} & \text{mutual termination} \\ M_n + M_m & \text{disproportionation} \end{cases}$$

The termination of chain propagation usually proceeds by standard radical coupling (mutual termination), when two growing radical chains combine, or disproportionation, when unsaturation exchanges through β-hydrogen migration involving ambient oxygen [97]. The rate of termination can be expressed as follows:

$$r_T = k_T[\bullet M]^2, \tag{7}$$

where k_T is the rate constant of the termination step.

Assuming that monomer radical is formed very quickly and its concentration is constant in the course of the chain reaction (steady-state approximation), one can show that in the case of termination by coupling, the average degree of polymerization $\langle p_n \rangle$ is a function of the concentrations of monomer $[M]$ and initiator $[PI]$, which can be controlled experimentally:

$$\begin{aligned} [\bullet M] &= \left(\frac{k_I}{k_T}\right)^{1/2} [PI]^{1/2} \\ \langle p_n \rangle &= 2\frac{\text{the rate of propagation}(r_P)}{\text{the rate of termination }(r_T)} = \frac{2k_P}{\sqrt{k_I k_T}}[M][PI]^{-1/2} \end{aligned} \tag{8}$$

Therefore, kinetics of chain reaction controls the polymerization as follows:

- The slower the initiation of the chain (the lower $[PI]$ and the smaller k_I), the higher is degree of polymerization

- The higher the concentration of monomer *[M]*, the longer is the average length of polymer chain
- The higher the concentration of initiator *[PI]*, the shorter is the average polymer chain

2.2.2. Time dependence of photopolymerization

In order to find the time course of the concentration of the polymer $[M_n]$, we begin with the kinetic equation for polymerization

$$-\frac{d[M]}{dt} = k_P \sqrt{\frac{r_I}{k_T}}[M]. \tag{9}$$

Separating variables and integrating with the assumption of no time dependence for all terms except the concentration of monomer *[M]* gives

$$[M] = [M]_0 e^{-k_P \sqrt{\frac{r_I}{k_T}} t}. \tag{10}$$

The concentration of the polymer $[M_n]$ can be obtained from the mass conservation law and Eq.(5) for the average degree of polymerization

$$\begin{aligned} &[M]_0 = [M] + [M_n]\langle p_n \rangle \\ &[M_n] = \frac{[M]_0 - [M]}{\langle p_n \rangle} = \frac{\sqrt{r_I k_T}}{2k_P}(1 - e^{-k_P \sqrt{\frac{r_I}{k_T}} t}) \end{aligned} \tag{11}$$

Expanding the exponential function into a series in powers of $x = k_P \sqrt{\frac{r_I}{k_T}} t$ and limiting ourselves with only two terms ($e^{-x} \approx 1-x$), the increase of polymer concentration with time can be estimated as follows:

$$[M_n] \approx \frac{r_I}{2} t. \tag{12}$$

2.2.3. Photochemical and processing parameters and the point of gelation

At the gel point, $[M_n] = [M_n]_c$, the critical concentration of the polymer represents a critical conversion point of a gel formation and corresponds to the dose of energy (E_{gel}) absorbed by the system during the time of exposure t_e. Our goal is to find the relation between E_{gel} and the photochemical and processing parameters. The initiation rate r_i (*mol/L/s*) is related to the photonic flux *I(z)* (= amount of photons per surface area per second, *Einstein/m*^{2}*/s* = $10^{-3}m \times mol/L/s$) [98,99]

$$r_I = \phi\varepsilon[PI]I(z), \tag{13}$$

where ϕ is the quantum yield of the photoinitiator, ε is the molar extinction coefficient (*L/mol/m*), and *I(z)* is the incident intensity of the electromagnetic field transmitted by EW at depth z. In the absence of a photoinitiator, the photonic flux, or intensity, at depth *z* decreases exponentially according to Eq. (4). However, in the presence of the molecules of photoinitiator additional absorption should be taken into account by Beer's Law factor:

$$I(z) = I_0 e^{-2z/d} e^{-\varepsilon \ln 10[PI]z} = I_0 e^{-z/d_{eff}}, \tag{14}$$

where I_0 is the incident intensity at the fiber core/solution interface (*z* = *0*), $d_{eff} = \dfrac{d}{2 + d\varepsilon'[PI]}$ is the effective penetration depth of the evanescent waves, and ε' stands for $\varepsilon ln10$. It seems reasonable that this effective depth of penetration of the evanescent field into the hydrogel forming solution is inversely related to the extinction coefficient and concentration of photoinitiator. Now, in the first approximation, the dose of energy received at depth *z* and time *t* is given by the following relationship:

$$E(z,t) = I_0 t e^{-z/d_{eff}}. \tag{15}$$

We use this equation to define the E_{gel} as the dose of energy (= amount of photons per surface area, $Einstein/m^2 = 10^{-3} m \times mol/L$) absorbed by the system to polymerize a layer with thickness l_p at exposure time t_e, i.e.:

$$E_{gel} = E(l_p, t_e) = I_0 t_e e^{-l_p/d_{eff}}. \tag{16}$$

Substitution of Eqs. (16) and (13) into Eq. (12) at $z = l_p$, $t = t_e$, and $[M_n] = [M_n]_c$ allows the following expression to be obtained for the critical dose of energy necessary to reach the gel point as a function of the concentration of photoinitiator

$$E_{gel} = \frac{2[M_n]_c}{\phi\varepsilon[PI]}, \tag{17}$$

Note that the less the concentration of the photoinitiator in the polymerization formulation, the higher dosage is necessary to polymerize a stable film on the fiber core. Combining Eqs. (17) and (16), one can predict the dependence of the thickness of polymerized hydrogel on the photochemical properties of the hydrogel forming solution (*[PI]*, *[M]*) and processing condition (ϕ, ε, $I_0 \times t_e$)

$$l_p = d_{eff} \ln\frac{I_0 \times t_e}{E_{gel}} = \frac{d}{2 + d\varepsilon'[PI]} \ln\frac{I_0 \times t_e}{2[M_p]_c / \phi\varepsilon}[PI]. \tag{18}$$

With all these assumptions in place, one can conclude that the presence of photoinitiator results in the decrease in the effective depth of penetration of evanescent light because of its absorption by the photoinitiator molecules. The second and main finding is that for a given concentration of photoinitiator *[PI]*, the thickness of polymerized film l_p remains zero until the dose of energy received by the system $I_0 \times t_e$ exceeds the critical value defined by the critical concentration of polymer $[M_n]_c$ corresponding to the point of gelation:

$$I_0 \times t_e > E_{gel} = \frac{2[M_p]_c}{\phi\varepsilon[PI]}$$

According to this model, if $I_0 \times t_e > E_{gel}$, the polymer thickness around the fiber core is expected to increase linearly with the logarithm of $I_0 \times t_e$, whereas the slope of this dependence is equal to the effective penetration depth of the evanescent wave.

3. Optical fiber evanescent wave spectroscopy (FEWS)

3.1. Real time recording of polymerization on the side surface of optical fiber core

The polymerization as well as observation and characterization of the nanofilm formation on the side surface of a fiber core, if any, are the-state-of-the-art and the "know-how". How to detect the formation of ultrathin films made of tiny cross-linked polymeric threads with optical properties similar to the solvent? This is the critical question. Experimental approaches to the question of hydrogel nanofilm monitoring usually fall into direct microscopic methods, such as optical microscopy (OM), atomic force microscopy (AFM), and electron microscopy (EM), which can be used at the step of film characterization. However, they are not convenient for routine measurements especially at the step of film formation. Spectroscopic methods are much practical in this case.

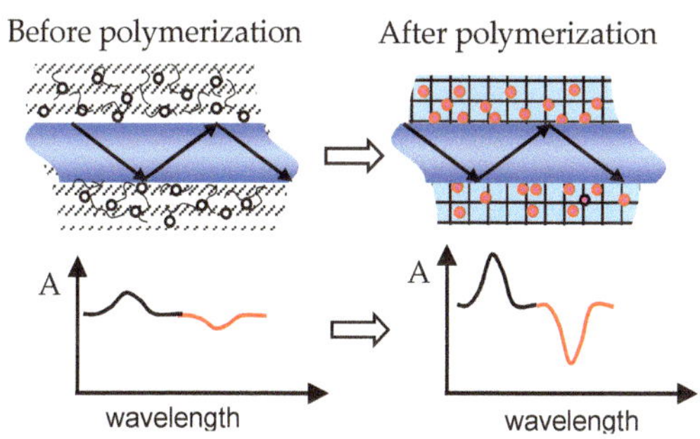

Figure 8. Spectroscopic detection of polymerization on the optical fiber core. The formation of polymer network changes the microenvironment of the fluorescent probe resulting in an increase in fluorescence emission (red line). Absorbance increases due to scattering on microheterogeneities of newly formed polymer network (black line).

Strikingly, the optical fiber arrangement for polymerization by evanescent waves (Figure 7) can be used for the real-time monitoring of hydrogel photopolymerization on the side surface of the fiber core. Indeed, in the course of initiation of polymerization, the intensity of light passing through the fiber core decreases due to excitation of molecules of photoinitiator, radical formation, and consumption of oxygen within the evanescence area. Furthermore, during polymerization the refractive index of the material changes, thus, inducing a modification of the refractivity of the core/solution interface. And finally, when the point of gelation is reached, the energy of evanescent wave scatters on the microheterogeneities of the cross-linked network.

The other way to record the polymerization on the side surface of a fiber core with time is to apply fluorescent probes. The fluorophores capable of changing the quantum yield of fluorescence depending on the microenvironment can be used (Figure 8). For example, 1-anilinonaphthalene-8-silfonate (ASN) may exhibit 100-fold emission intensity in the polymerized network in comparison to the pre-polymerization solution. Moreover, the maxima of both absorption and fluorescence picks shift with changes in the polarity of the microenvironment [100,101].

3.2. The ideas of hydrogel film detection by FEWS

Actually, the measured polymerization kinetics itself is the evidence for hydrogel formation. Nevertheless, a hydrogel film, which has already been polymerized on the fiber core, could be detected spectroscopically. Indeed, studying the mechanical and electrochemical responses of hydrogels in the course of swelling/de-swelling of polymer matrix [102-106], we have observed that being swollen in water hydrogels were soft, gentle, and transparent materials, whereas under dehydration or wet/dry transition, polymer networks revealed a drastic volume decrease (up to 1000 times) resulted in the transparency drop. Thus, the correspondence between variations in physical properties (volume, thickness, density, refractive index, etc.) of hydrogel films and changes in the spectral composition and total intensity of light passing through the fiber stands behind the ideas of hydrogel film detection on the fiber core.

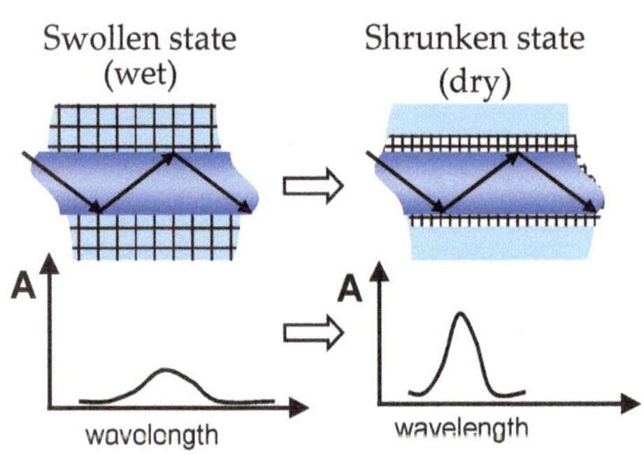

Figure 9. Spectroscopic detection of a polymer network on the optical fiber core: an increase in density (refractive index) of the polymer network in the course of dehydration of gel results in higher scattering/absorption of the evanescent light nearby the fiber core.

The first approach to the detection is illustrated in Figure 9. The scattering/absorption of evanescent light is greater in the dried polymer network than in its swollen wet state.

The second approach is based on using fluorescent probes (e.g. fluoresceins) with a capability of self-quenching at high concentrations [107-109]. Fluorescence resonance energy transfer (FRET) is a distance-dependent interaction between the electronic excited states of two dye molecules in which excitation is transferred from a donor molecule to an acceptor molecule without emission of a photon. Fluorescein/fluorescein is a donor/acceptor pair with 4.4–nm Förster radius, the distance at which ~50% of excited donors are deactivated by FRET. Fluorescence intensity is expected to decrease in the course of nanofilm shrinkage (the distance between fluorescent molecules becomes smaller) and to increase with hydrogel swelling (Figure 10).

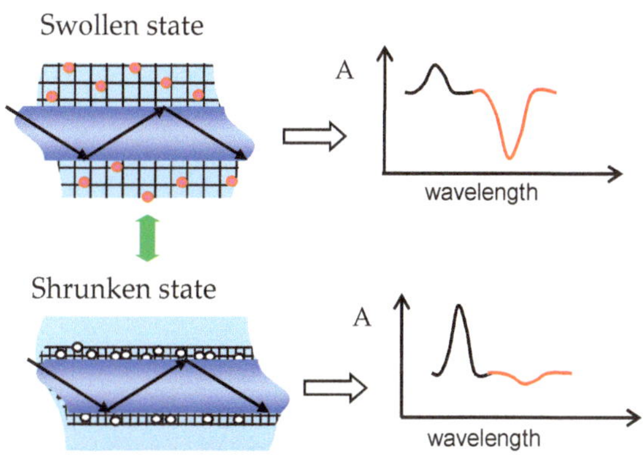

Figure 10. Spectroscopic detection of a polymer network on the optical fiber core: when fluorophores are loaded into the swollen hydrogel layer (red circles), the shrinkage of the polymer network will decrease the distance between these trapped molecules (white circles), resulting in a decrease in the fluorescence emission (red line). Herein, the absorption (black line) increases due to scattering of evanescent light on the denser polymer network.

3.3. Cylindrical hydrogel film characterization by FEWS

To the best of our knowledge, the first report on polymerization of hydrogel in a cylindrical configuration using the evanescent tail of light propagating in fiber was published in 2010 [95]. So long the hydrogel is on the fiber core, experimental techniques concerned with characterization of the extent of its volume change modality are in demand. The formed cylindrical polymer network is supported by, but not covalently bound to the side-surface of fiber core. It is a kind of "cylindrical sock" on a fiber core, and its thickness can vary with the photochemical and processing parameters discussed in Section 2.2.3. Understanding of the hydrogel functionality, specificity, and biochemical activity is required as well. It looks quite relevant to use the optical fiber evanescent wave spectroscopy (FEWS) as a method for characterization of hydrogels on the fiber core:

On the one hand, because hydrogels are the so-called soft matter materials with environmental response, i.e. relatively small external perturbations can cause dramatic structural changes in them followed by substantial physical expansion/contraction. Since the perturbed polymer networks with swelling/de-swelling ability in many cases, are not in equilibrium, it is of great importance to characterize the relaxation of their physical dimensions (especially, at the nanometer length scale) with time.

On the other hand, because the experimental scheme of FEWS allows one not only to record time-changes in integral intensity of light passing through the fiber, but also to provide a novel information on the spectral composition of light scattered/absorbed within the medium in the immediate vicinity of the fiber core surface, and how the spectrum changes with time at different external conditions, thereby taking advantage of time-resolved and nanometer-scaled measurements.

Thus, the hydrogel/fiber optics assemblage offers a method for studying the kinetics of swelling and shrinking spectroscopically with minor modifications of the same experimental scheme as used for polymerization and detection.

4. Experimental procedures

4.1. Photopolymerizable formulations

Even a brief review of physical and chemical properties of hydrogels present in Introduction allows the reader to realize that any hydrogels with relevant chemical composition and environmental responsiveness could be synthesized for numerous applications. The photopolymerizable formulations are specified by the desirable functionality and specificity of the hydrogel film on the fiber core. For example, in order to prepare a temperature sensitive hydrogel, *N*-isopropylacrylamide (NIPA) monomer can be used, since it forms polymers with volume phase transition temperature varied from 20 to 45 °C, i.e. in the physiologically relevant range [110]. Combined temperature and ionic sensitivity of the hydrogels can be achieved by co-polymerization of NIPA with corresponding cationic (e.g. vinylimidazole) or anionic (e.g. carboxyl) moieties [8-10]. Hydrogel materials with biological responsiveness built in can be achieved by introducing ligands that bind specific biomolecules into the the material matrix [111]. When exposed to a biological target (nutrient, growth factor, receptor, antibogy, enzyme, or whole cell), molecular recognition events in the bioresponsive hydrogels trigger macroscopic responses, such as swelling/de-swelling or solubility transitions.

The cross-linking density, which is the main effector of swelling ability, can be varied by changing the [monomer]/[cross-linker] ratio. As a standard cross-linker *N,N'*-methylenebisacrylamide (MBA) is used in many formulations. To optimize the composition of polymerizable formulations in terms of the final swelling/de-swelling ability, a hydrogel forming solutions of different composition should be tested on bulk gels by varying concentrations of monomer, co-monomer, and cross-linker (see Figure 4).

To prove the concept of photopolymerization by evanescent waves on the side surface of the optical fiber core, the composition of temperature sensitive PNIPA hydrogel has been optimized. The method for preparation of a bulk PNIPA hydrogel by polymerization of a hydrogel-forming medium induced by a UV-lamp (~365 nm, 100 W) has been described elsewhere [102,103]. From the homogeneity standpoint, it was found that during polymerization the temperature should be kept below the Lower Critical Solution Temperature (LCST, T_c) of the final hydrogel. Particularly, the polymerizable aqueous solution of NIPA (5 wt %) and MBA (0.5 wt %) yielded a clear temperature-responsive hydrogel with 10-fold volume change in water near T_c.

Successful photopolymerization relies on the effective initiation step, so that the photointiator (PI) is a critical component of a polymerizable formulation to reach the final cross-linked product. The vast majority of photoinitiating systems are based on free radicals as reactive species. Commercially available photoinitiating systems with different specific maximal absorption wavelengths (λ_{max}) and corresponding molar extinction coefficients (ε) were studied [96,112]. The typical examples of PIs sensitive in UV range are 2,2-diethoxyacetophenone (DEAP, 365 nm) and 2,2-dimethoxy-2-phenylacetophenone (DMPA, 365 nm). Sensitivity of the photoinitiating system can be extended to the visible range either by adding a suitable co-sensitizer (e.g. 4,4′-bis(dimethylamino)benzophenone, 300-500 nm) to one of the aforementioned PIs or by using the typical visible light photoinitiating systems such as camphorquinone, ethyl 4-N,N′-dimethylaminobenzoate, trethyl amine, isopropyl thioxanthone (470 nm). A considerable attention has been paid to the studies on the PIs sensitive in near infrared range (>700 nm). For example, cyanine dye-borate complexes have been studied [113] as promising PIs with λ_{max} = 780 nm. The near infrared sensitive PIs give a possibility to use cheap, compact and easy operational laser diodes and polymer fibers, which are less fragile and more robust and rugged than UV transparent ones.

Because of differences in chemistry for synthesis of hydrogels with the desired end properties, two or three different initiators are suggested to be tested. As an example, Figure 11 shows the photopolymerization kinetics for two hydrogel forming solutions with the same monomer (NIPA, 5 wt %) and cross-linker (MBA, 0.5 wt %) content, but different PIs – DEAP and DMPA. The time course of light intensity at λ = 365 nm and 391 nm scattered at a 90° angle from the polymerizable formulation shows that photoinitiating activity of DMPA is higher than that of DEAP under the same conditions. Indeed, polymerization started three times earlier when initiated by DMPA than DEAP (arrows).

The concentration of PI should be also optimized because of the following reasoning: on the one hand, the high concentration of PI provides a reliable polymerization with a high rate. On the other hand, too high rate and high level of cross-linking near the surface does not allow the light effectively penetrate farther inside the hydrogel forming solution. Moreover, limited water solubility of PI may lower its initiating efficiency and increase heterogeneity of the resultant polymer network. Solubility in water may be improved by dissolving the initiator in 70% ethanol solution before adding to the polymerizable formulation. The range of reasonable concentrations for DEAP was found to be from 0.01 to 1 %, whereas for DMPA it is 3 times lower due to a higher extinction coefficient.

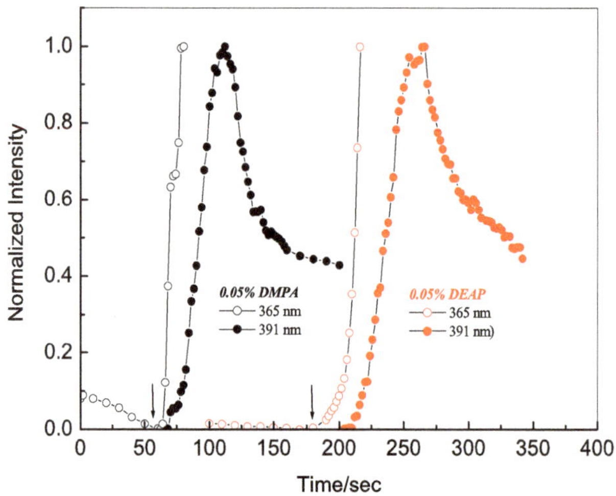

Figure 11. Polymerization kinetics of bulk PNIPA hydrogel with two different photoinitiators – 2,2′-diethoxyacetophenone (DEAP, red circles) and 2,2′-dimethoxy-2-phenylacetophenone (DMPA, black circles) measured by intensity of two spectral lines 365 nm and 391 nm) scattered perpendicular to the light initiating the polymerization. Diffuse radiation of Blak® Ray mercury lamp (365 nm, 100W) was used for the photopolymerization. Arrows show the instant when polymerization starts.

4.2. Optical fiber preparation and reaction chamber design

When selecting an optical fiber, it is important to work towards its transparency in a required spectral range, maximizing the light intensity collected into the fiber, and optimizing the length of the cladding-free area to provide a sufficient average path length within evanescence area for detectable level of an evanescent wave spectrum.

A plastic clad silica optical fiber with a high OH content is needed for UV and visible transmission. Since the light penetrates ~0.1λ into surrounding polymer, the depth of evanescent field penetration varies from ~30 to 70 nm for 300–700 nm wavelength range. The systematic studies show that the total path length for spectral measurements should be at least on the order of a micrometer. An appropriate path length necessary for an evanescent wave spectrum recording is determined by the portion of fiber with removed cladding. Herein, the critical angle of the total internal reflection and number of reflections depend on the core diameter (∅) and numerical aperture (NA) of a selected fiber: For example, for a fiber core with ∅600 μm and NA = 0.48, a 4-cm-long cladding free area provides a path length of ~1.6 μm for an evanescent wave with λ = 365 nm.

Based on this estimations the reaction chamber has been designed (Figure 12). A 50-cm length SMA-SMA ending fiber cord (BFH48-600, Thorlabs Inc., Newton, NJ) having a 600-μm core diameter and numerical aperture of 0.48 was used. This hard polymer clad multimode fiber with high OH content is transparent in the spectral range from 300 to 1200 nm. The evanescence area is prepared by removing a plastic clad in the middle of the cord. In general, there are three methods known for the removal of the cladding on different types of fibers: mechanical, thermal, or chemical. A chemical cladding-removal technique would be ideal, since it greatly reduces the possibility of mechanical damage to the surface of the fiber core. However, a 30-μm thickness polymer cladding of this fiber is too hard to be dissolved in any solvents in a reasonable time. Mechanical methods also are not reliable due to fragility of the fiber core. The fiber cladding was removed by a burning method using a propane or butane burning torch. When the fiber is stripped from the buffer, it should be held by an appropriate fixture in which fiber can rest and which protects that portion of the fiber core after cladding removal. The design of the reaction chamber (removable bottom cover) allows it to be used as the fiber fixture during the burning process and as a cleaning bath after. The surface of fiber core free of polymer cladding thoroughly cleaned with isopropyl alcohol, distilled water, and finally, compressed air to remove any carbonization. The quality of the surface was controlled under an optical microscope.

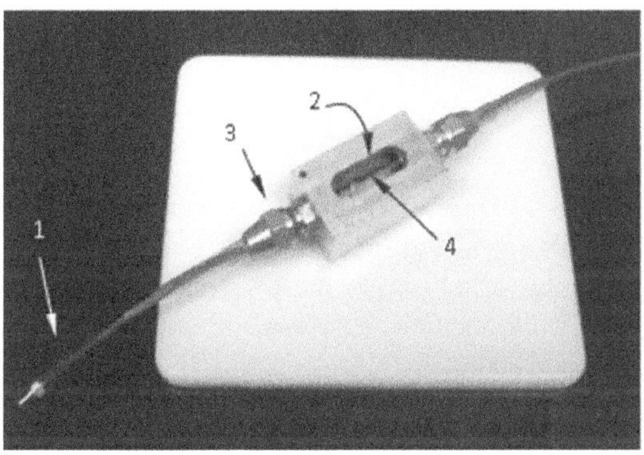

Figure 12. The reactor-container: (1) SMA ending fiber cord, (2) reservoir containing a polymerizable formulation (~7 mL), (3) sealing gaskets, (4) fiber core without cladding exposed to a hydrogel-forming solution.

The recent studies suggested [114] that evanescent field can be pronounced and sensitivity of FEWS can be drastically increased by decreasing the diameter of fiber core. Since the extra caution should be taken with fibers under 200-μm diameter, the evanescent wave cuvette of a new design was developed (Figure 13). Now, the so-called detection element is a movable module made of a cladding-free fiber core with the SMA compatible connectors. The distal ends of the detection element are polished to maximize the intensity of light passing through the fiber core. The modified reaction chamber offers the following advantages: maintains the integrity of the evanescence area, offers the feasibility of handing at all steps (polymerization, cleaning, detection, characterization), suits to a broad range of applications, such as a sensor with specifically modified functionality, and allows real-time watching the evanescence area during polymerization.

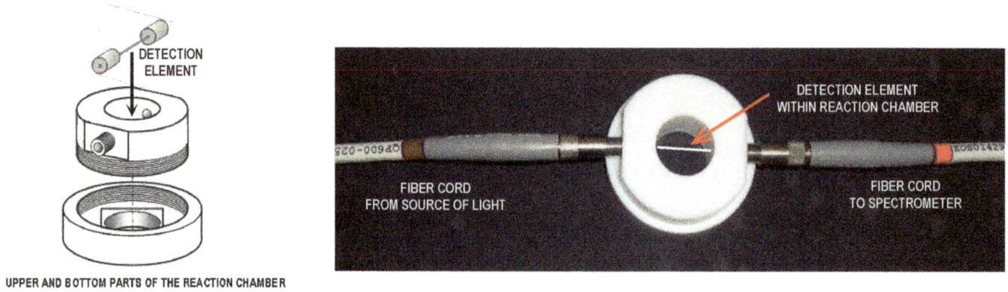

Figure 13. Schematic (a) and manufactured (b) the evanescent wave cuvette of a new design for polymerization on a side surface of the fiber core and recording the hydrogel swelling/de-swelling kinetics. The status of the hydrogel on fiber core can be controlled microscopically in the course of polymerization and swelling/de-swelling. Detection element is a removable part allowing for characterization by the other methods (e.g. atomic force microscopy).

4.3. Instrumentation

Light sources. Of equal importance to the PI, however, is the light source. Basically, two types of light sources are available: arc lamps, lasers, and light-emitting diodes. Recent advantages in light source technology are the development of lamps, lasers and/or diodes which offer increased light intensity at specific wavelengths. Regardless of the type of light source the emission spectra of the source must overlap the absorption spectrum of the chosen initiator. In this work, the Oriel Research Arc Source (Newport Corp., NJ) with a condenser, built-in ignitor, and 200 W Hg-Xe or 100 W Hg lamp was used for the purposes of PEW on optical fiber. A Blak Ray mercury lamp (365 nm, 100 W) was used for polymerization of a bulk hydrogels. A Deuterium-Tungsten halogen light source (DT-MINI-2) from Ocean Optics was used for spectroscopic detection and characterization of hydrogel films on the fiber core.

Fiber-optic spectrophotometers. Fiber-Optic CCD Array UV-Vis Spectrophotometer (Ocean Optics, Inc.) with a convenient fiber optics connectorization, high sensitivity, high recording speed (1 spectrum/sec), broad wavelength range (180–900 nm), and high spectral resolution (<1 nm) was used for routine spectroscopic measurements. SpectraSuite software was used to record, treat, and store radiation, transmission and absorption spectra.

Optical microscopy. Digital Optical Microscope (DC3 163 PH, Microscope World, CA) with Imaging Software (Motic Images 2000, Motic Group, Ltd.) was used to measure the diameter of the bare fiber core and thickness of the cylindrical hydrogel on the fiber core after washing. To detect and characterize a relatively thick hydrogel film (> 10 μm) on fiber core after polymerization or during swelling/de-swelling, an optical microscope with low magnification was used as well.

4.4. Experimental setup for photopolymerization on side-surface of optical fiber core

The effect of radiation intensity and exposition time on the hydrogel film formation by evanescent wave photopolymerization was examined using the 200 W Hg-Xe lamp. The UV light was focused with a condenser and collimating lenses on SMA905 ending fiber, which directs the light through the bandpass filter to the cladding-free portion of fiber core fixed within the chamber containing a polymerizable solution as shown in Figure 14. The other SMA end of the fiber cord was connected through an attenuator to a fiber optic spectrophotometer (USB4000-UV-VIS) with CCD array, which allowed for observing the spectrum of light transmitted through the whole pathway. The spectra recorded in the range from 180 to 900 nm were stored in a 1 sec time interval.

The reaction chamber was placed into a temperature insulating box to provide a temperature control (±0.1°C), A temperature sensor was dipped into the reactor-container nearby the fiber core.

When the hydrogel forming solution was poured into the reaction chamber in the amount necessary to cover the fiber core, light was on. Passing through the fiber core, it causes evanescent waves penetrating outside to initiate polymerization in the close vicinity to the fiber core. Any changes in the intensity of light transmitted through the fiber are recorded by the fiber optic spectrophotometer which serves as a light detector. The PNIPA hydrogel film synthesized on fiber core was washed in deionized water for 3 days at room temperature before further characterization. It was found that thickness of the hydrogel film could vary from nanometer scale (<1 μm, cannot be detected by optical microscopy) to hundreds of microns by varying the dose of energy – the product of power of the light source and polymerization time (see section 2.2.3).

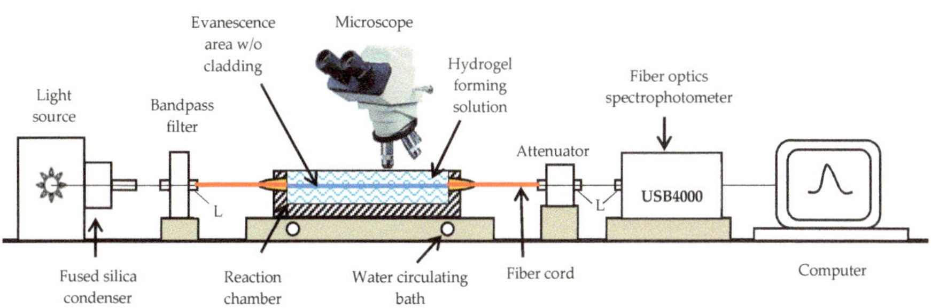

Figure 14. Optical scheme of the system for polymerization of cylindrical hydrogel on a side surface of the fiber core and recording the hydrogel swelling/de-swelling kinetics. L – lenses.

The same optical scheme (Figure 14) was used to examine the kinetics of hydrogel swelling/de-swelling in the course of film hydration and during the volume changes. The only difference was that a Deuterium–Tungsten Halogen light source (DT-MINI-2) provided continuous spectrum in the whole UV-VIS range from 180 to 900 nm. The time course of integral intensity and spectral composition of light passing through the fiber core with hydrogel film on it was measured at different temperatures. Temperature control inside the reaction container was carried out permanently in the course of swelling kinetics study. Transmission or absorption spectra were saved each second when temperature stabilized on a certain level within the experimental uncertainty of ±0.1°C.

5. Properties of hydrogel film on the fiber core

5.1. Hydrogel films of nanometer and micrometer thicknesses

5.1.1. Kinetics of photopolymerization of a thin hydrogel film on the fiber core

To polymerize a thin PNIPA hydrogel on a side surface of the optical fiber core, a mercury 100 W lamp without condenser was employed as a light source. The lamp was turned on after a 15 min equilibration of the hydrogel forming solution within the reaction chamber at room temperature. The polymerization reaction was followed *in situ* by monitoring the integral intensity of light passing through the fiber core and recording it every 0.2 sec (sample time) using the above described setup (Figure 14). A typical integral intensity versus exposure time kinetic curve is shown in Figure 15.

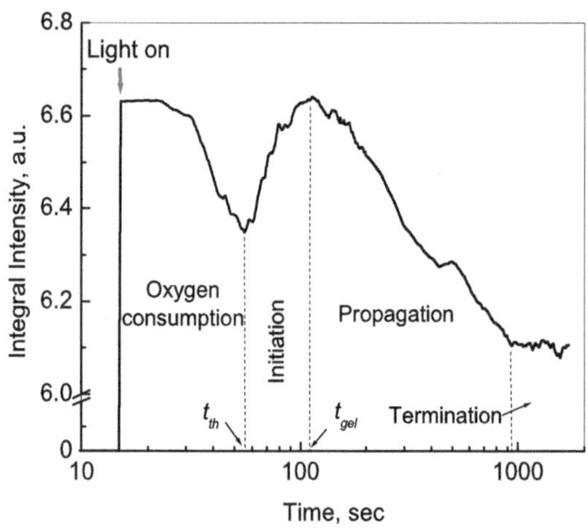

Figure 15. Time course of integral intensity during photopolymerization by evanescent waves on the fiber core surface: (UV lamp power = 100 W, composition of the hydrogel forming solution is 5 wt % NIPA, 0.5 wt % MBA, and 0.5 wt % DEAP).

As it has been explained in the section 2.2.1 on the photopolymerization process, there is a delay between the beginning of irradiation and the formation of gel because the absorbed dose of energy is used not only for the decomposition of the initiator molecules into radicals, but also for the consumption of oxygen which deactivates the formed radicals. The pronounced decrease in integral intensity at the very beginning of UV illumination (red arrow) can be assigned to the oxygen consumption. When the concentration of oxygen decreases enough, the reaction of radicals with monomers competes successfully with the deactivation processes and the intensity of light returns to a higher level. Therefore, the minimum in integral intensity located at about 60 seconds (t_{th}) corresponds to the threshold dose of energy necessary to consume a sufficient amount of oxygen. The radicals react with functional groups of monomer and cross-linker forming a polymer, but absorbed dose of energy is not sufficient yet to reach the gel point. This happens only after ~100 seconds (t_{gel}), when a significant portion of evanescent light scatters on the polymer network as it becomes denser to decrease the integral intensity – propagation of polymerization starts. A further decay of the integral intensity continues until the thickness of hydrogel reaches the depth of penetration. At this step (~15 min), the average level of integral intensity becomes constant indicating that radicals react with each other and polymerization is terminated. The total polymerization time was ~30 min. Thus, the three key steps of polymerization kinetics (initiation, propagation, termination) have been distinguished.

5.1.2. Hydrogel film of micrometer thickness on fiber core

Theoretically, the thickness a hydrogel film on the fiber core polymerized by evanescent waves can be varied within the depth of evanescent field penetration out of the fiber core surface (typically ~λ/10). To examine experimentally how thick the hydrogel film can be formed, the intensity of UV light collected into the fiber and exposition time were maximized. The power of the UV light was increased by using the Hg-Xe 200 W lamp and fused silica condenser along with focusing lenses (Figure 14).

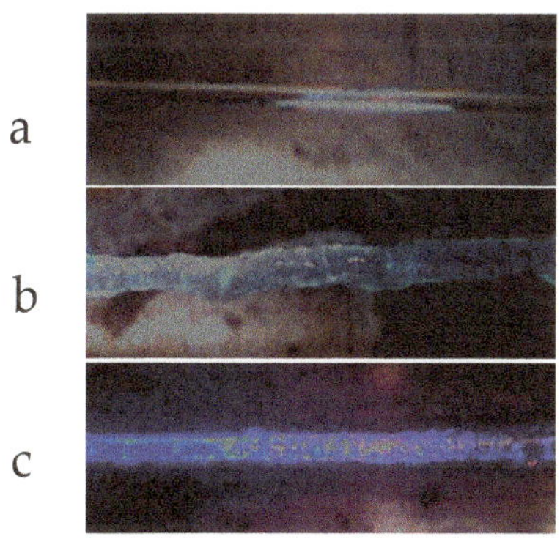

Figure 16. Optical micrographs (x10) of (a) bare fiber core on air, (b) hydrogel film on the fiber core on air after polymerization and washing, and (c) the same hydrogel film as for (b) dried overnight.

As in the previous experiment, before the UV light was allowed to pass through the fiber, thoroughly cleaned fiber core (Figure 16a) was completely covered with the hydrogel forming solution. The appearance of hydrogel was controlled by optical microscope and by measuring intensity of light passing though the fiber (polymerization kinetics) in real time. Figure 16b shows the cylindrical hydrogel film around the fiber core as a result of two hours of its exposition to UV light. It was a surprise that the average thickness of the film in wet state was ~175 μm, i.e. remarkably thicker than the depth of evanescent wave penetration. After overnight drying on air, the thickness decreased up to ~30 μm, still remaining thicker than the penetration depth (Figure 16c). In both cases, a blue glitter around the hydrogel cylinder is an indication that evanescent waves scatter within the hydrogel film. Herein, the intensity of light scattered inside the film in a wet state is much weaker than in the same film but dried overnight. Thus, the polymerization of hydrogel film of a macroscopic thickness, definitely, is the result of a successive scattering of evanescent waves on the newly formed polymer layers farther and farther from the surface of fiber core.

5.2. Properties of environmentally sensitive hydrogel film on the fiber core

5.2.1. Detecting the hydrogel film of nanometer thickness by dry-wet swelling

After measuring the polymerization kinetics as described in Section 5.1.1, the area exposed to polymerization was thoroughly and carefully washed in the reaction chamber by flowing distilled water at least for 48 hours. As expected, inspection of the fiber core surface under microscope did not allow us to prove the formation of a hydrogel film, since its thickness and the surface roughness was beyond the optical resolution. To detect the hydrogel film, the washed area of the fiber core was dehydrated by drying for 2 days on air. Then the measurements of the integral intensity of light passed through the fiber were carried out on air and after the addition of distilled water (see Sections 3.2 and 4.4 for details). Figure 17 shows three steps of changes: (i) an abrupt drop in intensity corresponds to the change in the refractivity of the fiber core/external medium interface, (ii) a very fast increase in the integral intensity due to immediate swelling of hydrogel in water, and (iii) a slow relaxation of the polymer network to equilibrium by smoothing up its microheterogeneities. Control experiments with bare fiber core did not show the later two steps. Together with polymerization kinetics data, these results unequivocally confirm the formation of a thin hydrogel film on the surface of fiber core.

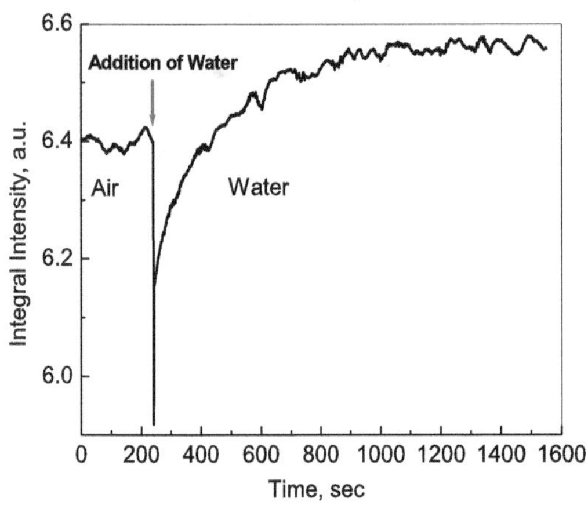

Figure 17. Time-dependence of the integral intensity of light passing through the fiber core during hydrogel film hydration on its surface.

5.2.2. Dry-wet transition in hydrogel film of micrometer thickness

The macroscopic PNIPA hydrogel film (Figure 16) was washed in deionized water for 3 days and dried on air for 120 hours at room temperature. The kinetics of hydrogel swelling in the course of film hydration is shown in Figure 18. The process is a multistep one. As for the thin film, the three steps can be clearly seen: (1) penetration of water inside the gel to reach the fiber core surface, (2) swelling of the hydrogel throughout the entire thickness, and (3) a slow relaxation of the polymer network to the equilibrium state. Totally, fiber transmittance increases 2-fold as a result of the hydrogel swelling (hydration).

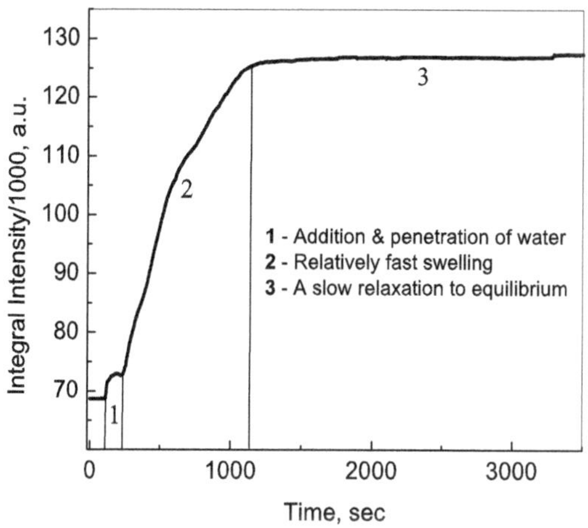

Figure 18. Integral intensity of light passed through the fiber core covered with dried micrometer PNIPA hydrogel film in the time course of its hydration.

5.2.3. Temperature sensitive volume transition of hydrogel micrometer film on the fiber core

The PNIPA hydrogel film on fiber core (Figure 16) was equilibrated at room temperature in the reaction chamber filled with distilled water. The hydrogel in water is in a swollen state. The integral absorbance (200–880 nm) in the fiber core was measured at different temperatures using the same scheme as described in previous section. Figure 19 clearly shows that absorption/scattering of evanescent light inside the PNIPA hydrogel film begins to increase at ~32 °C. This value is in good agreement with the transition temperature (T_c) known for PNIPA hydrogels from the literature [110] and our previous data [103] as the so-called Lower Critical Solution Temperature (LCST) at which a temperature sensitive hydrogel in aqueous medium collapses. As a result, the portion of evanescent wave scattered on the dense polymer network increases. Assuming the linear proportionality between absorbance and density, the temperature sensitive PNIPA hydrogel film underwent more than 4-fold volume decrease when temperature increased from 32 to 55 °C. Absorbance in the bare fiber core without any hydrogel has been measured for comparison, to illustrate that changes are attributed to the hydrogel layer collapsing.

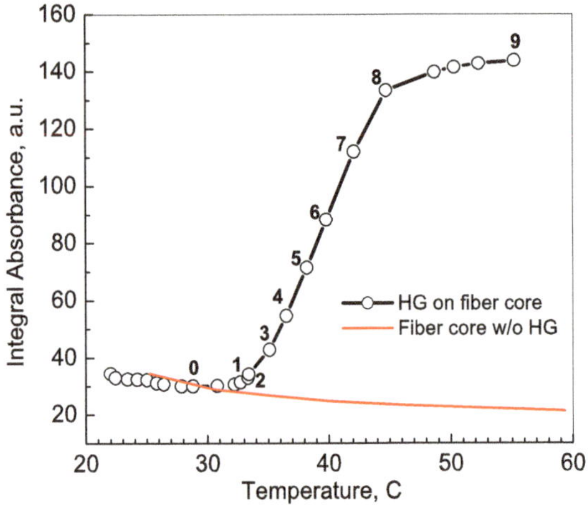

Figure 19. Integral absorbance (200 – 880 nm) in fiber core with and without PNIPA hydrogel film as a function of temperature.

Importantly, the continuous (without abrupt steps) changes in fiber transparency in the course of both dry-wet and temperature sensitive volume transitions indicated that the gel polymerized on the fiber core was tightly adhered to the surface without losing a contact at swelling/contraction. Applications of hydrogel spots on fiber distal ends and on a fiber core as chemical and biochemical sensors also confirm an excellent adhesion of hydrogel to the fiber surface.

5.2.4. New spectral properties of the temperature sensitive hydrogel film on the fiber core

The hydrogel/fiber core assembly proposes the experimental scheme which allows not only the integral intensity, but also the transmission (or absorption) spectra to be recorded in the whole range of wavelengths. For the first time, the spectral composition of light passing through the fiber with PNIPA hydrogel on it has been analyzed at different temperature in the course of the polymer network collapsing. The absorption spectrum at 28.8°C, below the volume transition temperature, was taken as a reference (0) with respect to which the spectra at higher temperatures are represented in Figure 20.

Strikingly, the spectral range of absorption depends on temperature, widening with the extent of temperature increase above T_c. Herein, maximum of absorbance shifts more and more to the longer wavelengths with temperature. Finally, at temperatures higher than 55°C, absorption was maximized in the whole range from 200 to 900 nm (in red) enveloping the absorption ranges at lower temperatures. This process was reversible, reflecting the reversibility of the volume transition.

The observed variations in the absorption range with temperature can be explained by scattering of evanescent waves [115] on growing microheterogeneities of the polymeric network when it collapses. At the volume transition temperature T_c, the size of heterogeneities is small, so that only light with short wavelengths considerably scatters, contributing to the absorption range of the recorded spectrum. In the course of temperature rising, the heterogeneities grow and new spectral components of evanescent light with longer wavelengths begin to scatter – the spectral range of absorbed light expands from ultraviolet to visible and further to the near infrared domain. Moreover, one can expect that concentration of growing heterogeneities also increases with density of collapsing polymer network. At some instant, the scattering features of evanescent light changes from Rayleigh to Mie scattering [15,116].

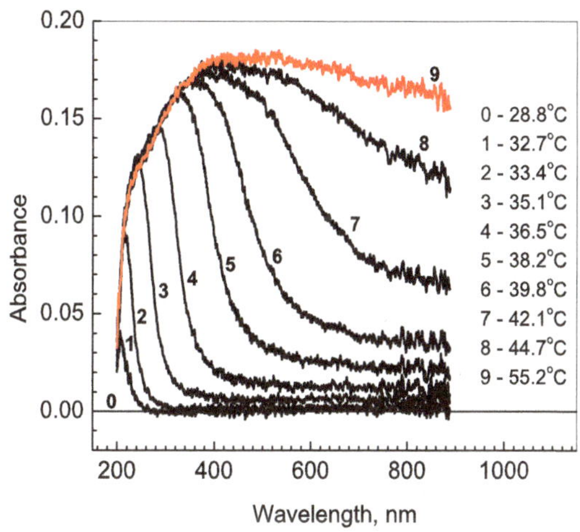

Figure 20. Absorption spectra of the fiber core covered with the cylindrical PNIPA hydrogel (PNIPA) at different temperatures indicated and numbered in accord with points labeled in Figure 19.

6. Future applications of hydrogel film/fiber core structures

6.1. Scientific and practical aspects of photopolymerization, detection, and characterization

The main reason of focusing on optical fiber core is that being a source of evanescent waves it provides the spectroscopic access to the surrounding medium. Connected to a spectrometer, an optical fiber with open core becomes an apparatus for polymerization and analytical device in one.

Since polymer networks of different chemical origin are increasingly studied for applications in biosensing, drug delivery and tissue engineering, their polymerization kinetics will be examined using the evanescent wave spectroscopy on the fiber core. The diversity of polymerizable formulations will initiate the development of new photoinitiating systems for reliable polymerization by visible and near infrared light.

Controllable formation of hydrogel nanoparts of different shapes and sizes on side surface of a fiber core is another direction of the future studies. Good quality and reproducibility of the nanoparts patterning on cylindrical support is a specific aim of the research on PEW. The development of technology for optical fibers with sub-wavelength diameter will drastically increase the evanescent field power and resolution of cylindrical nanopatterning.

Once hydrogel film is formed on a fiber core, its swelling/shrinking kinetics should be compared to theoretical predictions. However, quantitative kinetic model of swelling/de-swelling is still in demand, especially on a submicrometer scale. Despite some of the kinetic approaches to hydrogel swelling [117,118] exist, the changes in spectral composition of evanescent light scattered within polymer networks affected by external stimuli, chemical composition, dimensions, and load have never been able to measure.

Immobilization of chemical and biological specificity on the hydrogel cylindrical nanofilms is the other aspect of the future studies. A possible scheme for introduction of a polypeptide specific to biological moieties into polymer matrix of nanofilm developed in conjunction with bacterial spore interaction with liposomes [119] is shown in Figure 21. At Step 1, amine–groups (NH_2–) are incorporated into hydrogel film by *co*-polymerization of *N*-isopropylacrylamide with *N*-(3-aminopropyl)methacrylamide in the presence of *N,N'*-methylenebisacrylamide (cross-linker) and 2,2-diethoxyacetophenone (photoinitiator). Polypeptide-hydrogel conjugation is performed with heterobifunctional cross-linker [85,86], N-succinimidyl 3-(2-pyridyldithio)propionate (SPDP), as shown in Figure 21 (Steps 2 and 3).

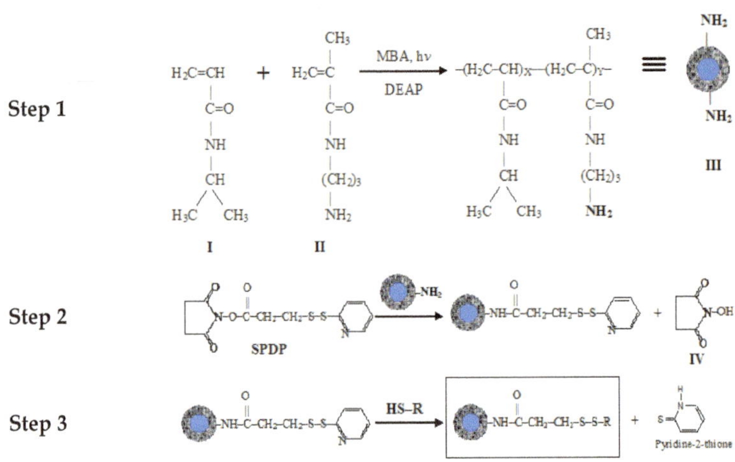

Figure 21. Polypeptide–hydrogel conjugation: (Step 1) Photo-co-polymerization of *N*-isopropylacrylamide (I) and *N*-(3-aminopropyl)methacryl-amide (II) on the surface of a fiber core. The amine-groups are everywhere in the bulk of hydrogel including the surface of the nanofilm. (Step 2) Preparation of pyridyl disulfide-activated surfaces by aminolysis leading to the release of *N*-hydroxysuccinimide (IV) and formation of an amide bond. (Step 3) Polypeptide (R) coupling to hydrogel nanofilm.

6.2. New materials: Quasi-two-dimensional network

The proposed cylindrical architecture of hydrogel film on side surface of fiber core may revolutionize conventional applications of supported polymer matrices. The use of evanescent waves for preparation of a cylindrical ultrathin hydrogel film on the side surface of a fiber core is a promising method for thickness control on nanometer scale. The dosage of energy (intensity times polymerization time) can be optimized in an attempt to get a new class of materials – quasi-2D-polymer webs (Figure 22).

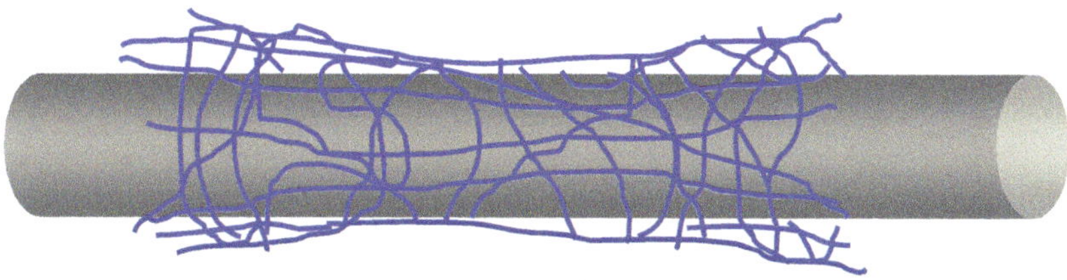

Figure 22. Cylindrical quasi-two-dimensional polymer web on optical fiber.

Future applications of the "ultrathin film/fiber core structures" originate from the unique properties of the ultrathin hydrogel film, namely: (1) the world's highest level of expansion and contraction, (2) the most sensitive externally induced mass transfer, and (3) the record fastest response to different stimuli. Therefore, the cross-linked polymer matrix of nanometer thickness on a fiber core is of great potential for a much broader spectrum of applications than just sensors. Generally speaking, the intersection of optical fiber technology and polymer nanotechnology offers numerous new science and technology opportunities. For example, the cylindrical sheet (Figure 22) prepared from a conductive polymer is expected to be stable, highly flexible, strong and remarkably conductive, so that electrons can travel with small scattering over submicron distances. Those cylindrical nanofabrics are important for making very-fast-switching transistors leading to biological computers which can use living cells and microorganisms as high memory density elements [120].

6.3. Modeling cell-like structures

Recent successful approaches to the development of new models of biological membranes by designing solid-supported phospholipid bilayers have demonstrated the importance of a water-filled space between the solid substrate and bilayer [121-125]. Up to now, only the planar architecture of the model has been considered. The proposed nanofilm/fiber structure might be an initial step in building a *cylindrical* model of biological membrane which is named "shish kebob structure" (Figure 23). We believe that the cylindrical architecture is a starting bottom-up polymer-chemistry-driven model of intracellular organization with the possibility to monitor biochemical processes spectroscopically within the cell, thereby revolutionizing cell biology.

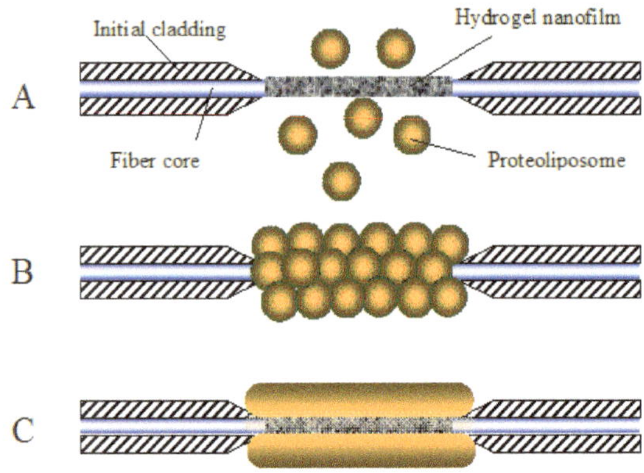

Figure 23. "Shish kebob structure" as a cylindrical polymer-driven model of a biological membrane: A) the preliminary prepared nanofilm/fiber structure is exposed to the proteoliposome suspension, B) the proteoliposomes adhere on the appropriately modified surface of the hydrogel ultrathin layer, C) spontaneous fusion of the proteoliposomes and formation of the cylindrical bilayer supported by the nanofilm/fiber structure.

6.4. New types of biosensors: From chemicals and biological agents to biofilms and single cell

The proposed cylindrical hydrogel/fiber core structure is a bicompartmental (optical and chemical) spectroscopic tool which can be specifically designed for accurate monitoring/detection of chemicals and living entities within "microscopic" environment, particularly, within a single cell or microorganism.

The optical part (fiber core) allows the monitoring of biochemical processes if not within the single cell, then within a biological community, and manipulating of cellular processes by transmission of information into biological systems, for example, using different wavelengths of electromagnetic radiation delivered by the fiber into the cell as shown in Figure 24.

The chemical part provides the following features: (i) hydrogel is a suitable semiwet 3D-environment for biological interactions, (ii) many hydrogels are inert surfaces without nonspecific adsorption of proteins (antifouling), (iii) biological molecules can be covalently incorporated into hydrogel using different chemistries (e.g. see Figure 21), (iv) mechanical properties of hydrogel materials are tunable by modifying cross-linking density, and (v) hydrogel can be responsive to external or internal stimuli (see Introduction).

Imagine Figure 23B where proteoliposomes are substituted for microorganisms. The proposed structure looks extremely potential for application as a microbial biofilm – microorganisms' culture growing on a supported polymeric matrix [126]. Cells within a biofilm are embedded in a matrix of the so-called extracellular polymeric substances (EPS), primarily produced by the microorganisms themselves. Little is known about this matrix.

The composition of EPS varies depending on the organisms present and environmental conditions. Characterization of EPS matrix is a daunting proposition.

The ability of bacterial spores to coordinate their alert sensory behavior and processes to respond to the germination effectors [119] are increasingly studied. However, several key features of sporulation and germination in bacteria are not understood at the molecular level. Presumably [127-129], under proper pre-germination circumstances, the spores can and will attempt to influence their surrounding chemical and physical environment by secreting specific biological macromolecules and solutes. The challenge is in determining when, why and how the bacterial spore population regulates external matrix composition and what the ultimate functional consequences are. It would be especially interesting to deploy spores on specifically modified hydrogel/fiber core structure.

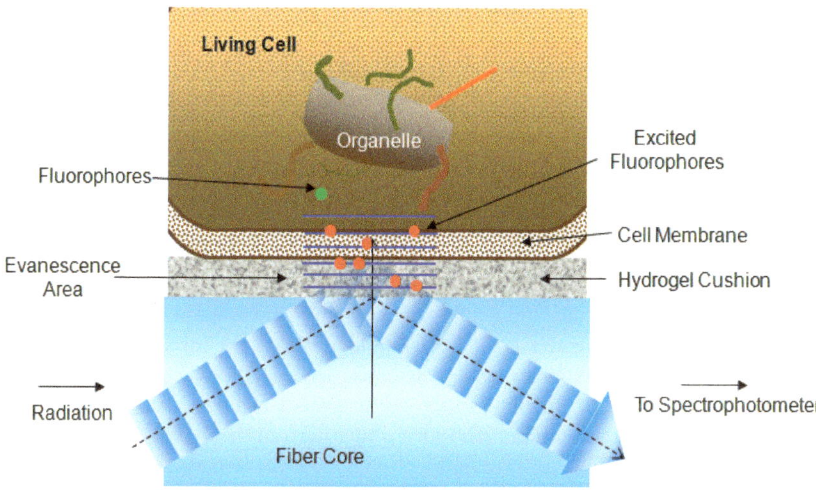

Figure 24. Single cell supported by cylindrical hydrogel film on optical fiber core.
Evanescent field penetrates through the hydrogel cushion and membrane into the living cell either for detection or initiation of biochemical processes inside. With fluorescent probe or without, spectroscopic signal can be recorded.

Both for dormant (spores) or living cells, fiber core can serve as an optical guide for recording spectroscopic information on the structural transformations and chemical changes in real time, whereas responsive hydrogel material can be modified to provide triple functionality: (i) the cell-adhesive ligands (e.g. polypeptides [119]) can be incorporated into the so-called *bioselective* hydrogel to adsorb a certain type of cells (spores), (ii) the *bioresponsive* hydrogels can be designed to contain bioactive ligands (peptides, polysaccharides) to direct cell behavior (germination, secretion, differentiation, division, etc.) [130-133], and (iii) the *biointeractive* hydrogel can be modified to contain receptors (proteins, antibodies, enzyme-sensitive substrates, enzymes) for biomolecules or cells that, when stimulated, trigger local or bulk changes in the hydrogel properties (swelling/collapse) [134-137].

6.5. Tissue engineering

Hydrogels have been widely used in tissue engineering [130-138]. Due to their biocompatibility, ability to simulate the properties of living tissue, and controlled diffusivity of nutrients and methabolites, hydrogels are versatile scaffolds which can support the implanted cells. The whole pool of recent study of 3D hydrogels scaffolds for tissue engineering puts forth a hypothesis that the thinner is the hydrogel scaffold, the more effective are the processes of cell differentiation and growth, presumably, because of the enhanced mass transfer in thin films. Thus, nanofilm/fiber structures may offer significant improvement over the current methods of tissue engineering providing the most sensitive biological, chemical, and physico-mechanical response of ultrathin hydrogel film. Furthermore, nanofilm/fiber structures offer the capability of spectroscopic controlling the properties of a functional tissue in the course of cell growth on the side surface of an optical fiber.

6.6. Drug delivery devices

The nanofilm/fiber structure with drug filled polymer matrix on the side surface of the fiber core can be inserted into the organ of interest. The battery of such kind of structures can be patched on the skin. Both configurations of the drug delivery devices can provide methods for triggering the drug release and for spectroscopic control of the delivery dose. Numerous techniques [2, pp. 95-126] have been used for immobilizing drugs on/within a hydrogel: physical entrapment, electrostatic attraction, physical adsorption, and/or chemical bonding. The polymerization of hydrogel by evanescent waves in the presence of a drug offers a new method for entrapping the drug into the ultrathin hydrogel film. Thus prepared nanofilm/fiber structures could be a key step in designing a combined drug delivery systems with precise quantification of a delivery dose: (1) various drugs entrapped in different structures can be simultaneously or in desired order released in the targeted organs in the body; (2) several nanofilm/fiber structures can be loaded with different pre-drug reagents (non-harmful for surrounding organs) which are allowed to react once being delivered to the targeted site, thereby the surrounding organs would be protected against the final form of the drug.

7. Closing remarks and conclusions

The increasing amount of studies has already demonstrated that environmentally responsive hydrogel materials could be successfully used for detection of chemicals, biological compounds, living cell, and microorganisms, for targeted delivery of drugs in response to disease specific biomolecules, and for triggering cell behavior and tissue repairing. Traditionally, only the planar design of these "smart" polymer networks has been considered. In this paper, cylindrical configuration of responsive hydrogels supported on side-surface (not distal end) of an optical fiber core was introduced. It was shown that absorption/scattering of the evanescent field penetrating out of the fiber core surface (typically, ~λ/10) can be feasibly used not only for the fabrication of the responsive hydrogel film around the cylindrical surface of fiber core, but also for recording the response of thus deposited soft material to environmental changes. Photolymerization by evanescent waves is a versatile and convenient method for synthesizing hydrogels films with thicknesses varied from nanometers to hundred micrometers. A simplistic model of PEW predicts that within the depth of evanescent field penetration out of the fiber the thickness is controlled by the dosage of absorbed energy (light intensity × polymerization time). The thicker hydrogel films are due to scattering of evanescent waves on the newly formed polymer microstructures and initiation of polymerization relatively far from the surface of fiber core.

The designed bicompartmental spectroscopic tool of cylindrical configuration allows one to use the same experimental scheme to polymerize, to study the kinetics of polymerization, to detect and characterize the deposited polymer network simultaneously. One major new option that was not previously available is the capability of real time analyzing the spectral composition of light absorbed/scattered within a hydrogel during polymerization, swelling/de-swelling, and other macroscopic changes in response to external stimuli or triggered internal interactions.

The option of spectral analysis was demonstrated on swelling kinetics of the temperature sensitive polymer network of poly(*N*-isopropylacylamide) synthesized around fiber core in the course of film hydration and the gel temperature sensitive volume transition. Particularly, the hydrogel supported by fiber core began to absorb/scatter light in ultraviolet region at the temperature of volume transition T_c. However, the spectral range of absorption continued to expand into the visible and farther to the near infrared domains while temperature was getting higher.

Polymer network formed is not covalently bound to fiber side-surface. It is a kind of "cylindrical sock" on the fiber core. Its structure and responsiveness depend on polymer chemistry whereas its thickness is a function of polymerization kinetics. Since the sensitivity of the method is expected to increase with fiber core of sub-micrometer diameter, the hydrogel/fiber core nanostructures will be especially useful for testing of nanometer-sized biological objects and spectroscopic study of an individual living cell. We believe that further research on optimization of the non-planar (cylindrical) design of supported bioresponsive and biointeractive hydrogel nanoparts will bring us to a stage where we better understand both potentials and limitations of the approach in order to compete with that induced by conventional flat nanopatterning in the market.

The approach is generic – any soft biomaterials (pigments, surfactants, liposomes, lipobeads, proteins, polysaccharides, biomembranes, blood, cells and their organelles, spores, and even organisms) could be supported on the side surface of an optical fiber core and the study of their structural changes using the evanescent wave spectroscopy would benefit from the development of the hydrogel/fiber core nanostructures.

This is not a complete review on possible applications. Nevertheless, the aforementioned examples allow one to conclude that the methods for hydrogel nanofilm preparation and support on a fiber core of sub-micrometer diameter are in demand.

Author details

Sergey V. Kazakov
Department of Chemistry & Physical Sciences, Pace University, Pleasantville, NY, USA

Acknowledgement

This work was financially supported in part by Pace University (Dyson College of Arts and Sciences, Summer and Scholarly Research Funds). Acknowledgement is made to the Donors of the American Chemical Society/Petroleum Research Fund for partial support of this research (PRF grant # 44161-GB5).

8. References

[1] Harland RS, Prud'homme RK (Eds.) (1992) *Polyelectrolyte Gels: Properties, Preparation, and Applications*. Washington, DC: American Chemical Society.
[2] Peppas NA (1987) *Hydrogels in Medicine and Pharmacy*. Boca Raton, FL: CRC Press.
[3] Galaev IY, Mattiasson B. (1999) Smart polymers and what they could do in biotechnology and medicine. *Trends Biotechnol.* 17: 335–340.
[4] Hennink WE, van Nostrum CF (2002) Novel cross-linking methods to design hydrogels. *Advanced Drug Delivery Reviews* 54: 13–36.
[5] Erman B, Flory PJ (1986) Critical phenomena and transitions in swollen polymer networks and in linear macromolecules. *Macromolecules* 19: 234-252.
[6] Pelton R (2000) Temperature-sensitive aqueous microgels. *Adv. Coll. Interface Sci.* 85: 1-33 and references therein.
[7] Tanaka T, Fillmore D, Sun ST, Nishio I, Suislow G, and Shah A (1980) Phase transitions in ionic gels. *Phys. Rev. Lett.* 45: 1636-1639.
[8] Eichenbaum GM, Kiser PF, Simon SA, Needham D (1998) pH and ion-triggered volume response of anionic hydrogel microspheres. *Macromolecules* 31: 5084-5093.
[9] Eichenbaum GM, Kiser PF, Dobrynin AV, Simon SA, Needham D (1999) Investigation of the swelling response and loading of ionic microgels with drugs and proteins: The dependence on cross-link density. *Macromolecules* 32: 4867-4878.
[10] Eichenbaum GM, Kiser PF, Shan D, Simon SA, Needham D (1999) Investigation of the swelling response and loading of ionic microgels with drugs and proteins: The dependence on functional group composition. *Macromolecules* 32: 8996-9006.
[11] Markland P, Zhang Y, Amidon GL, Yang VC (1999) A pH- and ionic strength-responsive polypeptide hydrogel: synthesis, characterization, and preliminary protein release studies. *J. Biomed. Mater. Res.* 47: 595-602.
[12] Tanaka T, Nishio I, Sun ST, Nishio SU (1982) Collapse of gels in electric field. *Science* 218: 467-469.
[13] Suzuki B, Tanaka T (1990) Phase transition in polymer gels induced by visible light. *Nature* 346: 345-347.

[14] Miyata T, Asami N, Uragami T (1999) A reversibly antigen-responsive hydrogel. *Nature* 399: 766-769.

[15] Kazakov S (2007) Phase transitions in smart polymer solutions and light scattering in biotechnology and bioprocessing. In: Galaev I, Mattiasson B, editors. *SMART POLYMERS: Application in biotechnology and biomedicine*, Taylor & Francis: Chapter 1, pp 1-52.

[16] Gehrke S (1993) Synthesis, equilibrium swelling, kinetics, permeability and applications of environmentally responsive gels. Adv. Polym. Sci. 110: 81-144.

[17] Peppas NA (2007) Drug delivery using smart polymers: Recent advances. In: Galaev I, Mattiasson B, editors. *SMART POLYMERS: Application in biotechnology and biomedicine*, Taylor & Francis: Chapter 11, pp. 331-358.

[18] Kabra BG, Akhtar MK, Gehrke SH (1992) Volume change of poly(*N*-isopropylacrylamide) gels. Polymer Int. 28: 29-36.

[19] Harsh D, Gehrke SJ (1991) Controlling the swelling characteristics of temperature-sensitive cellulose ether hydrogels, Journal of Controlled Release 17: 175-186.

[20] McPhee W, Tam KC, Pelton R J (1993) Poly(N-isopropylacrylamide) latexes prepared with SDS. J. Colloid Interface Sci. 156: 24-30.

[21] Hirotsu S, Hirokawa Y, Tanaka T (1987) Volume-phase transition of ionized *N*-isopropylacrylamaide. J. Chem. Phys. 87: 1392-1395.

[22] Zhou Sh, Chu B (1998) Synthesis and volume phase transition of poly(methacrylic acid-*co*-N-isopropylacrylamide) microgel particles in water. J. Phys. Chem. 102: 1364-1371.

[23] Shibayama M, Mizutani S, Nomura S (1996) Thermal Properties of Copolymer Gels Containing N-Isopropylacrylamide. Macromolecules 29: 2019-2024.

[24] Hirose T, Amiya T, Hirokawa Y, Tanaka T (1987) Phase transition of submicron gel beads. Macromolecules 20: 1342-1344.

[25] Kokufuta E, Wang B, Yoshida R, Khokhlov A, Hirata M (1998) Volume phase transition of polyelectrolyte gels with different charge distributions. Macromolecules 3: 6878-6884.

[26] Kazakov S, Kaholek M, Gazaryan I, Krasnikov B, Miller K, Levon K (2006) Ion concentration of external solution as a characteristic of micro- and nanogel ionic reservoirs. J. Phys. Chem. 110: 15107-15116

[27] Ogawa Y, Ogawa K, Wang B, Kokufuta E (2001) A biochemo-mechanical system consisting of polyampholyte gels with coimmobilized glucose oxidase and urease. Langmuir 17: 2670-2674.

[28] Eremeev NL, Kukhtin AV, Kazanskaya NF (1998) Enzyme-dependent responses of stimuli-sensitive systems. Biosystems 45: 141-149.

[29] Horkay F, Tasaki I, Basser PJ (2001) Effect of monovalent-divalent cation exchange on the swelling of polyacrylate hydrogels in physiological sait solutions. Biomacromolecules 2: 195-199.

[30] Horkay F, Tasaki I, Basser PJ (2000) Osmotic pressure of polyacrylate hydrogels in physiological salt solutions. Biomacromolecules 1: 84-90.

[31] Eichenbaum GM, Kiser PF, Shah D, Meuer WP, Needham D, Simon SA (2000) Alkali earth metal binding properties of ionic microgels. Macromolecules 33: 4087-4093.

[32] Amiya T, Tanaka T (1987) Phase transitions in cross-linked gels of natural polymers. Macromolecules 20: 1162-1164.

[33] Horkay F, Tasaki I, Basser PJ (2001) Ion-induced volume transition in synthetic and biopolymer gels. Polymer Preprint 42(1): 267-268.

[34] Iwasa K, Tasaki I (1980) Mechanical changes in squid giant axons associated with production of action potentials. Biochem. Biophys. Res. Commun. 95: 1328-1331.

[35] Tasaki I, Iwasa K (1982) Rapid pressure changes and surface displacements in the squid giant axon associated with production of action potentials. Jpn. J. Physiol. 32: 69-81.

[36] Tasaki I, Nakaye T, Byrne PM (1985) Rapid swelling of neurons during synaptic transmission in the bullfrog sympathetic ganglion. Brain Res. 331: 363-365.

[37] Tasaki I, Byrne PM (1988) Large mechanical changes in the bullfrog olfactory bulb evoked by afferent fiber stimulation. Brain Res. 475: 173-176.

[38] Tasaki I, Byrne PM (1992) Discontinuous volume transitions in ionic gels and their possible involvement in the nerve excitation process. Biopolymers 32: 1019-1023.

[39] Tasaki I, Byrne PM (1994) Discontinuous volume transitions induced by calcium-sodium ion exchange in anionic gels and their neurobiological implications. Biopolymers 34: 209-215.

[40] Tasaki I (1999) Rapid structural changes in nerve fibers and cells associated with their excitation processes. Jpn. J. Physiol. 49: 125-138.

[41] Kokufuta E, Zhang YQ, Tanaka T, Mamada A (1993) Effects of surfactants on the phase transition of poly(N-isopropylacrylamide) gel. *Macromolecules 26:* 1053-1059.

[42] Khokhlov AR, Kramarenko EYu, Makhaeva EE, Starodubtsev SG (1992) Collapse of Polyelectrolyte Networks Induced by Their Interaction with Oppositely Charged Surfactants *Macromolecules* 25: 4779-4783.

[43] Gong JP, Osada Y (1995) Theoretical analysis of the cross-linking effect on the polyelectrolyte-surfactant interaction. J. Phys. Chem. 99: 10971-10975.

[44] Safranj A, Yoshida M, Omichi H, Katakai R (1994) Effect of surfactants on the volume phase transition of cross-linked poly(acryloyl-l-proline alkyl esters). Langmuir 10: 2955-2959.

[45] Yoshida M, Asano M, Omichi H, Kamimura W, Kumakura M, Katakai R (1997) Dependence of volume phase transition temperature of poly(acryloyl-L-proline methyl ester) gel on hydrophobic tail length of anionic surfactants. Macromolecules 30: 2795-2796.

[46] Okuzaki H, Osada Y (1994) Effects of hydrophobic interaction on the cooperative binding of a surfactant to a polymer network. Macromolecules 27: 502-506.

[47] Murase Y, Onda T, Tsujii K, Tanaka T (1999) Discontinuous binding of surfactants to polymer gel resulting from a volume phase transition. Macromolecules 32: 8589-8594.

[48] Ogawa K, Ogawa Y, Kokufuta E (2002) Effect of charge inhomogeneity of polyelectrolyte gels on their swelling behavior. Colloids and Surfaces 209: 267-279.

[49] Okuzaki H, Osada Y (1994) Effects of hydrophobic interaction on the cooperative binding of a surfactant to a polymer network. Macromolecules 27: 502-506.

[50] Narita T, Gong JP, Osada Y (1998) Kinetic study of surfactant binding into polymer gel – Experimental and theoretical analysis. J. Phys. Chem. B. 102: 4566-4572.

[51] Akira Mamada A, Tanaka T, Kungwatchakun D, Irie M (1990) Photoinduced Phase Transition of Gels. Macromolecules 23: 1517-1519.

[52] Suzuki A, Tanaka T (1990) Phase transitions in polymer gels induced by visible light. Nature 346: 345-347.

[53] Juodkazis S, Mukai N, Wakaki R, Yamaguchi A, Matsuo Sh, Misawa H (2000) Reversible phase transitions in polymer gels induced by radiation forces. Nature 408: 178-181.

[54] Lendlein A, Jiang H, Junger O, Langer R (2005) Light-induced shape- memory polymers. Nature 434: 879-882.

[55] Kuhn W, Hargitay B, Katchalsky A, Eisenberg H (1950) Reversible Dilation and Contraction by Changing the State of Ionization of High-Polymer Acid Networks. Nature 165: 514-516.

[56] Steiberg IZ, Oplatka H, Katchalsky A (1966) Mechanochemical Engines. Nature 210: 568-571.

[57] Sussman MV, Katchalsky A (1970) Mechanochemical turbine: a new power cycle. Science 167: 45-47.

[58] Osada Y, Hasebe M (1985) Electrically activated mechanochemical devices using polyelectrolyte gels. Chem. Lett. 1285-1288.

[59] DeRossi DE, Chiarelli P, Buzzigoli G, Domenci C, Lazzeri L (1986) Contractile behaviour of electrically activated mecltanochemical polymer actuators. Trans. Am. Soc. Artif. Organs 32: 157-162.

[60] Tanaka T, Nishio I, Sun ST, Nishio SU (1982) Collapse of gels in an electric field. Science 218: 467- 469.

[61] Kishi R, Osada Y (1989) Reversible volume change of microparticles in an electric field. J. Chem. Soc., Faraday Trans. 1 85: 655-662

[62] Katchalsky A, Steinberg IZ, Kam A (1967) Mechanochemical engine. US Patent 3321908.

[63] Osada Y, Rossmurphy SB (1993) Intelligen gels. Scientific American 268: 82-87.

[64] Schreyer HB, Gebhart N, Kim KJ, Shahinpoor M (2000) Electrical activation of artificial muscles containing polyacrylonitride gel fibers. Biomacromolecules 1: 642 -647.

[65] Onoda M, Kato Y, Shonaka H, Tada K (2004) Artificial muscle using conducting polymers. Electrical Eng. Jpn. 149: 7-13.

[66] Grodzinsky AJ, Grimshaw PE (1990) Electrically and chemically controlled hydrogels for drug delivery. Pulsed and Self-regulated Drug Delivery pp. 47-64.

[67] Moorthy J (2007) Hydrogels in microfluidics. In: Galaev I, Mattiasson B, editors. *SMART POLYMERS: Application in biotechnology and biomedicine*, Taylor & Francis: Chapter 14, pp. 437-457.

[68] Lee KY, Mooney DJ (2001) Hydrogels for tissue engineering. Chemical Reviews 101: 1869-1879.

[69] Schwall CT, Banerjee IA (2009) Micro- and nanoscale hydrogel systems for drug delivery and tissue engineering. Materials 2: 577-612.

[70] Schmaljohann D, Nitschke M, Schulze R, Eing A, Werner C, Eichhorn K-J (2005) In situ study of the thermoresponsive behavior of micropatterned hydrogel films by imaging ellipsometry. Langmuir 21: 2317-2322.

[71] Green PG (1996) Iontophoretic delivery of peptide drugs. J. Control. Release 41: 33-48.

[72] Thompson LF, Wilson CG, Bowden MJ, Eds. (1994) Introduction to Microlithography, American Chemical Society.

[73] Levenson MD (1995) Extending optical lithography to the gigabit era. *Solid State Technol.* February: 57–66

[74] Pease RFW (1992) Nanolithography and its prospects as a manufacturing technology. *J. Vac. Sci. Technol. B* 10: 278–285

[75] Cerrina F, Marrian C (1996) A path to nanolithography. *Mater. Res. Soc. Bull.* XXI: 56–62

[76] Ecoffet C, Espanet A, Lougnot DJ (1998) Photopolymerization by evanescent waves: A new method to obtain nanoparts. Adv. Mater. 10: 411-414.

[77] Yariv A (1989) *Quantum Electronics.* 3rd ed.; John Wiley &Son.

[78] Mehrvar M, Bis C, Scharer JM, Moo-Young M, Luong JH (2000) Fiber-optics biosensors – Trends and advances. *Analytical Sciences* 16: 677-691.

[79] Wolfbeis OS (2004) Fiber-optic chemical sensors and biosensors. *Anal. Chem.* 76: 3269-3284.

[80] Yang J, Lee C-J, Wei C-H (2002) Fiber-optic chemical sensors: a general review. *J. Chin. Chem. Soc.* 49: 677-692.

[81] Steemers FJ, Walt DR (1999) Multi-analyte sensing: From site-selective deposition to randomly-ordered addressed optical sensors. *Microchimica Acta* 131: 99-105.

[82] Michie WC, Culshaw B, McKenzie I, Konstantakis M, Graham NB, Moran C, Bergquist E, Carlstrom B (1995) Distributed sensor for water and pH measurements using fiber optics and swellable polymeric systems. *Optics Letters* 20: 103-105.

[83] Rassell AP, Fletcher KS (1985) Optical sensor for the determination of moisture. *Anal. Chim. Acta* 170: 209-216.

[84] Hypszer R, Wierzba HJ (1997) Fiber optic technique for relative humidity sensors. *SPIE Proceedings* 3054: 145-150.

[85] Wyatt WA, Bright FV, Hieftje GM (1987) Characterization and comparison of three fiber-optic sensors for iodide determination based on dynamic fluorescence quenching of Rhodamine 6G. *Anal. Chem.* 59: 2272-2276.

[86] Shortreed M, Bakker E, Kopelman R (996) Miniature sodium-selective ion-exchange optode with fluorescent pH chromoionophores and tunable dynamic range. *Anal. Chem.* 68: 2656-2662.

[87] Mohr GJ, Wolfbeis (1996) Optical nitrite sensor based on a potential-sensitive dye and a nitrite-selective carrier. *Analyst* 121: 1489-1494.

[88] Hartman P, Trettnak W (1996) Lifetime imaging of luminescent oxygen sensors based on all-solid-state technology. *Anal. Chem.* 68: 4512-4514.

[89] Ferguson JA, Healey BG, Bronk KS, Bonrnard SM, Walt DR (1997) Simultaneous monitoring of pH, CO_2 and O_2 using an optical imaging fiber. *Anal. Chim. Acta* 340: 123-131.

[90] Li L, Walt DR (1995) Dual-analyte fiber-optic sensor for the simultaneous and continuous measurement of glucose and oxygen. *Anal. Chem.* 67: 3746-3752.

[91] Shakhsher ZM, Seitz WR (1990) Optical detection of cationic surfactants based on ion pairing with an environment-sensitive fluorophore. *Anal. Chem.* 62: 1758-1762.

[92] Barnard SM, Walt DR (1991) A fibre-optic chemical sensor with discrete sensing sites. *Nature* 353: 338-340.

[93] Healey BG, Walt DR (1995) Improved fiber-optic chemical sensor for penicillin. *Anal. Chem.* 67: 4471-4476.

[94] Ge Z, Brown CW, Sun L, Yang SC (1993) Fiber-optic pH sensor based on evanescent wave absorption spectroscopy. *Anal. Chem.* 65: 2335-2338.

[95] Kazakov S (2010) Evanescent wave spectroscopy for studying swelling/de-swelling kinetics of soft matter. *Soft Matter* 6: 1191-1196.

[96] Wang D, Elisseeff JH (2004) Photopolymerization. In Encyclopedia of Biomaterials and Biomedical Engineering (EBBE), Wnek, GE, Bowlin GL, Eds: Marcel Dekker, Inc.: pp 1212-1225.

[97] Belfield KD, Crivello JV (1996) Photoinitiated polymerization. American Chemical Society.

[98] Decker C, Jenkins AD (1985) Kinetic approach of O_2 inhibition in ultraviolet- and laser-induced polymerization. Macromolecules 18: 1241-1244.

[99] Lee JH, Prud'homme RK, Aksay IA (2001) Cure depth in photopolymerization: Experiments and theory. J. Mater. Res. 16: 3536-3544.

[100] Geuskens G, Soukrati A (2000) Investigation of polyacrylamide hydrogels using 1-anilinonaphthalene-8-sulfonate as fluorescent probe. *Eur. Polymer J.* 36: 1537-1546.

[101] Thomas TL, Mishra AK (2002) ANS fluorescence as a tool to monitor cross-linking polymerization of acrylamide. *Eur. Polymer J.* 38: 1805-1810.

[102] Kazakov S, Kaholek M, Kudasheva D, Teraoka I, Cowman KM, Levon K (2003) Poly(N-isopropylacrylamide-co-1-vinylimidazole) hydrogel nanoparticles prepared

and hydrophobically modified in liposome reactors: Atomic force microscopy and dynamic light scattering study. *Langmuir* 19: 8086-8093.

[103] Kazakov S, Kaholek M, Teraoka I, Levon K (2002) UV-induced gelation on nanometer scale using liposome reactor. *Macromolecules* 35: 1911-1920.

[104] Kazakov S, Kaholek M, Krasnikov B, Levon K (2003) Approach to study pH-responsive micro- and nanogels. *Polym. Preprint* 44: 186-187.

[105] Kazakov S, Kaholek M, Levon K (2002) Nanometer scale ionic reservoir based on ion-responsive hydrogels. *SPIE Proceedings* 4695: 42-51.

[106] Kazakov S, Kaholek M, Levon K (2002) Hydrogel nanoparticles compatible with phospholipid bilayer. *Polym. Preprint* 43: 381-382.

[107] Munkholm C, Parkinson D-R, Walt DR (1990) Intramolecular fluorescence self-quenching of fluoresceinamine. *J. Am. Chem. Soc.* 112: 2608-2612 and references therein.

[108] Berney C, Danuser G (2003) FRET or no FRET: A quantitative comparison. *Biophys. J.* 84: 3992–4010 and references within.

[109] Molecular Probes Publications online: www.invitrogen.com/handbook/

[110] Schild HG (1992) Poly(N-isopropilacrylamide): experiment, theory and application. Prog. Polym. Sci. 17: 163-249.

[111] Ulijn RV, Bibi N, Jayawarna V, Thornton PD, Todd SJ, Mart RJ, Smith AM, Gough JE (2007) Materials Today 10: 40-48.

[112] Aldrich Polymer Products Application and Reference Information, Aldrich Chemical Co. online Publication, 1-52, www.sigmaaldrich.com

[113] Zhang S, Li B, Tang L, Wang X, Lui D, Zhou Q (2001) Studies on the near infrared laser induced photopolymerization employing a cyanine dye-borate complex as a photoinitiator. *Polymer* 42: 7575-7582.

[114] Warken F, Vetsch E, Meschede D, Sokolowski M, Rauschenbeutel A (2007) Ultra-sensitive surface absorption spectroscopy using sub-wavelength diameter optical fiber. Optics Express 15: 11952-11958.

[115] Wu H-J, Shah S, Beckham R, Meissner KE, and Michael A. Bevan MA (2008) Resonant effects in evanescent wave scattering of polydisperse colloids. Langmuir 24: 13970-13795.

[116] Trutna R, Liu M, Chamberlin D, Hadley J (2008) Nanoparticle measurement by spectroscopic Mie scattering. NSTI-Nanotech 1: 834-837.

[117] De SK, Aluru NR, Johnson B, Crone WC, Beebe DJ, Moore J (2202) Equilibrium swelling and kinetics of pH-responsive hydrogels: Models, experiments, and simulations. *Journal of Microelectromechanical Systems* 11: 544-555.

[118] Plunkett KN, Kraft ML, Yu Q, Moore JS (2003) Swelling kinetics of disulfide cross-linked microgels. *Macromolecules* 36: 3960-3966.

[119] Kazakov S, Kaholek M, Ji T, Turnbough C, Levon K (2004) Interaction of modified liposomes with *Bacillus* spores. *Chem. Commun.* 4: 430-431.

[120] Simpson ML, Sayler GS, Fleming JT, Applegate B (2001) Whole-cell biocomputing. *Trends Biotech.* 19: 317-323.

[121] Wagner ML, Tamm LK (2000) Tethered polymer-supported planar lipid bilayers for reconstitution of integral membrane proteins: silane-polyethyleneglycol-lipid as a cushion and covalent linker. *Biophys. J.* 79: 1400-1414.

[122] Zhao J, Tamm LK (2003) FTIR and fluorescence studies of interactions of synaptic fusion proteins in polymer-supported bilayers. *Langmuir* 19: 1838-1846.

[123] Kiessling V, Tamm LK (2003) Measuring distances in supported bilayers by fluorescence interference-contrast microscopy: polymer supports and SNARE proteins. *Biophys. J.* 84: 408-418.

[124] Sackmann E, Tanaka M (2000) Supported membranes on soft polymer cushions: fabrication, characterization and applications. *Trends Biotech.* 18: 58-64.

[125] Heyse S, Ernst OP, Dienes Z, Hofmann KP, Vogel H (1998) Incorporation of rhodopsin in laterally structured supported membranes: observation of transducin activation with spatially and time-resolved surface plasmon resonance. *Biochemistry* 37: 507-522

[126] Parsek MR, Fuqua C (2004) Biofilms 2003: Emerging themes and challenges in studies of surface-associated microbial life. *J. Bacteriology* 186: 4427-4440.

[127] Foster SJ, Johnstone K (1989) The trigger mechanism of bacterial spore germination. – In book: *Regulation of prokaryotic development.* I. Smith, R.A. Slepecky, P. Setlow (Eds.), Ch. 4, ASM, Washington, DC.

[128] Swerdlow BM, Setlow B, Setlow P (1981) Level of H^+ and other monovalent cations in dormant and germinating spores of *Bacillus megaterium*. *J. Bacteriology* 148: 26-29.

[129] Makino S, Moriyama R (2002) Hydrolysis of cortex peptidoglycan during bacterial spore germination. *Med. Sci. Monit.* 8: RA119-127.

[130] Hersel U, Dahmen C, Kessler H (2003) RGD-modified polymers: biomaterials for stimulated cell adhesion and beyond. Biomaterials 24: 4385-4415.

[131] Zourob M, Gough JE, Ulijn RV (2006) A micro-patterned hydrogel platform for chemical synthesis and biological analysis. Adv. Mater. 18: 655-659.

[132] Silva GA, Czeisler C, Niece KL, Beniash E, Harrington DA, Kessler JA, Stupp SI (2004) Selective differentiation of neural progenitor cells by high-epitope density nanofibers. Science 303: 1352-1355.

[133] Rajangam K, Behanna HA, Hui MJ, Han X, Hulvat JF, Lomasney JW, Stupp SI (2006) Heparin Binding Nanostructures to Promote Growth of Blood Vessels. Nano Lett. 6: 2086-2090.

[134] Lutolf MP, Hubbell JA (2005) Synthetic biomaterials as instructive extracellular microenvironments for morphogenesis in tissue engineering. Nat. Biotechnol. 23: 47-55

[135] Stevens MM, George JH (2005) Exploring and engineering the cell surface interface. Science 310: 1135-1138.

[136] Langer R, Tirrell DA (2004) Designing materials for biology and medicine. Nature 428: 487-492.

[137] Duncan R (2003) The dawning era of polymer therapeutics. Nat. Rev. Drug Discov. 2: 347-360.

[138] Lee KY, Mooney DJ (2001) Hydrogels for tissue engineering. *Chem. Rev.* 101: 1869-1879.

Chapter 6

New Polymer Networks for PDLC Films Application

Ana Isabel Mouquinho, Krasimira Petrova,
Maria Teresa Barros and João Sotomayor

Additional information is available at the end of the chapter

1. Introduction

1.1. Polymer Dispersed Liquid Crystals

Polymer dispersed liquid crystal (PDLC) films are a mixed phase of nematic liquid crystals (LC) commonly dispersed as inclusions in a solid polymer[1]. They have remarkable electro-optical behaviour since they can be switched from an opaque to a transparent state simply by application of an electric field [2].

PDLCs have been prepared by two general methods: one in which the system remains heterogeneous during the process, and another in which the system becomes heterogeneous. The first case includes the solvent induced phase separation (SIPS): the LC is mechanically dispersed in solution of a polymer. After evaporation of the solvent, the composite structure obtained is stabilised because of the polymer morphology, but is poorly controlled due to a coalescence of droplets during the preparation process. The second case is called thermally induced phase separation (TIPS) or polymerisation induced phase separation (PIPS)[1].

The main advantage of preparation by the PIPS method is the possibility to obtain a composite directly between glass plates coated with conductive indium tin oxide (ITO) film without additional laminating procedures. Thus, the PDLC film is produced in one technological step, in which the phase separation of the initially homogeneous mixture and the polymerisation occur simultaneously. The PIPS process was particularly suitable for our purpose because it is quite simple and allows for high degree of control over the final properties of the PDLC films.[3]

In this work the PIPS method was performed by thermal or photochemical polymerisation of a homogeneous mixture of polymerisable monomers, initiator and liquid crystal. These types polymerisations occur by the mechanism of free radical polymerisation, as explained bellow.

1.2. Free radical polymerisation

Free radical polymerisation has been widely practiced method for chain polymerisation and has been used in this work. This type of polymerisation can be described in three steps: initiation, propagation and termination. It is initiated by radicals - chemical species characterised by unpaired electrons which subsequently initiate polymerisations. Radicals can be formed by monomers themselves without added initiators (self-polymerisation) but normally by the action of initiators which were added deliberately. The decomposition of initiators in radicals can be achieved electrochemically, thermally or photochemically. Depending on the polymerisation conditions, the same monomer can originate polymers that differ in configuration, molecular mass and therefore in properties. In this work more attention was given to the thermal and photochemical polymerisation.

In thermal polymerisation the N,N-azobisisobutyronitrile (AIBN) is one of the most common initiator which fragments mainly into isobutyronitrile radicals (figure 1).

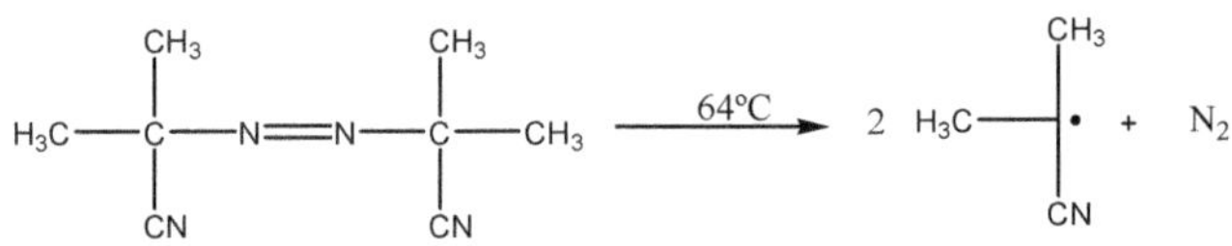

Figure 1. Reaction scheme for the thermal decomposition of N, N-azobisisobutyronitrile (AIBN) in two isobutyronitrile radicals.

Thermal initiator decomposition normally depends on temperature, but also on the solvent used and its solubility. AIBN decomposition normally occurs to an appreciable extent at temperatures higher than 64ºC.

In photochemical polymerisation, light sensitive compounds are required as initiators. These compounds absorb light in the ultraviolet / visible wavelength range and undergo fragmentation leading to the formation of radicals [4]. An example is the photochemical decomposition of 2,2-dimethoxy-2-phenylacetophenone (DMPA) which is illustrated in figure 2. The main advantage of polymerisation with UV radiation is the possibility to achieve high polymerisation rates in a fraction of second [5].

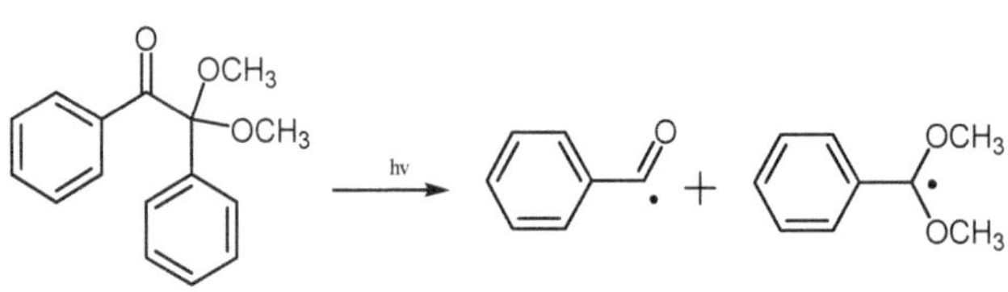

Figure 2. Reaction scheme for UV induced decomposition of 2,2-dimethoxy-2-phenylacetophenone (DMPA) into a benzoyl and an acetal radical fragments.

However, for thermal and photochemical initiators, it is necessary to take into account that some of the radicals formed may recombine in secondary reactions to form compounds that cannot decompose further into radicals and therefore only some radicals are useful to initiate a chain reaction.

The initiation step includes radical production (described above for the AIBN and DMPA) and the attack of these radicals to the monomer molecules. The overall mechanism for free radical chain polymerisation can be described as:

Initiation:

i. Generation of free radicals (R·) by homolytic dissociation of the initiator (I).

$$I \xrightarrow{k_d} 2R\bullet$$

(k_d- rate constant for the dissociation of the initiator)

ii. Reaction between radicals (R·) previously formed and monomer´s double bond (M_1) producing new active species (M_1·).

$$R\bullet + M_1 \xrightarrow{k_1} M_1\bullet$$

(k_1- rate constant for the initiation step)

Propagation:

i. Chain extension by successive addition of monomer molecules (M) to the monomer radical unities (M_n·) formed in the initiation step.

$$M_n\bullet + M_m\bullet \xrightarrow{k_{tc}} M_{n+m}$$

(k_p- rate constant for the propagation step)

Termination:

In last step, radicals combine or disproportionate to terminate the chain growth and form polymer molecules.

i. Combination (a simple interaction between two reactive species, M_n· and M_m·).

$$M_n\bullet + M_m\bullet \xrightarrow{k_{tc}} M_{n+m}$$

(k_{tc}- rate constant for the termination step by combination)

ii. Disproportionation (when hydrogen atom is transferred from one chain to another).

$$M_n\bullet + M_m\bullet \xrightarrow{k_{td}} M_n + M_m$$

(k_{td}- rate constant for the termination step by disproportionation)

1.3. The nematic liquid crystal mixture E7

The liquid crystal used in this work was thermotropic calamitic nematic, know as E7. It is a mixture composed by three different cyanobiphenyls and one cyanotriphenyl in different proportions. The molecular structures of the different components of the nematic liquid crystal E7 are shown in figure 3. The mass percentage and physical properties of E7 components are shown in table 1 [6,7].

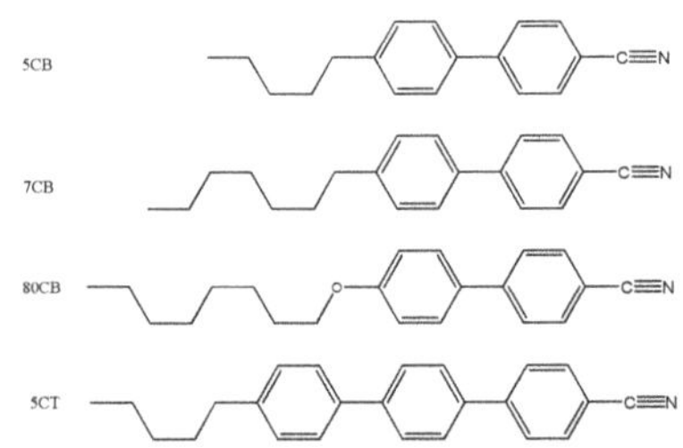

Figure 3. Molecular structures of the components of the nematic liquid crystal mixture E7.

E7 is widely used in polymer dispersed liquid crystals, and it was selected to be studied in this work, because it offers a wide range of operating temperatures in which it maintains anisotropic characteristics. The refractive indices of E7 at T=20ºC are given as: no=1.5183 (ordinary refractive index) and ne=1.7378 (extraordinary refractive index)[2]. It exhibits a nematic to isotropic transition at nearly T_{NI}=58ºC. At room temperature it still exhibits a nematic phase and no other transitions between 58 and -62ºC, where it shows a glass transition. Therefore, liquid crystalline properties are extended down to the glass transition[7]. These features are possibly due to the multicomponent nature of E7[6].

Designation	Molecular formula	IUPAC name	Composition (w/w)	T_{NI} (ºC)
5CB	$C_{18}H_{19}N$	4-cyano-4′-pentyl-1,1′-biphenyl	51%	35.3
7CB	$C_{20}H_{23}N$	4-n-heptyl-4′cyanobiphenyl	25%	42.8
8OCB	$C_{21}H_{25}NO$	4, 4′-n-octyloxycyanobiphenyl	16%	80
5CT	$C_{24}H_{23}N$	4′n-pentyl-4-cyanotriphenyl	8%	240

Table 1. Components and mass composition of the Merck E7 liquid crystal.

1.4. Microstructure of the polymer matrix

During polymerisation of what is initially a homogeneous solution of monomers and LC molecules, the polymeric components grow in molecular weight, and when the two components become sufficiently incompatible, there is a decrease of LC solubility which induces formation of separate phases. The polymer matrix acquires a particular morphology with liquid crystal dispersed in its clusters[1]. The polymerisation conditions[8], the chemical nature of the liquid crystal[9] and the polymerisable monomers[10] determine the morphology of polymer matrix. Conventional PDLCs have two main morphologies: Swiss cheese or polymer ball types each one with different characteristics. The observation of the microstructure of the polymer matrix was carried out by scanning electron microscopy (SEM). The dark areas in the SEM microphotographs reveal the absence of the material, which would have corresponded to the original liquid crystal domains.

The Swiss cheese morphology type (figure 4A) is characterised by liquid crystal randomly dispersed in a polymer matrix, as microdroplets. The size and shape of the LC microdroplets are in strong dependence on the parameters of preparation and the type of polymeric matrix [11, 12].

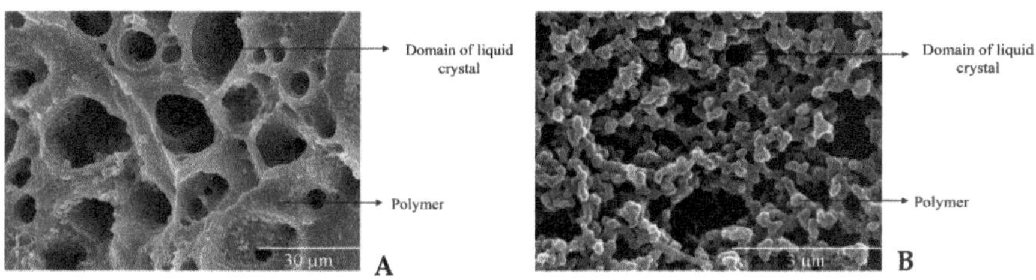

Figure 4. SEM micrograph for the microstructure of the polymer matrix with a A) Swiss cheese morphology type and B) polymer ball morphology type.

In PDLC morphologies with isolated LC microdroplets, two types of distribution of liquid crystal molecules are typically observed, as shown in figure 5.

The bipolar and radial configurations (figure 5) are related to droplet size and shape and depend on whether the liquid crystal prefers to align parallel or perpendicular to the polymer surface, respectively. Due to these different configurations, liquid crystal molecules would be randomly oriented within each droplet or from droplet to droplet, causing dispersion of the incident light.

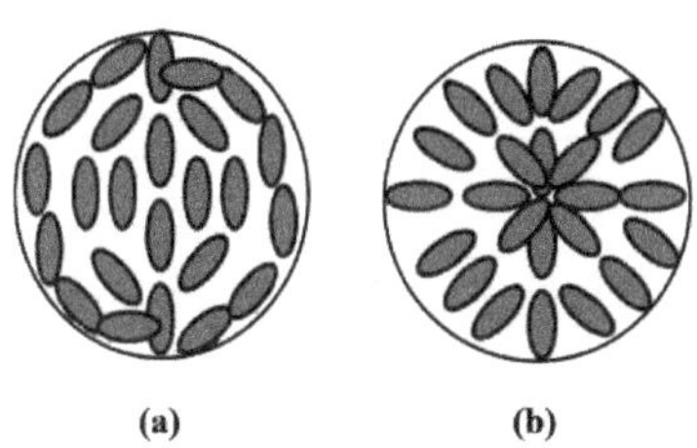

Figure 5. Schematic illustration of bipolar (A) and radial (B) director configurations of the liquid crystals inside spherical droplets.

In the polymer ball morphology type (figure 4B) the polymerisable monomers are phase separated from the continuous liquid crystals and form micro-sized polymer balls. These micro-sized polymer balls merge and form a large polymer network structure with diversely shaped voids in which LC exists. In this case, the LC is in a continuous phase and fills the irregular shaped voids of the polymer network, which are more or less interconnected[12].

1.5. PDLC films transmittance

As mentioned before, PDLC devices can be switched electrically from an opaque scattering state to a highly transparent state when a film of liquid crystal-polymer mixture is sandwiched between two conductive glass slides and the electric field is applied. The polymer matrix material is optically isotropic so it has a single refractive index (n_p). The liquid crystal within the micro-domains has an ordinary refractive index (n_o) and an extraordinary refractive index (n_e) when light ray travel through the liquid crystal.

When no electric field is applied although the liquid crystal molecules would be oriented within each droplet, this orientation changes from droplet to droplet and light propagation normal to the film surface will experiment a range of refractive indices between (n_o) and (n_e). Since the optical anisotropy of LC molecules used in PDLC is sufficiently large, the effective refractive index is not generally matched with the refractive index of the polymer (n_p), light will be scattered and the PDLC is opaque (OFF state). To maximize off-state scattering, the birefringence ($\Delta n= n_e-n_o$) should be as large as possible[9]. On the other hand, when an electric field with sufficient strength to overcome the interactions between polymer matrix and liquid crystal at interfaces of LC domains–polymer matrix, is applied across the film liquid crystal directors within each droplet become uniformly oriented parallel to the direction of the field. If refractive index of the liquid crystal matches the refractive index of the polymeric matrix (n_p) the film become transparent (ON state)[13].

Normally, when the applied electric field is removed, the nematic directors return to their random distribution. The film begins to appear opaque again. A schematic representation of a functional PDLC film is shown in figure 6.

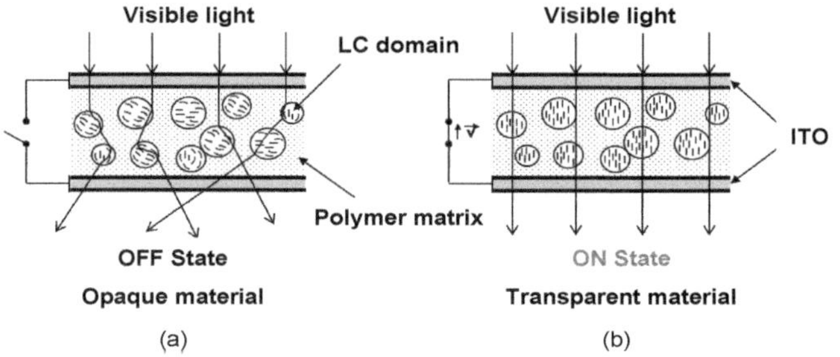

Figure 6. Schematic representation of the averaged molecular orientation of the liquid crystal within the microdroplets without (a) and with (b) an applied electric field.

1.6. Electro-optical properties of PDLCs

Three main types of electro-optical response can be observed (figure 7). The factors contributing to the different response are many and still poorly understood. Electro-optical response of PDLC is usually measured by ramping a PDLC up and down in voltage and comparing the optical response at each voltage[1]. The different types of change of transmittance of PDLC films as a function of the electric field are shown in figure 7.

One of the parameters used to evaluate the efficiency of electro optical response of PDLC is the electric field required to achieve 90% of the maximum transmittance and is designated as E_{90}. Thus, the lower the value of E_{90}, the more easily PDLC devices switch from the OFF state to ON state.

The most common electro-optical response reported in the literature for PDLC is when the increasing voltage curve is coincident to the decreasing voltage curve as shown in figure 7A. It was observed that when the electric field is removed liquid crystal molecules relax back, so that the long shaped molecules which were oriented in the same direction in each droplet return to their original random orientation.

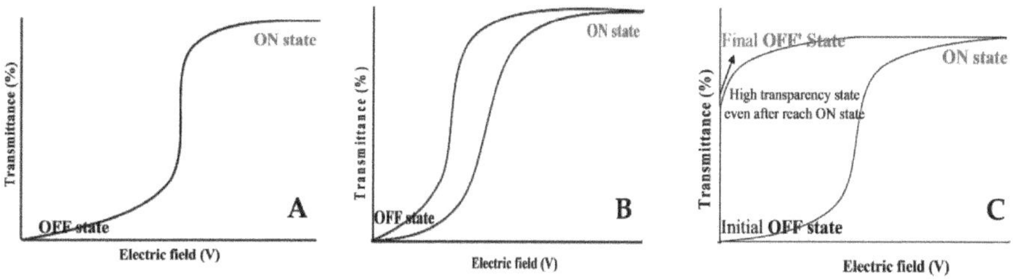

Figure 7. Example of electro-optical response of PDLC A) with no hysteresis, B) with hysteresis and C) with permanent memory effect.

When these curves are not coincident the PDLC shows electrical hysteresis. The transmission with increasing voltage is lower than the transmission when the electric field is decreased. This effect is illustrated on figure 7B and can be defined as the difference between the increasing voltage curve and the decreasing voltage curve.

This difference is commonly attributed in literature to the fact that liquid crystal molecules have a tendency to remain with a certain degree of alignment even after removal of the electric field. The PDLC film is more transparent under decreasing field voltage than with increasing field, because liquid crystals remain with certain degree of alignment caused by the increasing field voltage. However, at the end of measured electro-optical response, the value of the transmittance for the initial opaque state is coincident with that of the final opaque state[14].

In particular cases, not only the transmission with increasing voltage is lower than the transmission when the voltage is decreased but also a high transparency state is obtained for a long period of time at room temperature even after the applied voltage has been switched off, starting from an opaque state and later reaching a transparent state (figures 7C and 8)[1].

In figures 7C and 8, the initial OFF state corresponds to the transmittance of the initial opaque state (zero electric field), the ON state to the transmittance upon applying electric field and the final OFF' state is the transmittance after removal of the applied field. The PDLC film with this electro-optical response has a permanent memory effect and this still remains a poorly understood aspect of PDLC electro-optical behaviour.

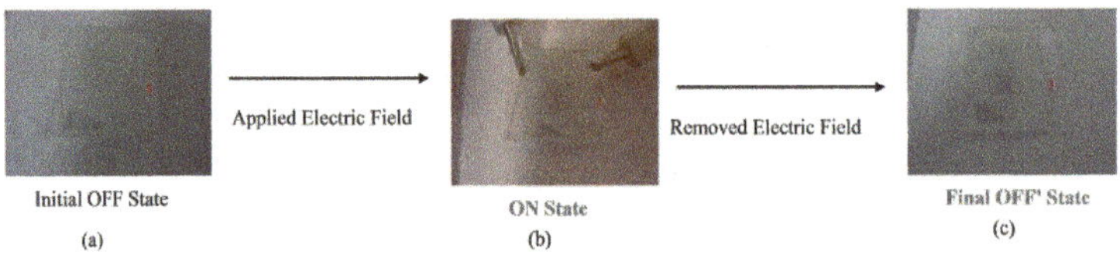

Figure 8. Images of PDLC devices with permanent memory effect: (a) initial OFF state, (b) upon applying electric field and (c) OFF state after removed electric field.

Various factors can influence the performance of a PDLC and therefore the permanent memory effect. However, the most frequently explanation mentioned in literature is related to the anchoring effect[6, 11, 15, 16]. This effect consists of the interaction between the liquid crystal molecules and polymer matrix at the interface of LC domains and polymer matrix. When an electric field is applied inducing the orientation of LC molecules, it opposes the anchoring effect. If this orientation remains even after the applied voltage has been switched off, the liquid crystal does not relax back completely but remains aligned with the electric field and a high transparency state is displayed without any more energy consumption. Therefore for this to happen there must be a weak force between LC molecules and the polymer surface.

The permanent memory effect is very revolutionary in the study of PDLCs because it allows a switch in transparency and the new state is kept without the need for energy to be spent. The only energy required is that needed to switch the PDLC from the OFF state to the ON state, therefore these devices become low power consumers and environmentally friendly.

1.7. Parameters that influence the performance of PDLC films

Despite the fact that PDLC films can be switched between a highly scattering opaque state and a clearly transparent state, these devices sometimes have disadvantages such as the high driving voltage (E_{90}) and the insufficient maximum transmittance (T_{MAX})[17]. The electro-optical properties of PDLC films prepared by the polymerisation induced phase separation (PIPS) method depend on a numbers of factors such as the type of liquid crystal, molecular structure of polymerisable monomers and polymeric conditions. These factors could seriously affect the microstructure of the polymer matrix, the size and the shapes of the LC domains and molecular interactions between the LC molecules and the polymer matrix (anchoring effect)[8, 18-20]. In general and in a simple way, in a nematic liquid crystal there is a relationship between the size and the shape of LC domains and the anchoring effect. When sizes decreased the anchoring effect increase. This effect is directly related with the voltage needed to align the liquid crystal molecules in the domain. Strong anchoring forces hinder the alignment of liquid crystal molecules. Therefore a high voltage is needed to align liquid crystal molecule along the electric field, and the reverse is valid for lower anchoring effects[18]. Therefore, an understanding of the relationships between preparation conditions and molecular structure of the monomers on electro-optical performance of PDLCs is crucial to control device properties.

1.8. Polymerisation conditions

The phase separation phenomenon between LC and polymeric matrix in PDLCs is a kinetic process where the transport parameters can play an important role in determining the domain size and amount of LC separated from the polymer matrix. The rate of polymerisation and also some physical parameters, such as the viscosity of the systems change the LC domain size. In general, the thermal polymerisation is slower (can take several hours), which combined with the effect of temperature that significantly decreases the viscosity of the medium may cause a promotion of the growth of LC domains by diffusion and coalescence. On the other hand, photochemical polymerisation produces a higher polymerisation rate. Smaller domain size can be achieved by higher viscosity of the systems and poor diffusion of free radicals of the polymerisable monomers during polymerisation and hence the reduced coalescence of LC domains. Therefore, small LC domains can be obtained[8,20].

1.9. Molecular structure of polymerisable monomers

The electro-optical response of PDLC films is greatly influenced by molecular structure of the monomers to be incorporated as polymeric matrix[10, 19, 20]. In literature reports there is not a great variety of monomers used in research for applications for PDLC films, besides the most common commercial ones. To the best of our knowledge, published results of the synthesis of monomers to be incorporated as polymer matrix, were mainly dedicated to the preparation of monomers bearing a cyanobiphenyl group[21, 22]. These monomers with a molecular structure similar to the E7 liquid crystal molecules, could lead to better miscibility and compatibility between the polymeric matrix and LCs during the preparation of PDLCs. This could lead to a uniform phase separation and control of the LC domains, which could enhance the performance of PDLCs. On the other hand, the low miscibility with the LC molecules can lead to a premature separation from the matrix. Therefore, one of the main objectives of this work was the design and synthesis of new photochemically and thermally polymerisable monomers, mimicking some structural elements of the E7 liquid crystal molecules. Besides the specific cyanobiphenyl groups, also aromatic systems in general with and without a linear chain spacer with five methylene units were synthesised. Thus, a number of aromatic mono- and dimethacrylates with and without spacer, as well as vinylic monomers, were synthesised under microwave irradiation. A linear chain spacer with five methylene units was introduced into some of the structures between the aromatic systems and the methacrylate group to mimic the structure of 5CB, which is the liquid crystal component present with the highest percentage in E7.

2. Synthesis and characterisation

As mentioned before, a number of photochemical and thermal polymerisable monomers, as well in some cases, their starting materials were prepared under microwave irradiation. The synthetic methods and the full characterisation of the compounds obtained have been described elsewhere[23]. All the compounds were synthesised using a microwave synthetic reactor under mild or solventless procedures. For some reactions, it was necessary to add a small amount of DMF (dimethylformamide) to obtain better homogenisation of the mixture. DMF was chosen because of its dipolar nature and the high ability to absorb microwave energy and convert it into heat, coupled with a comparatively high boiling point of 153ºC[24]. The protocol employed consisted in placing equivalent amounts of the corresponding reagents in an open quartz tube and then subjecting the mixture to microwave irradiation at 200W. The reaction time was optimised by following the reactions by thin layer chromatography (TLC) every minute and stopped when no starting material remained. The authors have described[23] the data of compounds characterisation by ^{1}H and ^{13}C NMR, FTIR spectroscopy, elemental analysis and melting points. Also, a comparison between the results described in the literature for the syntheses of the compounds under classical synthetic methods with the results obtained under microwave irradiation have been made. Part of the Gel Permeation Chromatography (GPC), Differential Scanning Calorimetry (DSC) and electro-optical studies presented here have been previously published[26].

2.1. List of synthesised monomers

The molecular structures of all synthesised monomers are shown in figure 9. Some of the synthesised monomers have been already mentioned in the literature, but were lacking full characterisation, which was provided in ref. 23. This was the case for compounds: 1[25], 2[27], 3[25],4 [28], 5[29], 6[30], 7[29], 9[31], 10, 11[32], 13[33], 14 [34],15 [35], 16[36], 25 [37], 26 [38]. To the best of our knowledge, the compounds 8, 12, 17, 18, 19, 20, 21, 22, 23 and 24 had not been synthesised before.

2.2. Selected monomers

Among all the synthesised monomers, the monomers containing cyano and/or spacer chain group (figure 10) were studied further due to higher chemical structural compatibility with E7 liquid crystal molecules. The monomers (with cyano and/or with spacer chains groups) properties will be compared to the analogous monomers without cyano group and/or spacer chain and the results will be presented later on.

2.3. Calorimetric characterisation

In order to determine the transition temperatures and their associated enthalpies differential scanning calorimetry (DSC) was employed. Figures 11-14 present the DSC thermograms collected on heating and on cooling run of cycle I. This cycle corresponds to the first heating and cooling runs and the cycle II to the second heating and cooling runs.

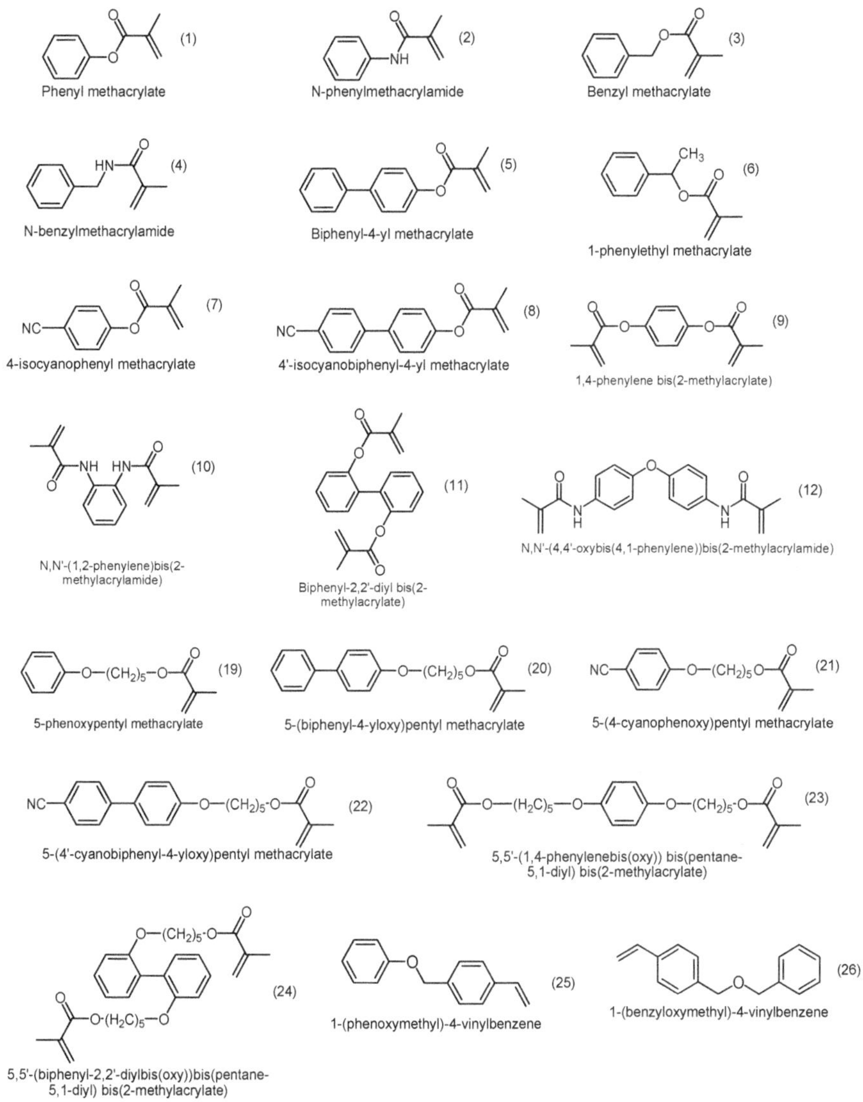

Figure 9. Molecular structures of synthesised monomers.

As mentioned before, the melting temperature (T_m) was obtained at the peak of the melting endothermic transition, and in analogues way the crystallisation temperature either for cold- (T_{cc}) or for melt- (T_c) crystallisation was obtained at the peak of the respective exothermic transition peak. The glass transition temperature (T_g) was taken at the inflection point of the specific heat capacity variation in the transition. The melting and crystallisation enthalpies

(ΔH_m) and (ΔH_{cr}), respectively, were determined from the areas under the curve that represents the respective transitions. The information obtained on the temperatures and enthalpies is summarised in table 2.

Figure 11A presents the DSC thermogram collected on cycle I for liquid monomer 19. The heat flux presents a discontinuity characteristic of the glass transition at T_g=-86.01ºC. At higher temperatures an exothermic peaks characteristic of cold crystallisation emerges at -40.19ºC with an enthalpy of -97.45 Jg^{-1} followed by endothermic peak due to melting at T_m=-3.62ºC with an enthalpy of 115.52 Jg^{-1}. During the first cooling rate no transitions were detected, so it is possible that the liquid remained in a supercooled state. During cycle II the same calorimetric behaviour as that obtained for cycle I was observed.

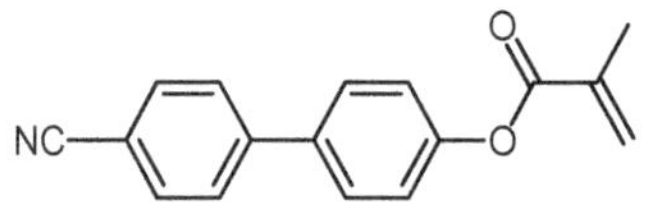

(8)

4'-isocyanobiphenyl-4-yl methacrylate

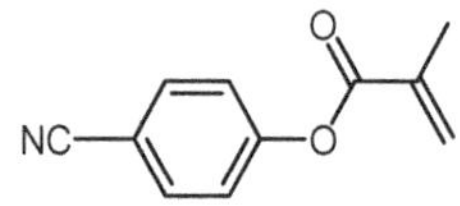

(7)

4-isocyanophenyl methacrylate

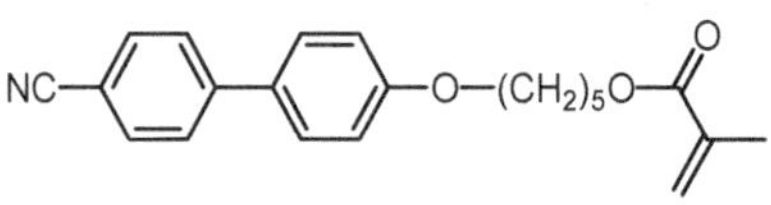

(22)

5-(4'-cyanobiphenyl-4-yloxy)pentyl methacrylate

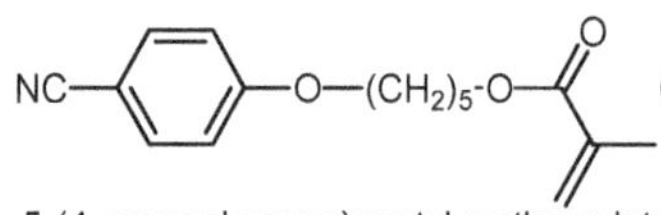

(21)

5-(4-cyanophenoxy)pentyl methacrylate

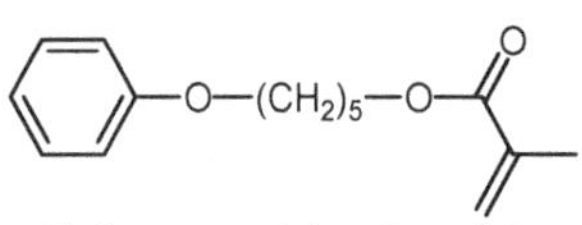

(19)

5-phenoxypentyl methacrylate

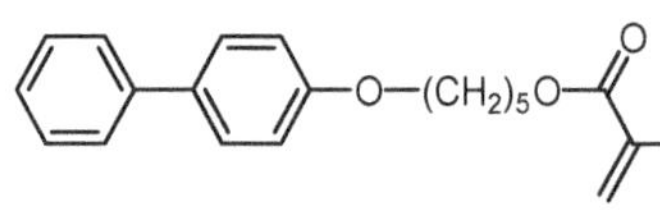

(20)

5-(biphenyl-4-yloxy)pentyl methacrylate

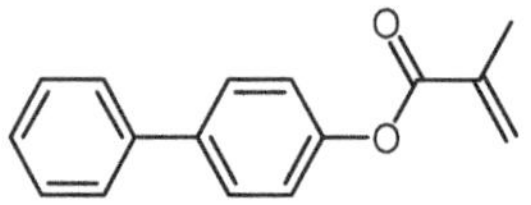

(5)

Biphenyl-4-yl methacrylate

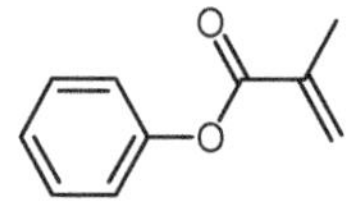

(1)

Phenyl methacrylate

Figure 10. Molecular structure of synthesised monomers with and without cyano and spacer chain group.

During cycle I for liquid monomer 1 (figure 11B) the thermogram showed endothermic peak at 18.86ºC with an enthalpy of 108.97 Jg^{-1} due to the melting of the crystalline form; on cooling rate no transitions were detected, thus the liquid could have remained in a supercooled state. However, during the cycle II the thermograms showed endothermic transitions assigned to a glass transitions at T_g=-89.02ºC and melting temperature at T_m=18.86ºC with an enthalpy of 107.67 Jg^{-1}. Cold crystallisation was also observed at T_{cc}=-55.13ºC with an enthalpy of -71.0 Jg^{-1}. These thermal transitions were very similar to those observed in the thermogram obtained for liquid monomer 19 (figure 11A).

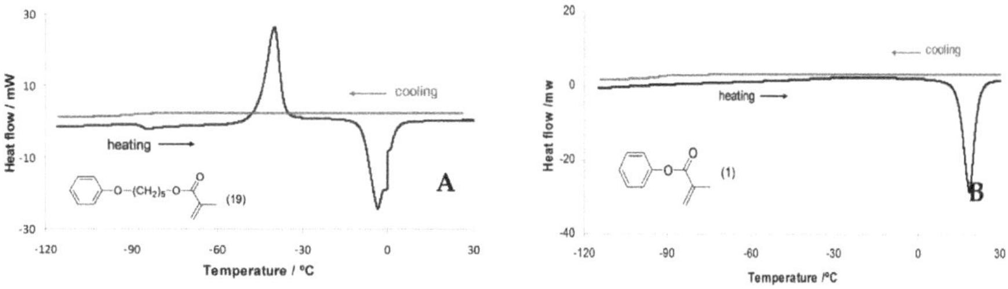

Figure 11. DSC measurement obtained for A) monomer 19 and B) monomer 1 during cycle I at 10ºC min-1.

The solid monomers, 7, 5, 22, 21 and 20 had different thermal behaviour compared with the liquid monomers (19 and 1) described above. The thermograms for monomers 7, 5, 22 and 21 in figures 12 and 13, showed reversible systems that on heating gave endothermic peaks characteristic of melting transition and on cooling rate - exothermic peaks of crystallisation. The second heating and cooling rates were characterised by the same thermal behaviour.

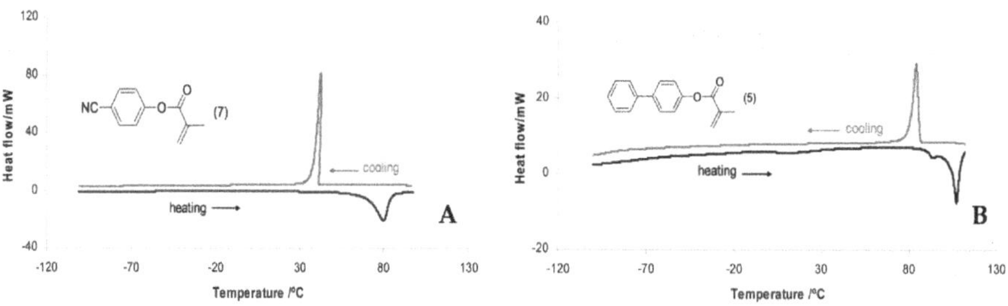

Figure 12. DSC measurement obtained for A) monomer 7 and B) monomer 5 during cycle I at 10ºC min-1.

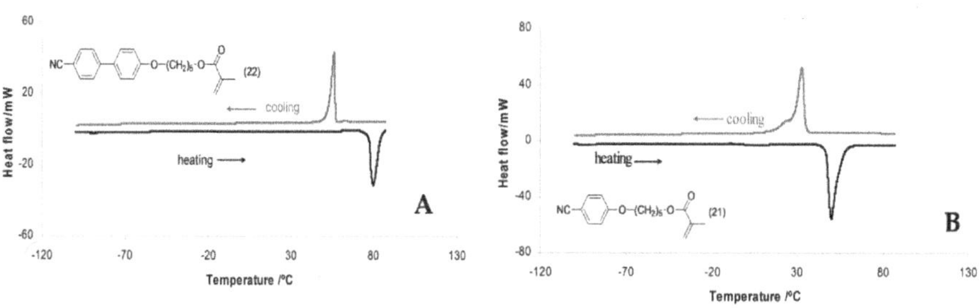

Figure 13. DSC measurement obtained for A) monomer 22 and B) monomer 21 during cycle I at 10ºC min-1.

Monomer 20 (figure 14A), despite having the same thermal behaviour as described above, it was not reversible. The thermogram of monomer 20 showed that the crystalline form melted and partially crystallised on cooling. During the first heating rate it exhibited endothermic peak due to melting transition with ΔHm=57.84 Jg-1 and during the cooling - exothermic peak due to melt-crystallisation but with ΔH_{cr}=-18.65 Jg^{-1}. During the second heating this crystalline solid formed melted with ΔHm=14.93 Jg^{-1} and partially crystallised again on cooling, but also with lower value of ΔHcr=-2.86 Jg^{-1}. The difference between the values of enthalpies associated with endothermic and exothermic transitions suggested that part of the material previously melted could have remained in supercooled state, degraded or polymerised, and only a small fraction of the monomer crystallised during cooling run. This possibility was supported by previous studies of DSC between 25 and 200ºC, which indicated that the temperature of 100ºC seemed to be enough to initiate polymerisation of the monomer.

All these monomers revealed high tendency to organise in crystalline form that melted during the heating run followed by melt-crystallisation during the cooling run. The only thermal event during the heating run was the melting transition that showed that until melting the material was in a crystalline phase.

The insertion of a spacer chain, in monomers 20, 21 and 22 decreased the melting temperatures, compared to the analogue monomers without the spacer chain, which have made the later structures more rigid.

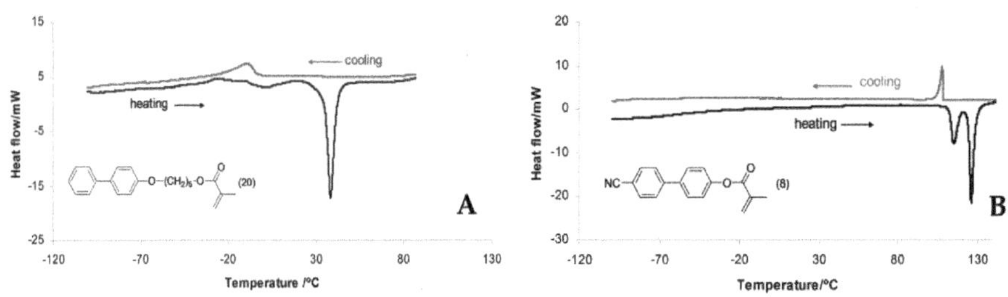

Figure 14. DSC measurement obtained for A) monomer 20 and B) monomer 8 during cycle I at 10ºC min^{-1}.

It can be seen in figure 14B that for monomer 8 on the heating run there were two endothermic peaks at 114.95 and 125.90ºC with enthalpies of 29.09 and 60.40 Jg^{-1}, respectively. However on the cooling run only one exothermic peak emerged at 107.72ºC with enthalpy of -14.74 Jg^{-1}. During the second heating run part of the amorphous material underwent cold crystallisation at T_{cc}=63.61ºC with an enthalpy of -10.43 Jg^{-1} followed by two endothermic transitions at 109.60 and 123.95 ºC with 7.48 and 12.13 Jg^{-1}, respectively. On the second cooling run no transitions were detected, so it was possible that the liquid have remained in supercooled state or that it had polymerised previously. This last possibility was supported by DSC studies between 25-200ºC. POM studies allowed better understanding of the thermal transitions, especially those involved in the thermal behaviour of monomer 8. Thus, the data from DSC and POM indicated that crystalline material transited to a mesophase state which could be assigned to the first endothermic peak, followed by transition to an isotropic state, assigned to the second endothermic peak. The peak at 114.95ºC could be assigned to solid to mesophase transition temperature (T_{SM}) and at 125.90ºC to mesophase to isotropic transition temperature (T_{MI}).

2.4. Polarised optical microscopy characterisation

The thermal properties of monomers were complemented with polarised optical microscopy (POM) studies. The polarised optical microscope equipped with a hot stage only allows the observation of transitions from 20 to 300ºC and therefore for the liquid monomers an analysis by POM with the same conditions used in DSC (-130 to 25ºC) was not possible. Optical observations allowed the determination of not only temperatures for each transition, but also the texture acquired by the sample. Furthermore, polarised optical microscopy can evaluate the possibility of the existence of liquid crystalline phases. An isotropic phase placed between crossed polarisers does not allow polarised light to pass through and what is observed is simply a black uniform image, unlike what is observed for a birefringent phase (optical anisotropy). The values of the thermal proprieties determined by POM are summarised in table 3.

		Heating run at 10ºC min^{-1}					Cooling run at 10ºC min^{-1}		
		glass transition	Cold-crystallisation		Melting		Melt-crystallisation		Glass transition
Monomer	**Cycle**	T_g/ºC	T_{cc}/ºC	ΔH_{cr}/ Jg^{-1}	T_m/ºC	ΔH_m/ Jg^{-1}	T_c/ºC	ΔH_{cr}/ Jg^{-1}	T_g/ºC
19	I	-86.01	-40.19	-97.45	-3.62	115.52	-	-	-
	II	-86.01	-39.83	-96.60	-3.58	114.80	-	-	-
1	I	-	-	-	18.86	108.97	-	-	-
	II	-89.02	-55.13	-71.00	18.86	107.67	-	-	-
7	I	-	-	-	81.53	56.03	44.28	91.92	-
	II	-	-	-	79.52	76.00	38.26	-90.85	-
5	I	-	-	-			83.99	-83.45	-
	II	-	-	-	106.04	71.35	82.72	-82.02	
8	I	-	-	-	114.95 125.90	29.09 60.40	107.72	-14.74	-
	II		63.61	-10.43	109.60 123.95	7.48 12.13	-	-	-
21	I	-	-	-	49.80	115.70	32.53	-105.98	-
	II	-	-	-	49.68	111.52	31.28	-91.19	
20	I	-	-	-	38.19	57.84	-9.24	-18.65	-
	II	-	-	-	16.27	26.48	-8.13	-2.87	-
22	I	-	-	-	79.77	85.21	55.90	-64.55	-
	II	-	-	-	78.14	69.36	53.71	-24.29	-

Table 2. Thermal properties of monomers obtained by DSC during different cooling/heating cycles

The texture for liquid monomers 19 and 1 which may be analysed in the optical microphotographs presented in figure 15, revealed an isotropic phase on heating and the cooling runs between 30 and 100ºC and no birefringent phase was observed. The spheres that appear in the images are spacers of ITO cell.

Optical microscopy pictures of monomers 7, 5, 22, 21 and 20 are present in figures 16-20. The analysis of these figures revealed a birefringent texture in the solid state. However, this property was lost with melting transition and the liquid state showed an optic isotropic structure that remained until the crystallisation temperature where again the monomers acquire a birefringent texture.

As mentioned before, in DSC analysis monomer 8 showed a particular thermal behaviour with two endothermic transitions one after the other in the heating run. The results of combined POM and DSC studies allow a better understanding of the type of transitions involved in all the monomers and in particular in monomer 8. The optical microscopy picture presented in figure 21 shows in the heating run that the birefringent solid material first transition to a mesophase at 114ºC (T_{SM}). This mesophase was assigned based on the observation of birefringence and fluidity. However this phase with the increase of temperature transited to an isotropic phase at 124ºC (T_{MI}). During the cooling run, at 102ºC the isotropic phase crystallised out, so that in the second heating run transit to an isotropic phase at 124ºC. This isotropic phase crystallised at 100ºC into a birefringent material. From POM studies it was possible to observe the onset of polymerisation at 140 ºC which is in agreement with the DSC analysis.

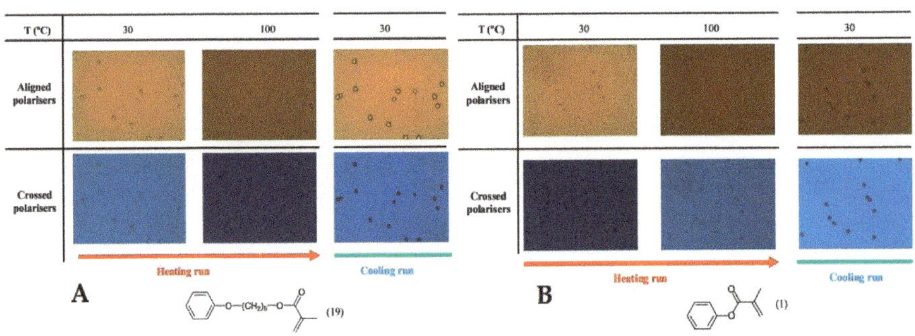

Figure 15. Optical micrographs of A) monomer 19 and B) monomer 1 at specific temperatures during the heating and cooling runs at 10ºC min^{-1}.

The melting temperature determined by DSC for all the monomers analysed seemed to be corroborated by the melting temperature determined by capillarity. However, the transition temperatures determined by POM showed small differences. Although the programme for the heating and the cooling runs was the same (10ºCmin^{-1}) it was difficult to keep exactly the same ramp throughout the POM analysis. During POM measurements the runs of heating and cooling were stopped momentarily to take optical micrographs.

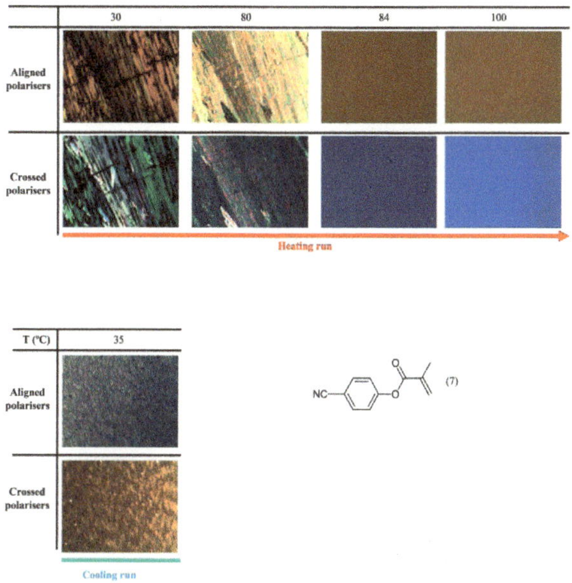

Figure 16. Optical micrographs of monomer 7 at specific temperature during the heating and cooling run at 10ºC min-1.

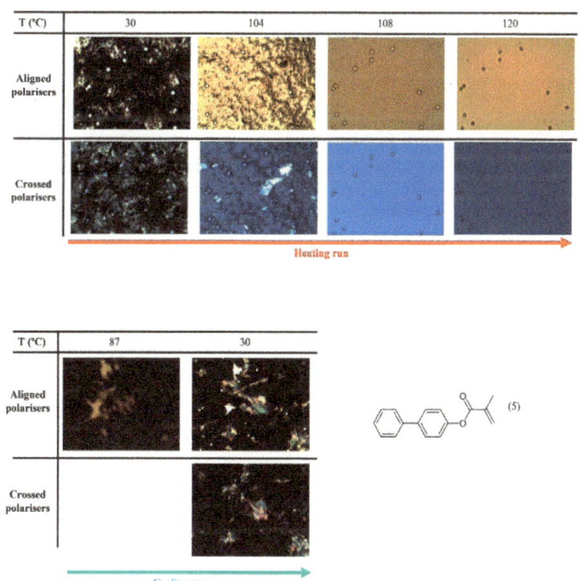

Figure 17. Optical micrographs of monomer 5 at specific temperatures during the heating and cooling runs at 10ºC min-1.

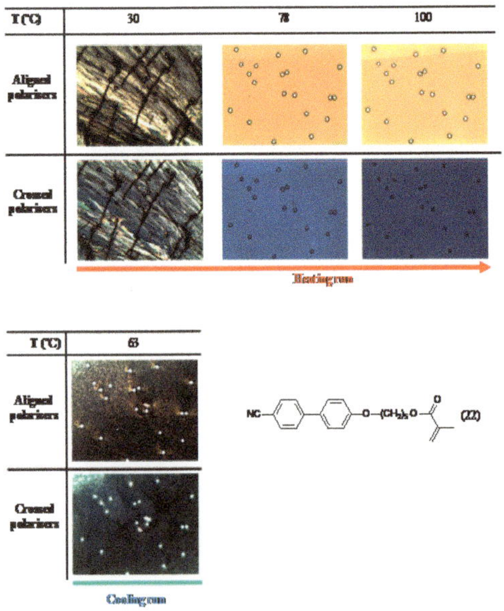

Figure 18. Optical micrographs of monomer 22 at specific temperatures during the heating and cooling runs at 10ºC min^{-1}.

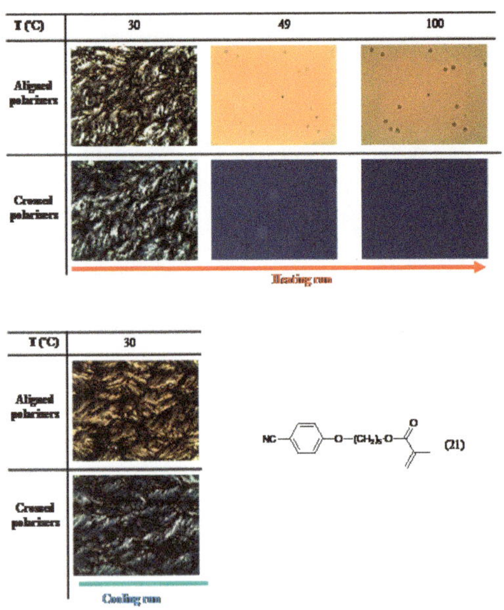

Figure 19. Optical micrographs of monomer 21 at specific temperatures during the heating and cooling run at 10ºC min^{-1}.

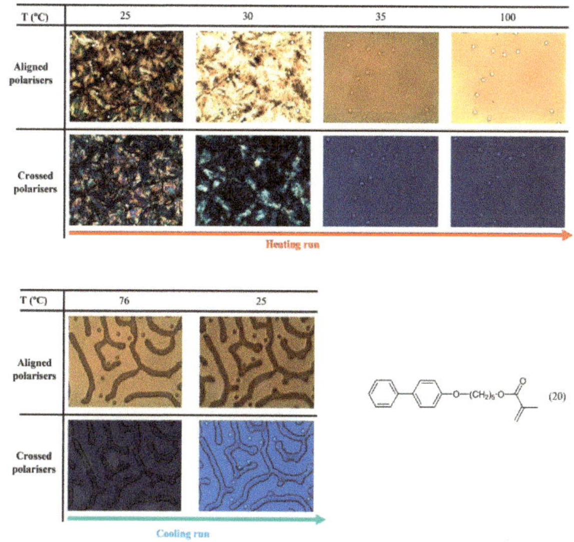

Figure 20. Optical micrographs of monomer 20 at specific temperatures during the heating and cooling runs at 10ºC min-1.

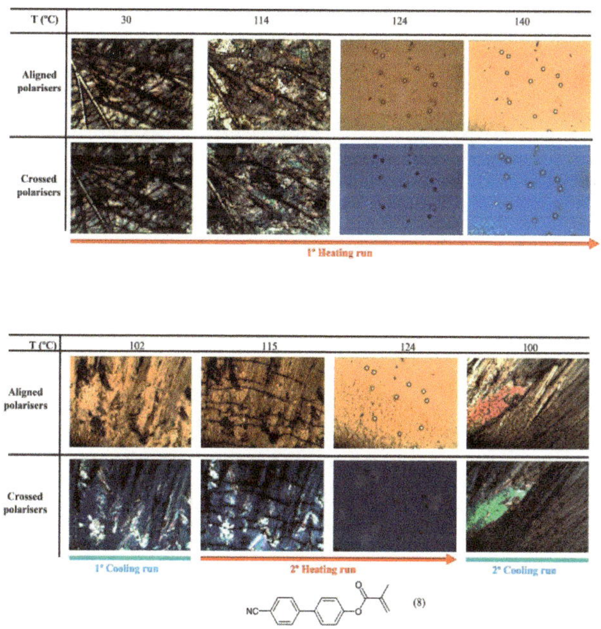

Figure 21. Optical micrographs of monomer 8 at specific temperatures during the first and second heating at 10ºC min-1.

		Heating run at 10ºC min^{-1}		Cooling run at 10ºC min^{-1}	
		Melting		Melt-crystallisation	
Monomer	Initial optical property	T/ºC	Optical property	T_c/ºC	Optical property
19	Isotropic	-	-	-	-
1	Isotropic	-	-	-	-
7	Birefringent	84(T_m)	Isotropic	35	Birefringent
5	Birefringent	108(T_m)	Isotropic	87	Birefringent
22	Birefringent	78(T_m)	Isotropic	63	Birefringent
21	Birefringent	49(T_m)	Isotropic	30	Birefringent
20	Birefringent	35(T_m)	Isotropic	-	-
8	Birefringent	114 (T_{SM}) 124 (T_{MI})	Birefringent Isotropic	102	Birefringent

Table 3. Thermal properties of monomers obtained by POM during heating/cooling runs.

3. Structural and electro-optical characterisation

3.1. Molecular weights and polymer structures

The molecular weight of the polymer matrix plays an important role in the morphologies of the matrix and therefore in the liquid crystal domain size and shape. If the polymer chain increases with an increase in molecular weight leading to a higher network density, the size of liquid crystal domains should decrease. This promotes the interaction between polymer and liquid crystal molecules; the opposite is valid for smaller chains. As previously mentioned, the size and the shapes of liquid crystal domains are correlated with the anchoring strength on the interface between the liquid crystal molecules and polymer molecules. The anchoring strength determines the electric field needed to achieve the transparent state in the PDLC. Generally, the strength of the field needed is inversely proportional to the liquid crystal domain size. A number of factors influence the electro-optical response of PDLCs and the investigation of these parameters allows control of the performance of these devices. So, it is important to understand and establish the effect of the molecular structure of the polymerisable monomers on the molecular weight of the polymer and the effect of molecular weight on the morphology and electro-optical properties.

3.2. Characterisation of molecular weights

Gel permeation chromatography (GPC) is a liquid chromatography technique that separates molecules according to sizes and not according to chemical affinities toward the porous substrate[39].Large polymer molecules with higher molecular weights can move into pores but since there are only few large pores available to them, they reside, on average, less time in pores than smaller molecules. Therefore, larger molecules are eluted first. On the other hand, smaller polymer molecules with lower molecular weights can fit into the small pores and penetrate into a larger number of pores. So, the elution time increases with decreasing molecular size[40]. This correlation between elution time and molecular weight was observed for the polymers analysed. The values of average molecular weights (M_n and M_w) and polydispersity index (PDI) for mixture of monomer/glycidyl methacrylate (50/50 w/w) after thermal and photochemical polymerisations are summarised in Tables 4 and 5, respectively. It was possible to observe differences between polymer matrixes prepared by thermal polymerisation and by photochemical polymerisation. For thermal polymerisation, the polymers had a number average molecular weight, Mn, around e^8 or e^{12}, and for photochemical polymerisation polydispersions were higher and Mn was around e^{12}. Generally, the polymers with lower molecular weight possessed lower polydispersion, indicating that there were chains with more uniform lengths (sizes), but no correlation was observed with the molecular structure of the respective monomers.

3.3. Microstructure of the polymer matrix

The microscopic images of the microstructure of the polymeric matrix were evaluated by scanning electronic microscopy (SEM). Some microscopic images can be assigned to the Swiss cheese (figure 22A) or polymer ball (figure 22B) types of morphology, but for some samples these morphologies were not uniform across the sample. For other samples it was not possible to identify any morphology. These results can suggest that the chemical affinity of polymerisable monomers with liquid crystal molecules due to their similarities in chemical structure produced smaller phase separation and therefore the E7 liquid crystal molecules

Monomers used in co-polymerisation	PDI	Peak/min	ln (M_w/gmol^{-1})	ln (M_n/gmol^{-1})
8	1.91	13.86	12.80	12.15
22	1.40	14.97	12.51	12.16
19	1.44	13.57	13.05	12.69
5	1.04	20.51	8.13	8.09
7	1.04	20.81	8.02	7.99
21	1.04	20.47	8.16	8.12
20	1.05	20.14	8.22	8.17

Table 4. Values of average molecular weights (M_n and M_w) and polydispersity index (PDI) for polymers prepared with a mixture of monomer/glycidyl methacrylate (50/50 w/w) by thermal polymerisation.

Monomers used in co-polymerisation	PDI	Peak/min	ln (M_w/gmol^{-1})	ln (M_n/gmol^{-1})
8	1.51	14.28	12.80	12.39
22	1.47	14.68	12.36	11.98
19	1.35	15.06	12.04	11.74
5	1.24	15.78	11.67	11.45
7	1.08	14.37	12.30	12.23
21	1.38	15.63	11.72	11.39
20	1.36	22.36	6.99	6.68

Table 5. Values of average molecular weights (M_n and M_w) and polydispersity index (PD) for polymers prepared with a mixture of monomer/glycidyl methacrylate (50/50 w/w) by photochemical polymerisation.

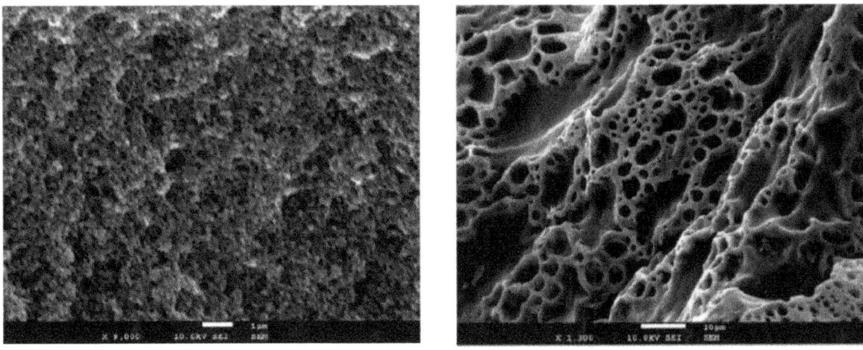

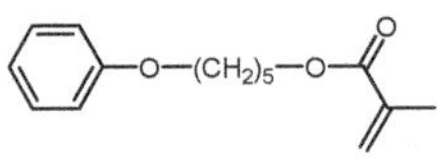

(19)

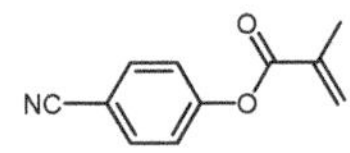

(7)

Figure 22. Examples of SEM micrographs for the microstructure of the polymer matrix prepared by thermal polymerisation of A) monomer 19 and B) monomer 7.

would be highly embedded in the matrix. It was also possible to observe that there could not be established a correlation between the molecular structure of the polymerisable monomers and polymerisation conditions. Although the samples for analysis by GPC and SEM were not prepared under the same conditions, it was not possible to establish a correlation between the molecular weight of polymers and the microstructure of the polymer matrix.

3.4. Electro-optical characterisation

The PDLCs prepared with the monomers synthesised showed poor electro-optical response (figure 23). In most of the studies the PDLC cells showed good opacity after polymerisation with transmittance of the initial opaque state TOFF◎0%. However, the electro-optical study did not reveal significant difference on the transmission of the sample with the application of electric field up to 400V (20V μm^{-1}), the maximum transmittance, TMAX, being less than 4%. The reason for this non-ideal behaviour is unclear, however, it was possible to suggest explications.

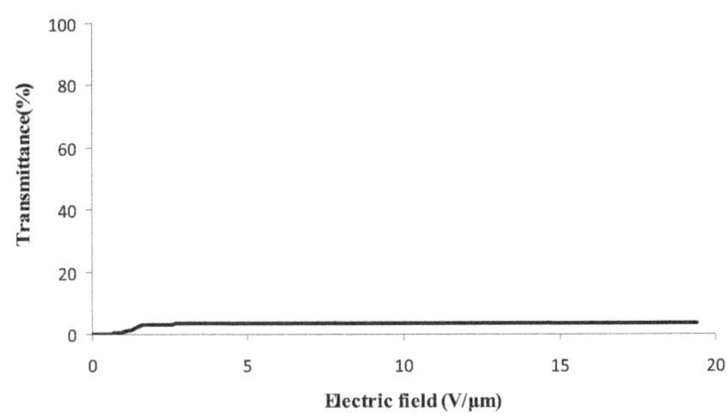

Figure 23. General electro-optical response for PDLC films prepared by the PIPS method with the monomers synthesised + E7 in a ratio of 30/70 (w/w).

As mentioned before, the electro-optical properties of PDLC films depend on numerous parameters, namely, the structure and the molecular weight of the polymeric matrix (size and shape of the liquid crystal domains). These factors control the dynamics between alignment and random distribution when the electric field is applied and after removal of the electric field, respectively. These dynamics are very much dependent on anchoring force. This force is related to the interactions between the liquid crystal molecules and the polymeric matrix at the interface of LC domains and polymer matrix. The electro-optical results for PDLC films studied in this work suggested that higher molecular affinity between the polymeric matrix and the liquid crystal could increase the anchoring effect.

The similar molecular structure between the molecules of monomer and the LC could have led to higher affinity between molecules of polymer and molecules of liquid crystal in the interfaces matrix–LC domains. In this way, the electric field applied could not overcome the interactions at the interface and the PDLC remained in an opaque state. This was due to the fact that if the liquid crystal molecules do not align themselves along the electric field, the PDLC films appear opaque. The higher affinity between molecular structure of polymer matrix and LC molecules caused an increase in anchoring strength, i.e. it hindered the alignment of the liquid crystal molecules and could be the cause of the low electro-optical response of PDLC films.

In conclusion, the chemical affinity between the polymerisable monomers and the liquid crystal molecules was able to provide better affinity between the two which created more homogeneous mixture with consequent phase separation, in which the liquid crystal was uniformly dispersed in the polymer matrix. However, this affinity could be high enough to increase the anchoring effect not allowing a good electro-optical response.

4. Conclusions

A series of new monomers with structurally diverse functionalisation was successfully synthesised under microwave irradiation. These results were compared with those obtained following classical methods. The comparison of these methodologies clearly indicated the considerable reduction in time and amount of solvent used in the reactions performed under microwave irradiation. Therefore, two important advantages, reduction in time and solvents, obtained by the use of microwave irradiation were demonstrated. Beyond that, the majority of monomers were synthesised in one step and in some cases in two steps, which provided simple and effective methods of synthesis. The GPC analysis of co-polymers (monomers synthesised and glicidyl methacrylate) revealed that it was not possible to establish a correlation between the molecular structure of the polymerisable monomers and the molecular weight of polymers. The microstructure of the polymer matrix was evaluated by SEM. From these studies, it was also not possible to establish a correlation between the molecular structure and molecular weight of the respective monomers. Nevertheless, the absence of a typical morphology on most microscopic images seemed to indicate that the liquid crystal remained highly impregnated in the polymer matrix and thus, their micro-domains could not be visualised. The phase transformations of the monomers were investigated by DSC and complemented by POM. In general, the thermal behaviour seems to be that of a reversible system and the monomers show a high tendency to melting followed by a melt-crystallisation. The spacer chain in the molecular structure of the monomers decreases the melting temperature in comparison with the respective monomers without this chain. The POM analyses confirmed the transition determined by DSC and allowed an observation of the texture acquired by the sample. The analysis of solid monomers showed a birefringent texture in the solid state that where lost in the melting transition where an isotropic phase is acquired. However, monomer 8 revealed a different thermal behaviour with a mesophase between 114-124ºC. On the other hand, the liquid monomers showed isotropic properties in all temperature ranges.

The electro-optical responses of all PDLCs prepared by the PIPS method with the composites of monomers synthesised–E7 exhibit a lower transmittance upon the application of an electric field. This behaviour can be interpreted as the result of a higher anchoring effect. Thus, even the application of a higher electric field could not orient the E7 molecules along the direction of the electric field and they scattered strongly the incident light. The similarity in molecular structure between the molecules of polymerisable monomers and LC should allow a good chemical affinity between them and a better homogenisation of the mixture and a uniform phase separation can be achieved but it should not be too high to avoid a higher anchoring effect which resulted in a low electro-optical response of the PDLCs.

Author details

Ana Isabel Mouquinho, Krasimira Petrova, Maria Teresa Barros and João Sotomayor
REQUIMTE, CQFB, Departamento de Química, Faculdade de Ciências e Tecnologia, Universidade Nova de Lisboa, Caparica, Portugal

Acknowledgement

This work was supported by Fundação para a Ciência e Tecnologia through Project PTDC/CTM/69145/2006.

5. References

[1] Drazaic, P. S., Liquid Crystal Dispersions World Scientific Publishing: Singapore, 1995.

[2] Benkhaled, L.; Coqueret, X.; Traisnel, A.; Maschke, U.; Mechernene, L.; Benmouna, M., A Comparative Study of UV and EB-Cured PDLC Films via Electro-Optical Properties. *Mol. Cryst. Liq. Cryst.* 2004, 412, 477/[2087]-483/[2093].

[3] Park, S.; Hong, W. J., Polymer dispersed liquid crystal film for variable-transparency glazing. *Thin Solid Films* 2009, 517, 3183-3186.

[4] Devdatt, L. K.; Nikolaos, A. P., Method of Determination of Initiator Efficiency: Application to UV Polymerizations Using 2,2-Dimethoxy-2-phenylacetophenone. *Macromolecules* 1994, 27, 733-738.

[5] Ates, S.; Aydogan, B.; Torun, L.; Yagci, Y., Synthesis and characterization of triptycene type cross-linker and its use in photoinduced curing applications. *Polymer* 2010, 51, 825-831.

[6] Bedjaoui, L.; Gogibus, N.; Ewen, B.; Pakula, T.; Coqueret, X.; Benmouna, M.; Maschke, U., Preferential Solvation of the eutectic mixture of liquid crystal E7 in a polysiloxane. *Polymer* 2004, 45, 6555-6560.

[7] Brás, A., R., E.; Henriques, S.; Casimiro, T.; Aguiar-Ricardo, A.; Sotomayor, J.; Caldeira, J.; Santos, C.; Dionísio, M., Characterization of a Nematic Mixture by Reversed-Phase HPLC and UV Spectroscopy: Application to Phase Behavior Studies in Liquid Crystal-CO_2 Systems. *Liq. Cryst.* 2007, 34, 591-597.

[8] Li, W.; Cao, Y.; Kashima, M.; Kong, L.; Yang, H., Control of the Microstructure of Polymer Network and Effects of the Structures of Polymerizable Monomers On the Electro-optical properties of UV- Cured Polymer Dispersed Liquid Crystal Films. *J. Polym. Sci., Part B: Polym. Phys.* 2008, 46, 1369-1375.

[9] Vaz, N. A.; Montgomery, G. P., Refractive indices of polymer-dispersed liquid crystal film materials: Epoxy based systems. *J. Appl. Phys.* 1987, 62, 3161-3172.

[10] Bulgakova, S. A.; Mashin, A. I.; Kazantseva, I. A.; Kashtanov, D. E.; Jones, M. M.; Tsepkov, G. S.; Korobkov, A. V.; Nezhdanov, A. V., Influence of the Composition of the Polymer Matrix on the Electrooptical Properties of Films with a Dispersed Liquid Crystal. *Russ. J. Appl. Chem.* 2008, 81 1446–1451.

[11] K.; Blaaderen, A., V.; Wiltzius, P., Morphology and electro-optical properties of polymer-dispersed liquid crystal films *Am. Phys. Soc.* 1997, 55, 1646-1654.

[12] Han, W. J., Morphological Studies of Polymer Dispersed Liquid Crystal Materials. *Korean J.Phys.Soc.* 2006, 49, 563-568.

[13] Nomura, H.; Suzuki, S.; Atarashi, Y., Electrooptical Properties of Polymer Films Containing Nematic Liquid Crystal Microdroplets. *Jpn. J. Appl. Phys.* 1990, 29, 522-528.

[14] Yokoyama, H., Surface Anchoring of Nematic Liquid Crystals. *Mol. Cryst. liq. Cryst.* 1988, 165, 265-316.

[15] Andy, F. Y. G.; Tsung, C. K.; Mo, H. L., Polymer Dispersed Liquid crystal Films with memory characteristics. *J. Appl.Phys.* 1992, 31, 3366-3369.

[16] Rumiko, Y.; Susumu, S., Highly transparent memory states by phase transition with a field in polymer dispersed liquid crystal films. *J. Appl. Phys.* 1992, 31, 254-256.

[17] Yan, B.; He, J.; Bao, R.; Bai, X.; Wang, S.; Zeng, Y.; Wang, Y., Modification of electro-optical properties of polymer dispersed liquid crystal films by iniferter polymerization. *Eur. Polym. J.* 2008, 44, 952-958.

[18] Wonsool, A.; Ha, K., Temperature effects on LC Droplets Formation of PDLC films with thermoplastic Matrix. *Korea Polym. J.* 1999, 7, 130-135.

[19] He, J.; Bin, Y.; Wang, X.; Yu, B.; Wang, Y., A novel Polymer dispersed Liquid Crystal film prepared by reversible addition fragmentation chain transfer polymerization. *Eur. Polym. J.* 2007, 43, 4037-4042.

[20] Ahmad, F.; Jamil, M.; Jeon, Y. J.; Woo, L. J.; Jung, J. E.; Jang, J. E.; Lee, G. H.; Park, J., Comparative study on the electrooptical properties of polymer-dispersed liquid crystal films with different mixtures of monomers and liquid crystals. *J. Appl.Polym. Sci.* 2011, 121, 1424-1430.

[21] Manni, A.; Gobbi, L.; Simoni, F., Novel PDLC Films Based On a Photoactive Polymeric Binder. *Mol. Cryst. Liq. Cryst.* 2003, 398, 281-291.

[22] Orzeszko, B.; Ksyta, M. D.; Orzeszko, A., An Efficient, Facile, and Fast Synthesis of 4-alkoxy-4′-hydroxybiphenyls. *Synth. Commun.* 2002, 32, 3425-3429.

[23] Barros, M. T.; Mouquinho, A.; Petrova, K.; Saavedra, M.; Sotomayor, J., Fast synthesis employing a microwave assisted neat protocol of new monomers potentially useful for the preparation of PDLC films. *Cent. Eur. J. Chem* 2011, 9, 557-566.
[24] Hayes, B. L., Microwave Synthesis: Chemistry at the Speed of Light CEM Publishing: Matthews, NC, 2002.
[25] http://www.sigmaaldrich.com/ (December,2011).
[26] Mouquinho, A.; Saavedra, M.; Maiau, A.; Petrova, K.; Barros, M. T.; Figueirinhas, J. L.; Sotomayor, J., Films Based on New Methacrylate Monomers: Synthesis, Characterisation and Electro-optical properties. *Mol. Cryst. Liq. Cryst.* 2011, 542, 132/[654]-140[662].
[27] Reddy, G. J.; Naidu, S. V.; Reddy, A. V., Synthesis and Characterization of Poly(N-phenyl methacrylamide-co-methyl methacrylate) and Reactivity Ratios Determination. *J. Appl. Polym. Sci.* 2003, 90, 2179.
[28] Casadei, M. A.; Cesa, S.; Inesi, A., Electrochemical Studies on Haloamides. Part XII. Electrosynthesis of Oxazolidine-2,4-diones. *Tetrahedron* 1995, 51, (20), 5891-5900.
[29] Vijayanand, P. S.; Kato, S.; Satokawa, S.; Kojima, T., Copolymerization of 4-cyanophenyl methacrylate with methyl methacrylate: Synthesis, characterization and determination of monomer reactivity ratios. *Polym. Bull* 2007, 59, 469-480.
[30] Chênevert, R.; Pelchat, N.; Morin, P., Lipase-mediated enantioselective acylation of alcohols with functionalized vinyl esters: acyl donor tolerance and applications. *Tetrahedron: Asymmetry* 2009, 20, 1191-1196.
[31] Lal, J.; Green, R., The Preparation of Some Esters of Methacrylic Acid. *J. Org. Chem* 1955, 20, 1030-1033.
[32] Pinazzi, C; Comptes Rendus des Seances de l'Academie des Sciences. *Serie C: Sciences Chimiques* 1972, 274, (44).
[33] Braun, J. V.; Steindorff, A., Ueber einige Verbindungen der Pentamethylenreihe. *Chem. Berichte (In German)* 1905, 38, 956-966.
[34] Slugovc, C.; Demel, S.; Riegler, S.; Hobisch, J.; Stelzer, F., Influence of functional groups on ring opening metathesis polymerisation and polymer properties. *J. Mol. Cat. A: Chem.* 2004, 213, 107-113.
[35] Berg, S.; Newbery, G., The Search for Chemotherapeutic Amidines. Part X . Substituted 4 : 4'-Diamidino-aw-diphenoxyalkanes and -diphenyl Ethers. *J. Chem. Soc.* 1949, 642-645.
[36] Yelamaggad, C. V.; Tamilenthi, V. P., Synthesis and thermal properties of liquid crystal trimers comprising cyanobiphenyl and salicylaldimine anisometric segments. *Tetrahedron* 2009, 65, 6403-6409.
[37] Dneprovskii, A. S.; Tuchkin, A. I., Radical-Anion Nucleophilic Substitution in p-Bromobenzyl Methyl and p-Bromobenzyl Phenyl Ethers. Competing Fragmentations of Radical Anions. *Russ. J. Org. Chem.* 1997, 33, 1601-1605.
[38] Ikeuchi, Y.; Taguchi, T.; Hanzawa, Y., Zirconocene-Mediated and/or Catalyzed Unprecedented Coupling Reactions of Alkoxymethyl-Substituted Styrene Derivatives. *J. Org. Chem* 2005, 11, 4354-4359.
[39] Elias, H.-G., An Introduction to Polymer Science. Wiley-VCH: New York, 1997.
[40] Braun, D.; Cherdron, H.; Rehahn, M.; Ritter, H.; Voit, B., Polymer Synthesis:Theory and practice. 4 ed.; Springer: 2005.

Section 2

Polymer Synthesis

Chapter 7

From Ruthenium Complexes to Novel Functional Nanocomposites: Development and Perspectives

Karen Segala and Angela S. Pereira

Additional information is available at the end of the chapter

1. Introduction

Nanotechnology is one of the most widespread topics for current development in fundamentally all research areas. Along with polymer science that is also one of the leading fields and their studies cover a broad range of topics. This would include nanoelectronics as the critical scale for modern devices is now below 100 nm. Other research areas may include polymer-based biomaterials, such as miniemulsion particles for drug delivery or tissue engineering, self-assembled layer-by-layer polymer films, electrospun nanofibers and imprint lithography.

Although the terminology "Nanotechnology" is quite recent, prior polymer science studies involving nanoscale dimensions were used. For example, phase separated polymer blends regularly acquire nanoscale phase dimensions; asymmetric membranes often have nanoscale void structure; miniemulsion particles can be synthesized below 100 nm and interfacial phenomena in blends and composites involve nanoscale dimensions. At times, amphiphilic macromolecules are used to modify the nanoparticles surface. The most common functionality of these types of polymers is a lipophilic part intercalated with the existing aliphatic chains at the nanoparticles surface, covering, or maybe even completely encapsulating the original passivated nanoparticle.[1, 2] This type of encapsulation has been described for copolymers [3, 4] and phospholipids,[5-7], *inter alia* and takes advantage of the hydrophobic colloidal stability of organically passivated nanoparticles (NPs) to prepare polymeric nanocomposites. In this chapter we plan on reviewing various nanocomposites using a polymer-based matrix with transition metal complexes, namely ruthenium based complexes, and a wide range of inorganic nanoparticles with varying optical, antimicrobial and magnetic properties. These nanocomposites have potential applications in biomedical implants as well as small devices for clinical applications.

2. Metallopolymers

Recent interest in functional polymer matrix based nanocomposites has emerged with great importance in many fields of scientific research as well as in industrial applications, which will increase greatly due to the variety of possible modifications in the chemical and physical properties of these materials.[8] This well-known interest in soft functional materials, led to the growing realization that the presence of metal centers in macromolecular materials can result in enhanced novel properties while maintaining the capacity of polymerizing through classic processing methods. Metal centers have been neglected as important structural components of synthetic polymers although they play a key role in determining the functions of many extended solids and biological macromolecules. The metallopolymer field has been delayed due to synthesis and characterization problems, such as what procedures to use, purification issues and a lack of commercially available starting reagents. Most of the classic polymerization procedures used in pure organic monomers originated inefficient or undesirable side reactions with the presence of the metal centers. These classic procedures originated low molar mass metallopolymers, contaminated by structural defects or missing conclusive characterization. The end of last century dictated the beginning of high molar mass metallopolymers with defined microstructures. Most of the synthetic strategies used are either compatible with the presence of coordinated metal centers or with pendant ligands that tolerate the consequent incorporation of metal ions.[9-11] This progress has been supported by the discovery and expansion of new characterization techniques, such as matrix-assisted laser desorption/ionization–time of flight spectroscopy or electrospray ionization mass spectrometry, as well as the rekindling of more traditional methods, such as analytical ultracentrifugation, leading to a vast collection of metal containing polymers that display an array of bonding modes and an assortment of structures. The ability to prepare metallopolymers with a variety of structures has fuelled efforts to improve the efficiency and control of synthetic procedures in order to create scaled-up, well-defined materials. These accomplishments lead to a range of potential applications which have now emerged, and the aim of this chapter is to illustrate selected recent advances of the use of metallopolymers in sync with inorganic nanoparticles to produce nanocomposites with varying antiseptic, optical and magnetic properties. These nanocomposites have potential applications in biomedical implants as well as small devices for clinical applications.

Network polymers based on styrene, prepared by well-developed suspension polymerization processes, retain many of their qualities for a microcomposite host matrix. The many advantages include facile preparation, controlled surface area and porous structure, good thermal resistance and reasonable chemical inertness.[12] Styrene-divinylbenzene copolymers have been the subject of several studies. Mahdavian *et al.* described an easy and efficient method for preparing polymer-supported chiral copolymers of styrene-divinylbenzene, which can be used in chromatographic columns packing.[13] Crosslinked copolymers prepared from styrene-divinylbenzene are of particular interest when used in the preparation of ion exchange resins by the introduction of various functional groups (polar or not) as alkyl, benzoyl, acetyl, hydroxymethyl and sulfonic derivatives.[14] Their main applicability is in catalytic supports and solid-phase extraction. The efficiency of sulfonate agents and their effects on the morphology of styrene-divinylbenzene beads have been investigated.[15] The authors found that the use of acetyl sulfate as a sulfonate agent preserved the spherical morphology of the beads.

Polymers based on the application of coordination forces have been prepared recently. The first direct synthesis of a polymer containing a transition metal complex gave rise to a coordination polymer based on ruthenium(II)-terpyridine complexes.[16,17] The combination of conventional polymers to metal centers generates intrinsic properties of the metal, retaining the properties of the conventional polymer. A new property acquired due to the metal center, is the electroactivity amongst all the transition metals, and the ruthenium has been widely used by coordination chemists mainly due to its ease of forming extremely stable complexes, especially with pyridine ligands and their derivatives.[18]

3. Ruthenium (II) complexes: a brief history

Ruthenium based metal complexes have shown a plethora of uses and applications. Namely, complex structures in the presence of a metal center containing ruthenium have shown good results in biological applications.[19-21] Coordination compounds of formula *trans*-$[RuX_2(L)_4]$, where (X) = Cl, Br, and I, (L) one monodentated ligand (Figure 1) are extremely well known in literature and many compounds containing pyridine groups and their derivatives have been prepared and characterized.[22-24]

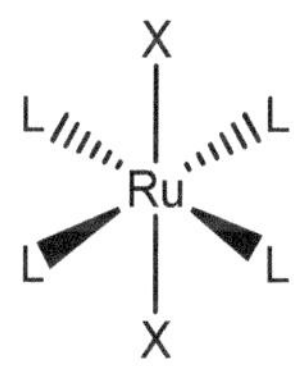

Figure 1. Molecular structure for ruthenium (II) metal complexes, where L is the different ligands monodentate and X, the halogens Cl, Br, I.[25]

The first studies were those with the synthesis of coordination complexes with four pyrrole group substituents. In 1996, progress was made in the synthesis and characterization of copper (I) complexes containing four tetranuclear pyrrole substituents attached to the pyridine ligand. The aim of the complex synthesis was to electropolymerize in the form of a copolymer, but the results were not encouraging and the authors remained focused only on the synthesis of the ruthenium complexes.[26] The preparation of a series of new electropolymerizable complexes with the general formula *trans*-$[RuX_2(L)_4]$ was achieved.[27, 28] The preparation and full characterization of poly-*trans*-$[RuCl_2(pmp)_4]$ (pmp=3-(pyrrol-1-ylmethyl) pyridine) was reported and its applicability demonstrated in the preparation of modified electrodes.[27, 29]

In order to fully understand the electropolymerization process and to determine the optimized electrochemical parameters for the achievement of a good film, electrochemical studies were made on *trans*-[RuCl$_2$(pmp)$_4$]. In 1997, the preparation of electrodes modified by electropolymerization of poly-*trans*[RuCl$_2$(pmp)$_4$] using galvanostatic and potenciometric techniques was accomplished.[28] Studies show that water-soluble ruthenium complexes are of high therapeutic value due to their capacity to interfere with nitric oxide (NO) in biological systems.[30] On the other hand, the complex *trans*-[RuCl$_2$(vpy)$_4$] (vpy=4-vinylpyridine),[31] composed of four groups, which are easily polymerized producing electroactive adherent films, compared to films made from the linkers vinylpyridine and vinylterpyridine, were also prepared.[32, 33] Previous studies have shown that the complex *trans*-[RuCl$_2$(vpy)$_4$] can be electropolymerized on different substrates, including inert materials such as Pt and Pd,[31] ferrous alloys[34, 35] and stainless steel.[36] Studies of cyclic voltammetry, electrochemical crystal microbalance quartz and Raman spectroscopy revealed the results of electropolymerization of the complex *trans*-[RuCl$_2$(vpy)$_4$] in Au and Pt electrodes in particular the effects of the polymerization potential on the characteristics of the film formed, and especially the understanding of the key processes involved in electropolymerization.[37]

Much of the recent work on polymer substrates has been focused on the structure of the polymeric matrix for subsequent complex formation with a metal ion. Antony *et al.* studied the synthesis and catalytic properties of poly-(styrene-divinylbenzene) supported by ruthenium (III complexes.[38] Styrene-divinylbenzene copolymers, with its reactive groups, have often been used as ligands with chelating properties. When reacted together with styrene (sty), divinylbenzene (DVB) can be used as a reactive monomer in polyester resins. The resulting cross-linked polymer is mainly used for the production of ion exchange resin. The authors showed that by a simple method, the new polymer of ruthenium catalyst was effective in the epoxidation of styrene and cyclic olefins. The facile preparation and recyclability of these catalysts should provide a useful strategy for oxidation of organic substrates under controlled experimental conditions.

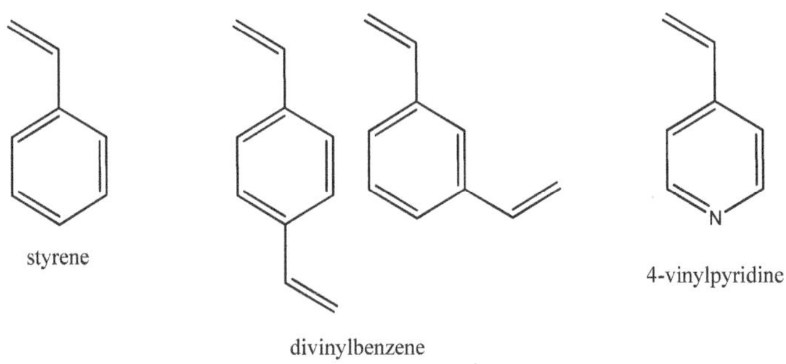

Figure 2. Chemical structures of monomers commonly used along with ruthenium complexes to form metallo-co-poymers.

However, few thermal studies were found in the literature involving copolymers of vinylpyridine and divinylbenzene. The vinylpyridine can be chemically modified by oxidation, producing vinylpyridine N-oxide. These modified materials can be used as catalyst supports and ion exchange resins to remove heavy metals due to the presence of oxidized vinylpyridine groups.[39] The covalent incorporation of the transition metal complexes in the polymer network should be accompanied by benefits in biological systems.[40] There is great interest in biomedicine and bio-nanotechnology in studies involving supported metal complexes as enzyme inhibitors.[41] Metals such as ruthenium, osmium, and iridium are capable of forming stable complexes which are not involved in biological environments, thereby expanding the coordination modes limited to carbon. This provides new opportunities for the construction of small molecular structures that cannot be occupied by purely organic compounds.[42] Recent studies have shown the importance of the presence of ruthenium structures in biological research involving cancer treatment.[43] Lately, many studies are being conducted using ruthenium complexes in nanocomposites and biomaterials. Such studies have shown that the nanocomposite containing the complex can be used as nanodrug, for targeting cancer, and intracellular labeling.[44] Among the various research groups involved in this issue, Zhang *et al.* showed that the effect of nano-scale carborane ruthenium (II)-arene complexes, showed significant results in treating lung cancer.[45] The authors suggest that this complex containing Ru(II) may be a strong candidate for lung cancer chemotherapy. The results provide evidence that tumor growth can be inhibited by anti-cancer ruthenium (II) arene complex *in vivo*. Clearly, a better understanding of the mechanism of action and possible chemotherapeutic activity of the complex of Ru(II)-arene could certainly benefit the future clinical treatment of this cancer.

Although one may propose alternatives of the possible reaction mechanisms between the monomers: the *trans*-$[RuCl_2(vpy)_4]$ complex, styrene, divinylbenzene and 4-vinylpyridine, it is not possible to safely affirm what structures will be formed by a combination of these monomers (Figure 2). However, the authors suggest that the formation of crosslinking between the complex combination with the monomers participating in the structure (Figure 3) is highly probable, where the polymerization and the polymeric chain propagation could occur in both, by double bond cleavage of the ligands of the metal complex (vpy) and the interpolation of the monomers (sty, DVB, vpy) among themselves. Thus, there is still no concrete answer on the exact formulation of co- and terpolymers synthesized. Efforts in this direction have been intensified by several researchers in order to have an idea of the structural composition of these polymers and how, or even if, these bonds significantly influence the formation of these new functional polymers nanocomposites.

In this chapter we plan on reviewing various nanocomposites using a polymer-based matrix with transition metal complexes and a wide range of inorganic nanoparticles with varying optical, antiseptic and magnetic properties, namely ZnO nanoparticles which have interesting optical properties; silver nanoparticles (AgNPs) which have antimicrobial properties as well as known optical properties and magnetic nanoparticles ($CoPt_3$) which have well-defined magnetic properties.

4. Inorganic nanoparticle incorporation

The incorporation of nanoparticles as a filler in polymer matrices is interesting mainly due to the different properties resulting therefrom and compared to their bulk analogues.

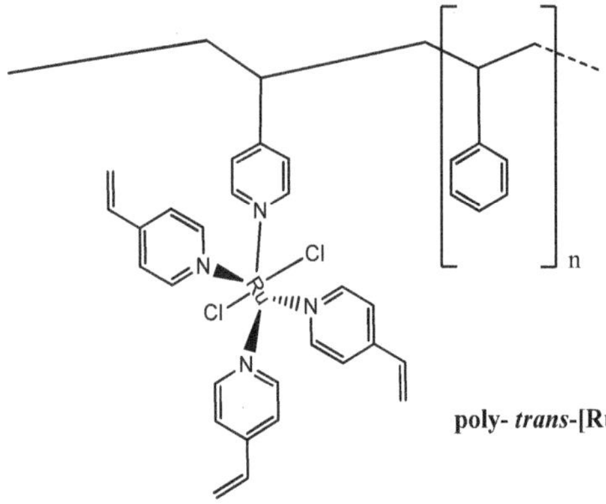

poly- *trans*-[RuCl$_2$(vpy)$_4$]-sty}

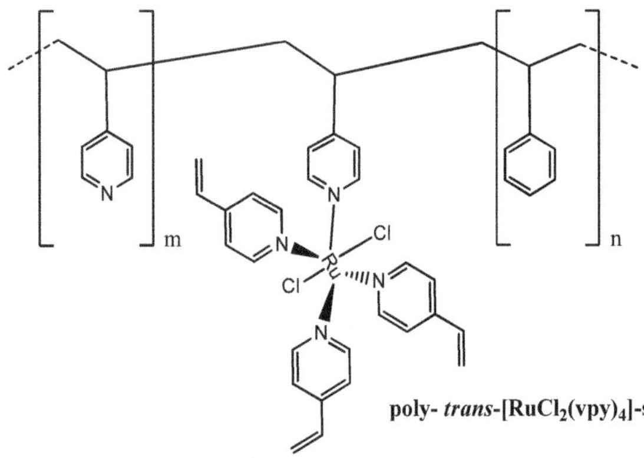

poly- *trans*-[RuCl$_2$(vpy)$_4$]-sty-4-vpy}

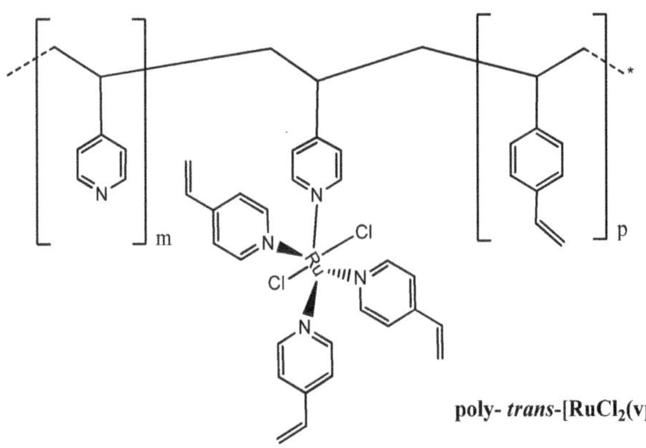

poly- *trans*-[RuCl$_2$(vpy)$_4$]-4-vpy-DVB}

Figure 3. Proposed polymeric structure containing monomers: styrene, divinylbenzene, 4-vinylbenzene and ruthenium complex.

Several authors have described changes in optical, electrical and mechanical properties when the size of a certain material reduces to nanoscale dimensions.[46, 47] This type of interaction can influence the molecular dynamics of a polymer hence resulting in significant changes in their physical, thermal behavior and mechanical properties. Bullions *et al.* investigated a hybrid material consisting of NPs embedded in SiO_2 poly(imide).[48] The authors concluded that the formation of chemical bonds between the NPs and certain terminal groups of poly(imide), confines the space accessible to the polymer chains in the molecular dynamics influencing polymer, which can result in changes in thermal stability and consequently its glass transition temperature. A controlled polymer synthesis in the presence of inorganic NPs, allows rigorous control over the matrix's physical/chemical properties. Nonetheless, the chemical affinity between nanoparticles and polymer is a major aspect in nanocomposite preparation. The compatibility of the inorganic material with the polymer matrix can be improved through the chemical surface modification of the particles. At times, surface modification may be necessary to promote chemical compatibility between components. For example, in the case of trioctylphosphine oxide (TOPO)-prepared semiconductor nanoparticles, these can be used directly in the synthesis of an organic matrix-based nanocomposite, once these are passivated with molecules of hydrophilic nature.

For the preparation of nanocomposites there is a diversity of polymer synthesis techniques, the most common being: dispersion polymerization, mass polymerization and emulsion polymerization. Recently the development of new radical polymerization methods, currently denominated by controlled/"living" polymerization, came to ease new methodologies that are prepared in a controlled manner.[18] This method was used for various inorganic materials, such as TiO_2[49-51] and SiO_2,[52] where these were directly dispersed in the monomer. More recently the NP encapsulation by an *in situ* miniemulsion polymerization was demonstrated with poly-styrene,[53-55] poly-butyl acrylate[53, 56] and a poly-styrene - poly-methylmetacrylate copolymer.[57] However changes in the crystallization of semicrystalline polymer matrices in nanocomposites has also been described.[58] For example, a significant influence of NPs morphology on the crystallization kinetics of polyamide in a hybrid material consisting of an array of poly(tetramethylene terephthalamide) fibers or containing SiO_2 spheres with nanometric dimensions.[59] When preparing Bi_2S_3/Nylon nanocomposites, the authors found that the Bi_2S_3 nanofibers act as nucleation sites, resulting in a larger number of spherulites in the nanocomposite when compared to the pure polymer.

Depending on the envisioned application, various types of fillers may be used for loading the nanocomposite which may differ, for example, their morphological or other properties, such as heat resistance or chemical reactivity. Studies have shown that for the incorporation of metal and semiconductor NPs in this polymeric matrix, these nanoparticles can be dispersed among the crosslinking of the polymeric matrix, as shown in the Figure 4. Amongst the most common fillers for polymer-based nanocomposites are carbonates, sulfates, alumino-silicates and metal oxides (Al_2O_3, Fe_2O_3, ZnO).[60] Semiconductor NPs such as CdS, ZnS, or CdSe have also been widely used due to their optical properties.[61, 62] Some of the examples of inorganic particles cited in the literature are the use of TiO_2 NPs in

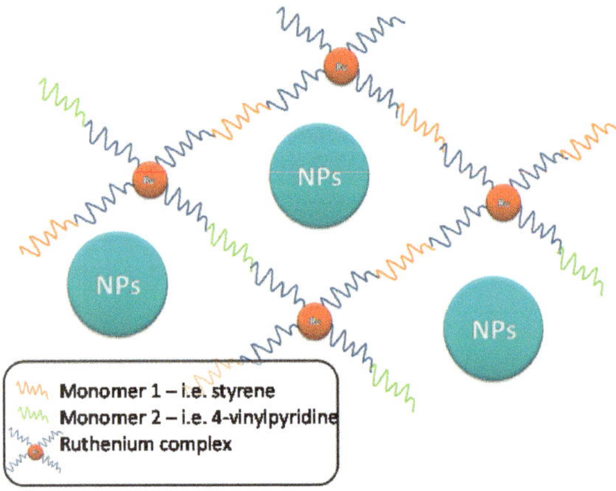

Figure 4. Proposed crosslinked structure of prepared metallpolymers nanocomposite.

the preparation of composites for paints and coatings[46, 63] and SiO_2 NPs, which can give added strength or flame retardant characteristics to nanocomposites.[64, 65]

5. Ruthenium complex based copolymers with incorporated inorganic nanoparticles

The following section summarizes relevant properties of functional ruthenium based nanocomposites when doped with diverse inorganic nanoparticles (magnetic, antimicrobial and luminescent nanoparticles).

5.1. ZnO NP doped Ru complex/styrene copolymer

Semiconductor NPs show size-dependent optical properties and have been widely explored as new optical materials.[47] New optically active inorganic/polymer nanocomposite have been prepared when using semiconductor NPs together with polymers. As with all nanocomposites the intrinsic properties of the starting materials maintain but new properties due to synergistic effects arising from the combination of the inorganic and organic components may arise.[66] The homogeneous distribution of inorganic NPs within the polymer matrix and the strong interface adhesion between the matrix and nano-fillers are important aspects to be considered because of their influence in the nanocomposites' performance.[67] Segala *et al*. reported the preparation of nanocomposites containing ZnO NPs dispersed in the co-polymer poly-{*trans*-[$RuCl_2(vpy)_4$]-sty}.[68] The polymer was prepared through reaction of the complex *trans*-[$RuCl_2(vpy)_4$] with styrene.[69]

The ZnO NPs were previously prepared and their optical properties monitored by UV-visible spectrocsopy and size through high resolution transmission electron microscopy (HR-TEM) (Figure 5). A quantum mechanical effect occurring in the semiconductor NPs is the increase in band gap, with the reduction in particle size. The quantum confinement effect is clearly seen in Figure 5 inset. The literature shows that bulk ZnO exhibit a band gap of 3.37 eV (or about 368 nm), whereas the NPs band gap has a characteristic optical feature at 357 nm. Theoretical calculations based on the effective mass approximation for spherical particles[70, 71] provides a size estimate of 5.2 nm diameter of the ZnO NPs. The size of the NPs was confirmed through HR-TEM in which the approximate size is in the order of 5 nm.

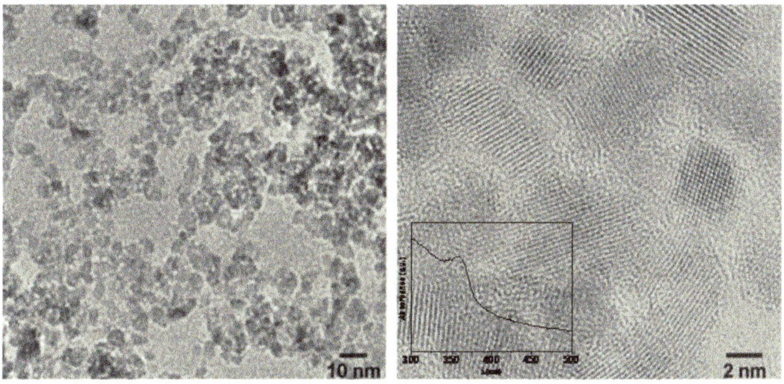

Figure 5. HR-TEM images of ZnO NPs used in the preparation of the nanocomposite. Inset: Optical spectrum of coloidal suspension of ZnO NPs.

When incorporating the semiconductor NPs into the copolymer matrix, the properties of the initial NPs were maintained, such as optical and structural properties.[68] Scanning electron microscopy (SEM) of the nanocomposite and energy-dispersive X-ray spectroscopy (EDS), clearly demonstrate the homogeneous presence of Zn and Ru throughout the whole sample. (Figure 6).

Nanocomposites containing ZnO and poly-{*trans*-[$RuCl_2(vpy)_4$]-sty} have been prepared and the results point to a homogeneous nanocomposite.[68] This study allows further development of this synthetic strategy to obtain homogeneous nanocomposites of variable composition that can find practical interest in ruthenium-based sensor devices.

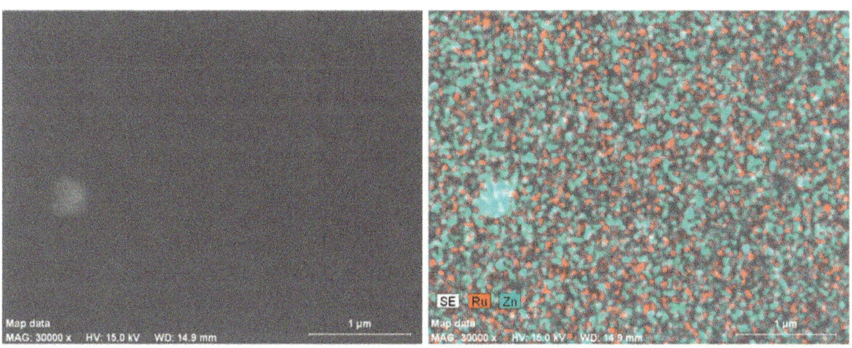

Figure 6. SEM image and EDS mapping of ZnO NPs dispersed in the co-polymer poly-{*trans*-[$RuCl_2(vpy)_4$]-sty}

5.2. Ag NP doped Ru complex/styrene/divinylbenzene copolymer

The bactericidal effects of ionic silver are well-known and have been used since ancient times. The precious metal was originally used as an effective antimicrobial agent and is relatively free of side effects. However, with the development of modern antibiotics to treat infectious disease, the use of silver in the clinical setting has been limited predominantly to topical cream based silver sulfadiazine to prevent infections associated with treatment of burns.[72-74] However recent studies revealed that Ag NPs have improved antibacterial properties when compared with bulk silver and that even nanomolar concentrations of Ag NPs can be more effective than micromolar concentrations of silver ions,[75] thus leading to an increased interest on materials containing nano-silver. Silver ions are known to be extremely reactive and could readily bind with an assortment of negatively charged molecules such as inorganic anions, proteins, RNA and DNA. Consequently, the antibacterial property of silver ions has been credited to its interaction with thiols, along with other groups such as carboxylates, phosphates, hydroxyls, imidazoles, indoles and amines. Roe and coworkers developed a method to coat plastic catheters with bioactive Ag NPs. The catheters are non-toxic devices capable of targeted and sustained release of bactericidal silver at the implantation site and were useful in preventing infectious complications in patients.[76] Bowler *et al.* assessed the antimicrobial properties of an absorbent topical wound dressing containing ionic silver, samples were tested with a diversity of known burn wound pathogens in a simulated wound fluid model. and verified that the silver-containing dressing is likely to provide a barrier to infection in controlling partial-thickness burns.[72] A relevant factor in the evaluation of biocide action of different materials is their behavior against the bactericidal dose-dependent effect on the different strains of bacteria. Because of the complexity in determining the actual bioavailability of Ag, values of minimum inhibitory concentration of about 100 strains of *Staphylococcus aureus* was reported, with a range 8-80 mg/L when using $AgNO_3$.[77] Similarly, minimum inhibitory concentration values for Ag studied with approximately 100 strains of *Pseudomonas aeruginosa* showed a range of 8 to 70 mg/L. These studies showed that some strains are resistant to Ag, and in fact, resistant strains of bacteria were isolated more than 70 years ago and its mechanism of resistance to Ag has been investigated.[77, 78]

Dutra *et al.* reported the preparation and characterization of the nanocomposite between a poly-{*trans*-[$RuCl_2(vpy)_4$]-styrene-divinylbenzene} or styrene-divinylbenzene-4-vinylpiridine matrix containing Ag NPs.[79] Non-aqueous polymerization using benzoyl peroxide as the initiator was the method used to prepare these materials. The Ag NPs were obtained through the chemical reduction using $NaBH_4$ as reducing agent and sodium citrate as stabilizer.[80] The Ag NPs optical properties were monitored by UV-visible spectroscopy and size and morphology confirmed through TEM. The nanocomposites morphology was characterized by Field Emission Gun Scanning Electron Microscope (FEG-SEM) and the thermal properties were studied.

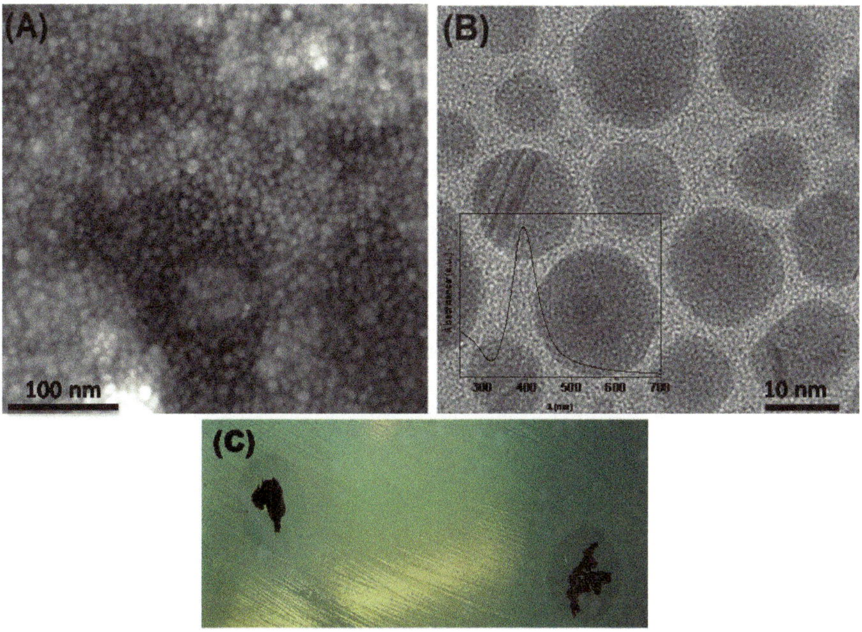

Figure 7. a) FEG-SEM image and b) HR-TEM image of silver nanoparticles. Inset: optical spectrum of AgNPs and c) Photograph of microculture plate containing the microorganism *Staphylococcus aureus* in the presence of poly-{*trans*-[$RuCl_2(vpy)_4$]-styrene–divinylbenzene}/AgNPs nanocomposite.

The antimicrobial action of the nanocomposite was evaluated using various microorganisms: *Staphylococcus aureus* and *Escherichia coli* and was confirmed by the presence of an inhibition halo of the bacterial growth in seeded culture media (Figure 7c). This growth inhibition caused by the presence of silver nanoparticles in the polymer reveals the antiseptic properties of the material. Authors suggested that the absence of ruthenium complex in the polymeric matrix interfered with the release of silver ions into the media, not conferring this material antimicrobial activity. The antimicrobial activity of the silver in the polymeric matrix can also be important for the prevention of proliferation of adherent bacteria in the surface of the polymeric materials, increasing like this the materials' lifetime. Polymeric matrix impregnated with Ag NPs ensures new surface properties that perform in the prevention of microorganism's adherence on the surface and deposits of proteins.

5.3. Magnetic NP doped Ru complex/styrene/divinylbenzene copolymer

Magnetic NPs are of great interest to researchers and cover a wide subject area, including magnetic fluids, catalysts, biotechnology, magnetic resonance imaging, prevention of the environment, among others.[81] Although a reasonable amount of methods have been developed for the synthesis of magnetic NPs with a wide variety of different compositions, the successful application of these magnetic NPs depend on their stability under different conditions. In most applications, the particles properties are enhanced when present in the 10-20 nm size range. However, depending on the application, an inevitable problem associated with the particle size is its intrinsic instability over time has to be regarded. Some smaller particles tend to form agglomerates to reduce the energy associated with large surface area relative to the volume of nanometric particles. For most applications it is crucial to develop strategies to protect the chemical stability of the NPs in relation to the magnetic degradation during and after the synthesis. These strategies include coating with organic species, including surfactants or polymers, or a coating with an inorganic layer such as silica or carbon. It is notable that in many cases, protection of the particles or coating not only stabilizes NPs, but may also be used for additional functionality, for example, with NP or other ligands, depending on the desired application.[81] Several syntheses of Fe, Co, FePt and $Fe_xCo_yPt_{100-x-y}$ NPs have recently been developed to produce particles of controlled size and narrow size distributions. They are currently the focus of scientific research aimed at understanding the formation mechanism of monodisperse NPs and the influence of size and shape in their magnetic behaviour.[82] The literature reports the preparation of polydisperse 2 nm-sized alloy of Co-Pt prepared from organometallic precursors in the presence of a polymer.[83-85]

On the other hand, researchers at Ohio State University reported the development of magnetic NPs that promises rapid detection and elimination of the bacteria *Escherichia coli (E.coli),* anthrax *(Bacillus anthracis)* and other bacteria harmful to human health.[86] These studies have demonstrated the potential of the magnetic NPs studied and their aid in the rapid detection of *E.coli* strains in approximately five minutes, and removing 88% of the bacteria by providing a very attractive field for the decontamination of pathogenic microbes and diagnostic applications.[86] These studies were pioneers in using magnetic NPs to detect, quantify and differentiate *E.coli* cells.

A promising application of magnetic NPs is as drug carriers as well as in drug release, initially proposed by Widder *et al.* in the 70's.[87] The magnetic drug carriers have the potential to carry a large amount of drug to a certain body part, thus avoiding toxicity and adverse effects of excessive use of a large dose of drugs in the body. Although to date, there has been a breakthrough in *in vivo* applications, clinical studies in real situations are still problematic. Many fundamental issues in systems for controlled drug release need to be resolved, such as size-controlled synthesis and stability of the magnetic NPs, the biocompatibility of the coating layers (polymer or silica), links between drug-particle, among other physiological parameters.[81]

The literature describes various synthesis methods of interest to develop new bio-magnetic nanocomposite applications.[56, 88] However, some of these proposed methods suggest a post-synthetic step in order to prevent the liberation of the magnetic NPs. To investigate this authors have proposed an *in situ* synthetic method to incorporate the magnetic NPs during the synthesis of the nanocomposite. This section will present some results obtained in the preparation and characterization of magnetic $CoPt_3$ NPs and *in situ* incorporation of these particles in the poly-{*trans*-[$RuCl_2(vpy)_4$]-sty} polymer matrix. Although these nanocomposites are not to be directly applied in Nanomedicine they may provide useful clues in the formulation of other materials.

The magnetic $CoPt_3$ NPs were characterized by UV-vis spectroscopy (Figure 8-inset).The optical spectrum of the magnetic NPs presents two characteristic bands, at 319 nm and 407 nm. The observation of the UV/vis spectra obtained from the starting materials, ie, salts of Co (II) and Pt (VI) in ethylene glycol showed an intense band at 522 nm for the Co (II) and two minor bands intensity for Pt (VI), 363 nm and 458 nm. However, the optical spectrum obtained for this alloy shows distinct bands of the starting materials, which indicates the presence of magnetic material, $CoPt_3$.[82] The morphology of the $CoPt_3$ NPs was analyzed by SEM and EDX confirmed the presence of metals in the magnetic alloy, and its respective mean ratio, Co:Pt, 1:3 (Figure 8). The presence of spherical agglomerates was observed by coalescence of smaller particles. However, the nanoscale of the $CoPt_3$ NPs was confirmed by TEM. Figure 8(right) shows TEM images obtained from a sample of these magnetic particles.

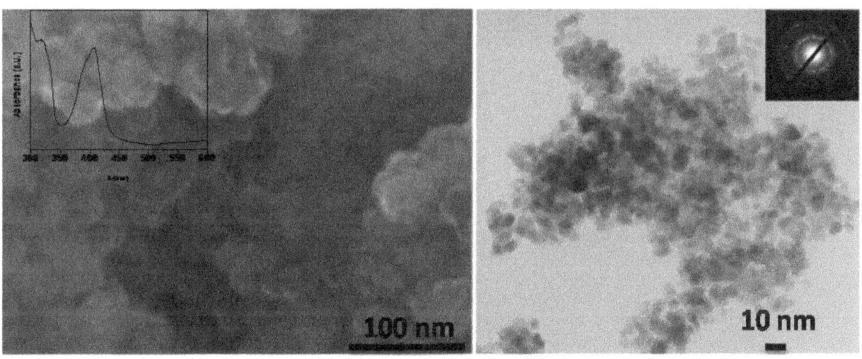

Figure 8. SEM image (left) with inset optical spectrum and TEM image (right) of $CoPt_3$ NPs.

The nanocomposite $CoPt_3$/poly-{*trans*-[$RuCl_2(vpy)_4$]-sty} was synthesized using the *in situ* method with a nominal percentage of 0.5%(w/w) $CoPt_3$ in relation with the styrene monomer and the complex *trans*-[$RuCl_2(vpy)_4$]-sty] and characterized by UV-vis spectroscopy. The optical spectrum of the magnetic nanocomposite present bands in the region of 320 to 410 nm (Figure 9b). A shoulder and band broadening observed in the 450 nm region of the optical spectrum, may suggest the overlap of the bands that characterize the complex *trans*-[$RuCl_2(vpy)_4$] with the magnetic NPs characteristic band. The morphology of the magnetic nanocomposite was analyzed simultaneously by SEM and TEM and the presence of the nanoparticles homogeneously dispersed in the polymeric matrix was observed. However, there are dots and small beads over the whole surface of the nanocomposite examined by SEM, suggesting an agglomeration of magnetic NPs incorporated in the copolymer matrix. The average size of the NPs is in the 10 nm range, showing that the average size of NP was not significantly altered during the synthesis of the nanocomposite. Larger particles are probably caused by coalescence of smaller particles.

EDX mapping was made on the SEM image of the nanocomposite and showed the homogenous presence of the $CoPt_3$ alloy over the entire surface of the polymer matrix.

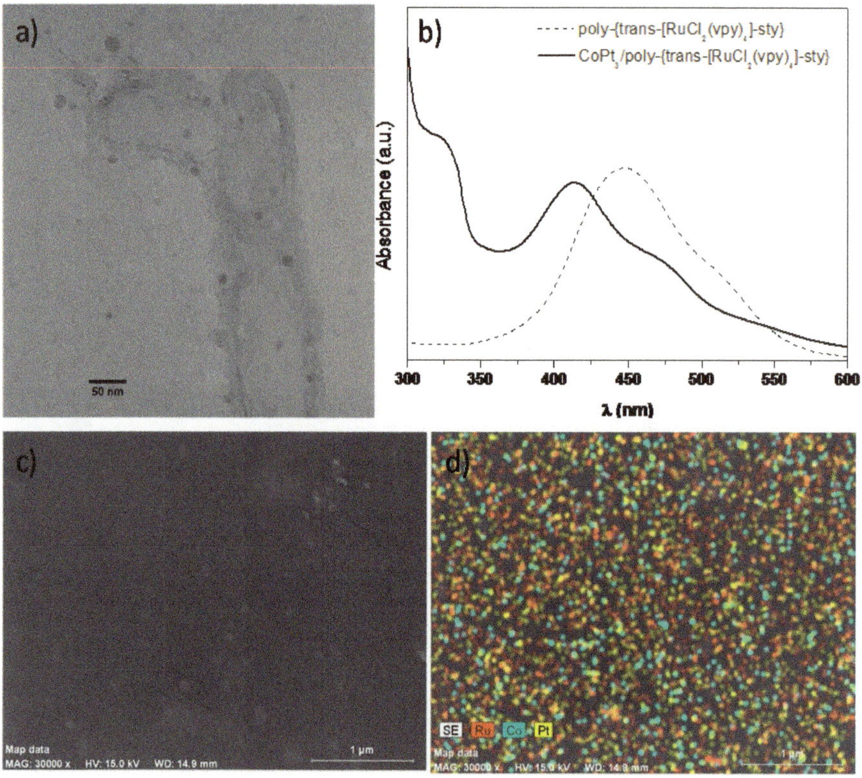

Figure 9. a) TEM image; b) optical spectra; c) SEM image and d) EDS mapping of the nanocomposite $CoPt_3$/poly-*{trans-* [$RuCl_2(vpy)_4$]-sty} 0.5%.

6. Concluding remarks and future trends

In this chapter we have discussed studies involving nanocomposites prepared from the complex *trans*-[$RuCl_2(vpy)_4$], with the monomers styrene, divinylbenzene and 4-vinylpyridine, with the incorporation of functional nanoparticles. To each type of nanoparticles, different properties may be attributed: (1) antimicrobial properties in the presence of Ag NPs, (2) luminescent properties in the presence of ZnO NPs and (3) magnetic properties of the alloy in the presence of $CoPt_3$. The Ag NPs precursors of antibacterial nanocomposites were prepared and characterized. The synthesis of the copolymers were carried out in these emulsions in the presence of complex *trans*-[$RuCl_2(vpy)_4$] monomer and other monomers such sty, DVB and vpy, which will form the polymeric matrices. These nanocomposites were characterized by optical, structural and morphological techniques through UV-visible spectroscopy, SEM and TEM analysis. The antiseptic property of these new nanocomposite materials was confirmed by microbiological testing using the *Gram-positive, S. aureus* (ATCC-25923) and *Gram-negative, E. coli* (ATCC-25922) for Ag NPs impregnated with the polymer. The nanocomposites containing ZnO NPs with luminescent properties incorporated in the copolymers poly-{*trans*-[$RuCl_2(vpy)_4$]-sty showed that the synthesis method using *in situ* promotes better dispersion and a better distribution of the ZnO NPs along the polymeric matrix compared with the nanocomposite prepared by the analogous *ex situ* method. ZnO/poly-{*trans*-[$RuCl_2(vpy)_4$]-sty}nanocomposites prepared were fully characterized at the optical, structural and morphological level, through the techniques of UV-vis spectroscopy, analysis by SEM and TEM. $CoPt_3$ NPs, with superparamagnetic properties, were incorporated into a magnetic poly-{*trans*-[$RuCl_2(vpy)_4$]-sty} nanocomposite prepared and characterized. TEM images show that $CoPt_3$ NPs remained unchanged and are homogeneously embedded through the polymeric matrix of the magnetic nanocomposite The development of highly technological materials with antiseptic, optical and magnetic properties present in ruthenium complexes with biological activity are promising therapeutic applications with a high possibility of success if applied in the medical field and a variety of technological applications.

Author details

Karen Segala
Department of Bioprocesses and Materials Engineering (DEMBio), State University of Campinas, UNICAMP, Brazil

Angela S. Pereira
CICECO, University of Aveiro, Portugal

Acknowledgement

K.S. is so grateful to CNPq and Capes (PNPD 02625/09-1-linha MEC/CAPES) for financially supporting this project and for the research (post-doctoral fellowship), Faculdade de Engenharia Química, FEQ/UNICAMP, Brasil, and LNLS/LNNano (Brazilian Synchrotron Light Laboratory) for HR-TEM and SEM/FEG images. A.S.P acknowledges Fundação para a Ciência e Tecnologia for her post-doctoral fellowship (SFRH/BPD/44398/2008). The authors would also like to acknowledge the financial support from the Universidade de Aveiro and FCT (PTDC/EQU-EQU/098617/2008).

7. References

[1] Hezinger AFE, Tessmar J, and Gopferich A, (2008)Polymer coating of quantum dots - A powerful tool toward diagnostics and sensorics. Eur. J. Pharm. Biopharm. 68: p. 138.

[2] Jorge P, Martins MA, Trindade T, Santos JL, and Farahi F, (2007)Optical fiber sensing using quantum dots. Sensors. 7: p. 3489.

[3] Gao XH, Cui YY, Levenson RM, Chung LWK, and Nie SM, (2004)*In vivo* cancer targeting and imaging with semiconductor quantum dots. Nat. Biotechnol. 22(8): p. 969.

[4] Pellegrino T, Manna L, Kudera S, Liedl T, Koktysh D, Rogach AL, Keller S, Radler J, Natile G, and Parak WJ, (2004)Hydrophobic nanocrystals coated with an amphiphilic polymer shell: A general route to water soluble nanocrystals. Nano Lett. 4(4): p. 703.

[5] Carion O, Mahler B, Pons T, and Dubertret B, (2007)Synthesis, encapsulation, purification and coupling of single quantum dots in phospholipid micelles for their use in cellular and in vivo imaging. Nat. Protoc. 2: p. 2383.

[6] Dubertret B, Skourides P, Norris DJ, Noireaux V, Brivanlou AH, and Libchaber A, (2002)In vivo imaging of quantum dots encapsulated in phospholipid micelles. Science. 298(5599): p. 1759.

[7] Liu SW, Lee CM, Wang SN, and Lu DR, (2006)A new bioimaging carrier for fluorescent quantum dots: Phospholipid nanoemulsion mimicking natural lipoprotein core. Drug Delivery. 13(2): p. 159.

[8] Struck M and Widdecke H, (1996)Surface functionalization of polymer networks: Sulfonation of the internal surface of macroporous styrene-divinylbenzene copolymers. Angew. Makromol. Chem. 235: p. 131.

[9] Williams KA, Boydston AJ, and Bielawski CW, (2007)Main-chain organometallic polymers: synthetic strategies, applications, and perspectives. Chem. Soc. Rev. 36(5): p. 729.

[10] Schubert US and Eschbaumer C, (2002)Macromolecules containing bipyridine and terpyridine metal complexes: towards metallosupramolecular polymers. Angew. Chem., Int. Ed. Engl. 41(16): p. 2892.

[11] Abd-El-Aziz AS, Shipman PO, Boden BN, and McNeil WS, (2010) Synthetic methodologies and properties of organometallic and coordination macromolecules. Prog. Polym. Sci. 35(6): p. 714.

[12] Maria L, Leite M, Costa MAS, Ribeiro JMS, Senna LF, and Silva MR, (2004)Characterization of magnetic microspheres based on network styrene and divinylbenzene copolymers. Mater. Lett. 58(24): p. 3001.

[13] Mahdavian AR and Khoee S, (2002)A facile and efficient method for preparation of chiral supported poly(styrene-divinylbenzene) copolymers. React. Funct. Polym. 50(3): p. 217.

[14] Leon-Gonzalez ME and Perez-Arribas LV, (2000)Chemically modified polymeric sorbents for sample preconcentration. J. Chromatogr., A. 902(1): p. 3.

[15] de Oliveira AJB, de Aguiar AP, de Aguiar M, and Maria LCD, (2005)How to maintain the morphology of styrene-divinylbenzene copolymer beads during the sulfonation reaction. Mater. Lett. 59(8-9): p. 1089.

[16] Franco CV, Marques da Silva Paula M, Goulart G, de Lima LFCP, Noda LK, and Gonçalves NS, (2006)Thermal analysis, Raman spectroscopy and scanning electron microscopy of new polymeric material containing in-chain ruthenium complex: Poly-

{*trans*-[$RuCl_2(vpy)_4$]-co-styrene} and poly-{*trans*-[$RuCl_2(vpy)_4$]-4 vinylpyridine-styrene}. Mater. Lett. 60(21-22): p. 2549.

[17] Kelch S and Rehahn M, (1999)Synthesis and properties in solution of rodlike, 2,2′:6′,2″-terpyridine-based ruthenium(II) coordination polymers. Macromolecules. 32(18): p. 5818.

[18] Asua JM, (2002)Miniemulsion polymerization. Prog. Polym. Sci. 27(7): p. 1283.

[19] Friese VA and Kurth DG, (2009)From coordination complexes to coordination polymers through self-assembly. Current Opinion in Colloid & Interface Science. 14(2): p. 81.

[20] Islam A, Singh SP, and Han L, (2011)Synthesis and Application of New Ruthenium Complexes Containing β-Diketonato Ligands as Sensitizers for Nanocrystalline TiO_2 Solar Cells. International Journal of Photoenergy. 2011.

[21] Moucheron C, (2009)From cisplatin to photoreactive Ru complexes: targeting DNA for biomedical applications. New J. Chem. 33(2): p. 235.

[22] Abel EW, Bennett MA, and Wilkinson G, (1959)Norbornadiene-metal complexes and some related compounds. J. Chem. Soc. p. 3178.

[23] Lewis J, Mabbs FE, and Walton RA, (1967)The preparation, infrared spectra, and magnetic properties of some ruthenium(II), ruthenium(III), and osmium(III) complexes. J. Chem. Soc. A. p. 1366.

[24] Gilbert JD, Rose D, and Wilkinson G, (1970)Preparative use of blue solutions of ruthenium(II): ruthenium-(II) and -(III) complexes with amines, nitriles, phosphines, etc. J. Chem. Soc. A. p. 2765.

[25] Lewis J, Mabbs FE, and Walton RA, (1967)The preparation, infrared spectra, and magnetic properties of some ruthenium(II), ruthenium(III), and osmium(III) complexes. J. Chem. Soc. A.

[26] Sugahara E, Paula MMS, Vencato I, and Franco CV, (1996)Synthesis, crystal structure and properties of tetranuclear copper(I) complex containing 3-(pyrrol-1-ylmethyl)pyridine as ligand. J. Coord. Chem. 39(1): p. 59.

[27] Paula MMS and Franco CV, (1996)Synthesis, properties and an electroactive film of ruthenium(II) complexes with the pyridine derivative ligand: trans- $RuCl_2(pmp)_4$ center dot(pmp=3-(pyrrol-1-ylmethyl)pyridine). J. Coord. Chem. 40(1-2): p. 71.

[28] Franco CV, Prates PB, deMoraes VN, and Paula MMS, (1997)Oxidative formation of electroactive film from polypyridinyl complexes of ruthenium(II)-containing 3-(pyrrol-1-ylmethyl)pyridine. Synth. Met. 90(2): p. 81.

[29] Paula MMS and Franco CV, (1995)Electrochemical studies of the bimetallic complex formed by ruthenium(II) polypyridine and manganese(III) porphyrin. J. Coord. Chem. 36(3): p. 247.

[30] Fricker SP, Slade E, Powell NA, Vaughan OJ, Henderson GR, Murrer BA, Megson IL, Bisland SK, and Flitney FW, (1997)Ruthenium complexes as nitric oxide scavengers: a potential therapeutic approach to nitric oxide-mediated diseases. Br. J. Pharmacol. 122(7): p. 1441.

[31] Paula MMS, de Moraes VN, Mocellin F, and Franco CV, (1998)Electroreductive deposition of polymeric coatings having Ru-II redox centers on Pt, Pd and sintered Fe-(5-10)%Ni electrodes. J. Mater. Chem. 8(9): p. 2049.

[32] Calvert JM, Schmehl RH, Sullivan BP, Facci JS, Meyer TJ, and Murray RW, (1983)Synthetic and mechanistic investigations of the reductive electrochemical polymerization of vinyl-containing complexes of iron(II), ruthenium(II), and osmium(II). Inorg. Chem. 22(15): p. 2151.

[33] Denisevich P, Abruna HD, Leidner CR, Meyer TJ, and Murray RW, (1982)Electropolymerization of vinylpyridine and vinylbipyridine complexes of iron and ruthenium - homopolymers, copolymers, reactive polymers. Inorg. Chem. 21(6): p. 2153.

[34] Bandeira MCE, Prochnow FD, Costa I, and Franco CV, *Corrosion characterization of electrodeposited organometallic films on Nd-Fe-B permanent magnets*, in *Advanced Powder Technology II*, L. Salgado, F.A. Filho, and R. Muccilo, Editors. 2001. p. 673.

[35] Bandeira MCE, Prochnow FD, Noda LK, Goncalves NS, Costa I, de Melo HG, Crayston JA, and Franco CV, (2004)Electroreductive polymerization of *trans*- $RuCl_2(vpy)_4$ on Nd-Fe-B magnets: electrochemical impedance spectroscopy interpretation, Raman spectroscopy, X-ray photoelectron spectroscopy and scanning electron microscopy analysis. J. Solid State Electrochem. 8(4): p. 244.

[36] Sobral AVC, Ristow W, Domenech SC, and Franco CV, (2000)Characterization and corrosion behavior of injection molded 17-4 PH steel electrochemically coated with poly trans-dichloro(4-vinylpyridine)ruthenium. J. Solid State Electrochem. 4(7): p. 417.

[37] Bandeira MCE, Crayston JA, Concalves NS, Noda LK, Glidle A, and Franco CV, (2007)Electropolymerization of *trans*- $RuCl_2(vpy)_4$ complex-EQCM and Raman studies. J. Solid State Electrochem. 11(2): p. 231.

[38] Antony R, Tembe GL, Ravindranathan M, and Ram RN, (2000)Synthesis and catalytic property of poly(styrene-co-divinylbenzene) supported ruthenium(III)-2-aminopyridyl complexes. Eur. Polym. J. 36(8): p. 1579.

[39] Jandrey AC, de Aguiar AP, de Aguiar M, and de santa Maria LC, (2004)Thermodegradation of poly(2-vinylpyridine-co-styrene-co-divinylbenzene) and N-oxide derivatives. Thermochim. Acta. 424(1-2): p. 63.

[40] Creczynski-Pasa TB, Bonetti VR, Beirith A, Ckless K, Konzen M, Seifriz I, Paula MS, Franco CV, Wilhelm D, and Calixto JB, (2001)Complexes trans- $RuCl_2(nic)_4$ and trans-$RuCl_2(i\text{-}nic)_4$ as free radical scavengers. J. Inorg. Biochem. 86(2-3): p. 587.

[41] Meggers E, Atilla-Gokcumen GE, Bregman H, Maksimoska J, Mulcahy SP, Pagano N, and Williams DS, (2007)Exploring chemical space with organometallics: Ruthenium complexes as protein kinase inhibitors. Synlett. (8): p. 1177.

[42] Meggers E, (2009)Targeting proteins with metal complexes. Chem. Commun. (Cambridge, U. K.). (9): p. 1001.

[43] Atilla-Gokcumen GE, Di Costanzo L, and Meggers E, (2011)Structure of anticancer ruthenium half-sandwich complex bound to glycogen synthase kinase 3 beta. J. Biol. Inorg. Chem. 16(1): p. 45.

[44] Wu CH, Wu DH, Liu X, Guoyiqibayi G, Guo DD, Lv G, Wang XM, Yan H, Jiang H, and Lu ZH, (2009)Ligand-Based Neutral Ruthenium(II) Arene Complex: Selective Anticancer Action. Inorg. Chem. 48(6): p. 2352.

[45] Zhang G, Wu CH, Ye HD, Yan H, and Wang XM, (2011)Nanoscaled carborane ruthenium(II)-arene complex inducing lung cancer cells apoptosis. J. Nanobiotechnology. 9.

[46] Sertchook H and Avnir D, (2003)Submicron silica/polystyrene composite particles prepared by a one-step sol-gel process. Chem. Mater. 15(8): p. 1690.

[47] Trindade T, O'Brien P, and Pickett NL, (2001)Nanocrystalline semiconductors: Synthesis, properties, and perspectives. Chem. Mater. 13(11): p. 3843.

[48] Bullions TA, Wei M, Porbeni FE, Gerber MJ, Peet J, Balik M, White JL, and Tonelli AE, (2002)Molecular dynamics in nanostructured polyimide-silica hybrid materials and their thermal stability. J. Polym. Sci., Part B: Polym. Phys. 40(10): p. 1056.

[49] Erdem B, Sudol ED, Dimonie VL, and El-Aasser MS, (2000)Encapsulation of inorganic particles via miniemulsion polymerization. I. Dispersion of titanium dioxide particles in organic media using OLOA 370 as stabilizer. J. Polym. Sci., Part A: Polym. Chem. 38(24): p. 4419.

[50] Erdem B, Sudol ED, Dimonie VL, and El-Aasser MS, (2000)Encapsulation of inorganic particles via miniemulsion polymerization. II. Preparation and characterization of styrene miniemulsion droplets containing TiO_2 particles. J. Polym. Sci., Part A: Polym. Chem. 38(24): p. 4431.

[51] Erdem B, Sudol ED, Dimonie VL, and El-Aasser MS, (2000)Encapsulation of inorganic particles via miniemulsion polymerization. III. Characterization of encapsulation. J. Polym. Sci., Part A: Polym. Chem. 38(24): p. 4441.

[52] Tiarks F, Landfester K, and Antonietti M, (2001)Silica nanoparticles as surfactants and fillers for latexes made by miniemulsion polymerization. Langmuir. 17(19): p. 5775.

[53] Esteves ACC, Barros-Timmons A, Monteiro I, and Trindade T, (2005)Polymer encapsulation of CdE (E = S, Se) quantum dot ensembles via in-situ radical polymerization in miniemulsion. J. Nanosci. Nanotechnol. 5(5): p. 766.

[54] Joumaa N, Lansalot M, Theretz A, and Elaissari A, (2006)Synthesis of quantum dot-tagged submicrometer polystyrene particles by miniemulsion polymerization. Langmuir. 22(4): p. 1810.

[55] Pereira AS, Rauwel P, Reis MS, Silva NJO, Barros-Timmons A, and Trindade T, (2008)Polymer encapsulation effects on the magnetism of EuS nanocrystals. J. Mater. Chem. 18(38): p. 4572.

[56] Martins MA, Neves MC, Esteves ACC, Girginova PI, Guiomar AJ, Amaral VS, and Trindade T, (2007)Biofunctionalized ferromagnetic $CoPt_3$/polymer nanocomposites. Nanotechnology. 18(21).

[57] Fleischhaker F and Zentel R, (2005)Photonic crystals from core-shell colloids with incorporated highly fluorescent quantum dots. Chem. Mater. 17(6): p. 1346.

[58] Fornes TD and Paul DR, (2003)Crystallization behavior of nylon 6 nanocomposites. Polymer. 44(14): p. 3945.

[59] Esteves ACC, Barros-Timmons AM, Martins JA, Zhang W, Cruz-Pinto J, and Trindade T, (2005)Crystallization behaviour of new poly(tetramethyleneterephthalamide) nanocomposites containing SiO_2 fillers with distinct morphologies. Compos Part B - Eng. 36(1): p. 51.

[60] Oriakhi CO, (2000)Polymer Nanocomposition Approach to Advanced Materials. J. Chem. Educ. 77(9): p. 1138.

[61] Rajeshwar K, De Tacconi NR, and Chenthamarakshan CR, (2001)Semiconductor-based composite materials: Preparation, properties, and performance. Chem. Mater. 13(9): p. 2765.

[62] Steigerwald ML and Brus LE, (1990)Semiconductor crystallites: A class of large molecules. Acc. Chem. Res. 23(6): p. 183.

[63] Fleming MS, Mandal TK, and Walt DR, (2001)Nanosphere-microsphere assembly: Methods for core-shell materials preparation. Chem. Mater. 13(6): p. 2210.

[64] Gilman JW, Jackson CL, Morgan AB, Harris R, Manias E, Giannelis EP, Wuthenow M, Hilton D, and Phillips SH, (2000)Flammability properties of polymer - Layered-silicate nanocomposites. Polypropylene and polystyrene nanocomposites. Chem. Mater. 12(7): p. 1866.

[65] Hsiue GH, Liu YL, and Liao HH, (2001)Flame-retardant epoxy resins: An approach from organic-inorganic hybrid nanocomposites. J. Polym. Sci., Part A: Polym. Chem. 39(7): p. 986.

[66] Sanchez C, Julian B, Belleville P, and Popall M, (2005)Applications of hybrid organic-inorganic nanocomposites. J. Mater. Chem. 15(35-36): p. 3559.

[67] Zhang F, Zhang H, and Su Z, (2008)*In situ* preparation of $Mg(OH)_2$/polystyrene nanocomposite via soapless emulsion polymerization. Polym. Bull. (Berlin). 60(2): p. 251.

[68] Segala K, Dutra RL, Franco CV, Pereira AS, and Trindade T, (2010)*In Situ* and *Ex Situ* Preparations of ZnO/Poly-{*trans*- $RuCl_2(vpy)_4$/styrene} Nanocomposites. J. Braz. Chem. Soc. 21(10): p. 1986.

[69] Segala K, Dutra RL, de Oliveira EN, Rossi LM, Matos JR, Paula MMS, and Franco CV, (2006)Characterization of poly-{*trans*- $RuCl_2(vpy)_4$ -styrene-4-vinylpyridine} impregnated with silver nanoparticles in non aqueous medium. J. Braz. Chem. Soc. 17(8): p. 1679.

[70] Brus LE, (1984)Electron-electron and electron-hole interactions in small semiconductor crystallites - The size dependence of the lowest excited electronic state. J. Chem. Phys. 80(9): p. 4403.

[71] Pesika NS, Stebe KJ, and Searson PC, (2003)Determination of the particle size distribution of quantum nanocrystals from absorbance spectra. Adv. Mater. 15(15): p. 1289.

[72] Bowler PG, (2004)Microbicidal Properties of a Silver-Containing Hydrofiber" Dressing Against a Variety of Burn Wound Pathogens. J. Burn Care Rehabil. 25(2): p. 192.

[73] Uygur F, Oncul O, Evinc R, Diktas H, Acar A, and Ulkuer E, (2009)Effects of three different topical antibacterial dressings on Acinetobacter baumannii-contaminated full-thickness burns in rats. Burns. 35(2): p. 270.

[74] Lalueza P, Monzon M, Arruebo M, and Santamaria J, (2011)Bactericidal effects of different silver-containing materials. Mater. Res. Bull. 46(11): p. 2070.

[75] Kong H and Jang J, (2008)Antibacterial properties of novel poly(methyl methacrylate) nanofiber containing silver nanoparticles. Langmuir. 24(5): p. 2051.

[76] Roe D, Karandikar B, Bonn-Savage N, Gibbins B, and Roullet J-B, (2008)Antimicrobial surface functionalization of plastic catheters by silver nanoparticles. J. Antimicrob. Chemother. 61(4): p. 869.

[77] Chopra I, (2007)The increasing use of silver-based products as antimicrobial agents: a useful development or a cause for concern? J. Antimicrob. Chemother. 59(4): p. 587.

[78] Franke S, Grass G, and Nies DH, (2001)The product of the ybdE gene of the Escherichia coli chromosome is involved in detoxification of silver ions. Microbiol. 147: p. 965.

[79] Dutra RL, Segala K, de Oliveira EMN, de Souza EP, Rossi LM, Matos JR, Noda LK, Paula MMS, and Franco CV, (2008)Preparation and characterization of the novels terpolymers of poly-{*trans*- $RuCl_2(vpy)_4$-styrene-divinylbenzene} and styrene-divinylbenzene-vinylpiridine impregnated with silver nanoparticles. Polym. Bull. (Berlin). 60(6): p. 809.

[80] Creighton JA, Blatchford CG, and Albrecht MG, (1979)Plasma resonance enhancement of raman-scattering by pyridine adsorbed on silver or gold sol particles of size comparable to the excitation wavelength. J. Chem. Soc., Faraday Trans. 75: p. 790.

[81] Lu AH, Salabas EL, and Schuth F, (2007)Magnetic nanoparticles: Synthesis, protection, functionalization, and application. Angew. Chem., Int. Ed. 46(8): p. 1222.

[82] Shevchenko EV, Talapin DV, Rogach AL, Kornowski A, Haase M, and Weller H, (2002)Colloidal synthesis and self-assembly of $COPt_3$ nanocrystals. J. Am. Chem. Soc. 124(38): p. 11480.

[83] Ely TO, Pan C, Amiens C, Chaudret B, Dassenoy F, Lecante P, Casanove MJ, Mosset A, Respaud M, and Broto JM, (2000)Nanoscale Bimetallic Co_xPt_{1-x} Particles Dispersed in Poly(vinylpyrrolidone): Synthesis from Organometallic Precursors and Characterization. J. Phys. Chem. B. 104(4): p. 695.

[84] Kim JH, Kim J, Baek KH, Im DH, Kim CK, and Yoon CS, (2007)Fabrication of CoPt nanoparticles with high coercivity on a polymer film. Colloids Surf., A. 301(1-3): p. 419.

[85] Frommen C, Malik S, Wurfel JU, Rosner H, and Didschies C, (2004)Synthesis and magnetic properties of $CoPt_3$ nanoparticle assemblies containing copper. Mater. Lett. 58(6): p. 953.

[86] El-Boubbou K, Gruden C, and Huang X, (2007)Magnetic glyco-nanoparticles: A unique tool for rapid pathogen detection, decontamination, and strain differentiation. J. Am. Chem. Soc. 129(44): p. 13392.
[87] Widder KJ, Senyei AE, Ovadia H, and Paterson PY, (1979)Magnetic protein-A microspheres - Rapid method for cell-separation. Clin. Immunol. Immunopathol. 14(3): p. 395.
[88] Fievet F, Lagier JP, Blin B, Beaudoin B, and Figlarz M, (1989)Homogeneous and heterogeneous nucleations in the polyol process for the preparation of micron and sub-micron size metal particles. Solid State Ionics. 32-3: p. 198.

Chapter 8

Oxidative Polymerization of Aniline: Molecular Synthesis of Polyaniline and the Formation of Supramolecular Structures

I.Yu. Sapurina and M.A. Shishov

Additional information is available at the end of the chapter

1. Introduction

The growing interest in nanostructured conducting polymers is caused by expanding area of their practical application. Conducting polymers can be used in new electronic devices (transistors, displays, sensors, energy-storage and memory cells), materials for shielding electromagnetic irradiation, polymer nanolithography, inhibition of corrosion, membrane constructing, catalysis and medicine. Miniaturization of devices, increasing their efficiency and lowering cost prices require development of new materials. The studies of polyaniline (PANI) and other conducting polymers (polyacetylene, polythiophene, polypyrrole, poly-*p*-(phenylenevinylene) have shown that they possess semiconducting properties. These polymers can exist in different oxidation states and, in common with inorganic semiconductors, respond to external influences by changing some characteristics (conductivity, color, density, magnetic properties, hydrophilicity or hydrophobicity, permeability to gases and liquids). Therefore, conducting polymers (which are sometimes called "smart polymers") may serve as an alternative to inorganic semiconductors (Trivedi, 1997). In 2000, the scientists who have discovered and studied conducting polymers were awarded Nobel Prize in chemistry.

Due to its high stability and unique complex of properties, PANI was the first among conducting polymers to be used in practice as an antistatic coating, electrode material for batteries and condensers, as a corrosion inhibitor and detecting material for sensors. Polyaniline possesses controlled conductivity within the 10^{-10} – 10^{1} $S{\cdot}cm^{-1}$ range combined with ionic and proton conductivity, redox activity, electro- and solvatochromism, non-linear optical properties and paramagnetism. In addition, the polymer is nontoxic, stable in aggressive chemical environments, has high thermal stability and low manufacturing cost.

PANI properties are determined by the regular structure of polymer chains. The polymer consists almost entirely of *para*-substituted monomer units coupled in the "head-to-tail" manner. The formation of organized supramolecular structures is also of great importance. Both these factors are responsible for the existence of elongated polyconjugated system and high conductivity of macroscopic sample (Skotheim & Reynolds, 2007).

A great number of publications are devoted to the synthesis and study of nanostructured PANI forms (Zhang & Wang, 2006; Stejskal et al. 2010; Wan, 2009; Tran et al. 2009). As a rule, nanostructured materials possess high homogeneity, unique surface properties and high charge carrier mobility. Due to the development of nanotechnology, organized supramolecular structures are considered, first of all, as materials for molecular electronics. Nanostructured forms of PANI show a striking diversity. The polymer chains can form one-dimensional structures (nanofibers, nanorods and nanotubes), planar two-dimensional objects (e.g., the so-called, ribbons, nanobelts and nanoplates) and three-dimensional particles (microspheres, nanospheres and granules). These main architectural elements may serve as a base for more complicated hierarchical formations such as flowers, urchins, branches, corals, "micromats", complex geometric figures.

Polyaniline is synthesized by oxidative polymerization of aniline (Higashimura & Kobayashi, 2004). All the above-listed supramolecular structures are also formed in the process of oxidative polymerization. PANI synthesis includes two interrelated processes. In the course of polymerization, monomer undergoes a chain reaction with the formation of regular macromolecules, and growing chains are simultaneously organized into complex supramolecular structures. As a result, conducting polymer containing stable supramolecular structures with various morphologies is formed. Most of these structures cannot be dissolved or melted without destruction of polymer chains and changing properties of the polymer. Therefore, synthesis is the only unique instrument allowing us to control structure of polymer chains and obtain nanostructured forms of the conducting polymer and its composites with other materials. In this connection, it is very important to develop methods of controlled synthesis of PANI giving polymers with predetermined properties and supramolecular structure. Experimentalists should have the capability of making deliberate choice of aniline polymerization parameters based on understanding the molecular mechanism of chain assembly and self-organization.

The present publication is devoted to developing concepts of mechanism of PANI synthesis and ways of formation of supramolecular structures. The authors attempt to answer the following fundamental questions:

- What is responsible for regular assembly of PANI chains during oxidative polymerization of aniline?
- Which synthesis parameters and why have the most pronounced influence on properties and morphology of polymerization products?
- In which manner is the assembly of all the types of PANI supramolecular structures realized and how does this process correlate with molecular stages of the synthesis?

- What are the experimental conditions for obtaining the main types of supramolecular structures: one-, two- and three-dimensional ones?
- How are supramolecular PANI particles organized and in which manner is the macroscopic structure of conducting polymer (possessing the properties of disorganized metal) formed?

First we should give a brief description of conducting PANI, structure of polymer chains, forms and properties of the polymer.

2. Polyaniline: Structure, forms and their properties

The chains of conducting PANI have ordered structure; they contain regularly alternating phenyl rings and nitrogen-containing groups. This structure provides for polyconjugation: polymer chain forms a zigzag lying in one plane, and π-electron clouds overlap above and below this plane. The lone electron pair of nitrogen performs the same function as π-electrons and assures polyconjugation. Polyconjugated system is a transport path providing charge carrier mobility. It is formed as a result of strictly regular assembly of monomer units. The chain of conducting PANI contains more than 95% of *para*-substituted aniline fragments linked in a "head-to-tail" configuration (Hagiwara et al. 1987). Defects in this regular structure arising during copolymerization of aniline with other monomers or after introduction of aniline units with another configuration (*ortho-* or *meta*-substituted fragments) lead to dramatic decrease in conductivity.

Charge carriers are formed in polymer during its oxidation. Nitrogen atoms of PANI serve as oxidation centers. During oxidation, i.e., removal of electron, positive polaron is generated in the chain; it deforms chain structure considerably. Content of oxidized nitrogen atoms in the polymer can change from zero (in its reduced form, leucoemeraldine) to nearly one (in the fully oxidized state, pernigraniline), see Scheme 1. The most stable form of PANI is emeraldine, in which every second nitrogen atom is oxidized, and polymer chain contains equal number of oxidized and reduced units. In the absence of external potential, the fully reduced and fully oxidized forms of the polymer simultaneously transform into this oxidation state. In the case of leucoemeraldine, the transformation into emeraldine takes place at the expense of slow oxidation of amino groups by air oxygen. This reaction is reversible. Pernigraniline also tends to lower its oxidation state and turn into emeraldine. When the degree of oxidation of PANI exceeds 0.7, the polymer becomes unstable, and irreversible transformations of macromolecules begin. Crosslinking and intrachain oxidative cyclization are the most common processes. Destruction of chains and formation of quinones accelerated in the presence of electron donor agents are also possible.

In all the conducting polymers, charge carriers are formed under the action of oxidizing or reducing agent. Charge carriers are located in the main chain and compensated by counterions. Removal of counterion results in changes in the oxidation state of polymer and disappearance of polaron. PANI demonstrates a special feature: removal of counterion does not necessary changes oxidation state of polymer chain. After withdrawal of counterion, the benzene ring adjacent to nitrogen atom "accepts" a fraction of positive charge. With that, benzoid structure is transformed into quinoid one, the ring leaves polyconjugation plane, and, therefore, conductivity of the chain is disturbed. Although the oxidation state remains unchanged, PANI loses its conductivity due to decrease in polyconjugation chain length.

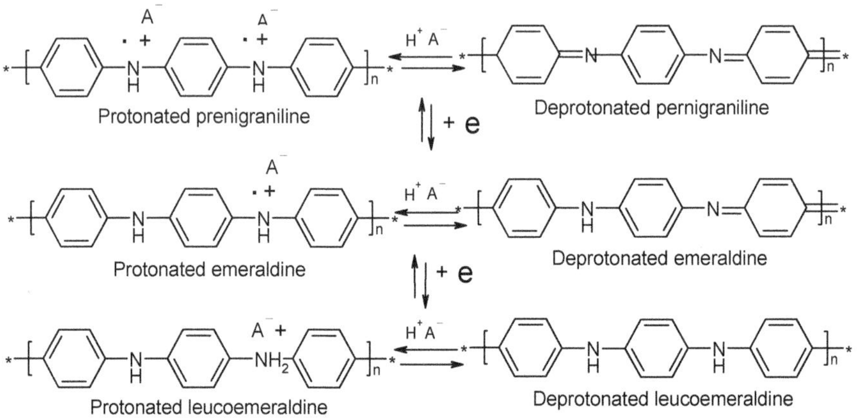

Scheme 1. Polyaniline forms

The best charge stabilizing agents for PANI are strong acids. Acids are sometimes called "doping agents", because they "introduce" charge carriers into polymer chain and impart conductivity to the polymer. However, this term is not correct, since acid molecule does not create positive polaron, and only stabilizes the polaron generated during oxidation. Another term, "protonation", does not adequately describe interaction between PANI and acid either. It was introduced from ammonia and amine chemistry. Protonation of amines by acids means binding a non-oxidized nitrogen atom with proton, with the formed ammonium cation being stabilized by acid anion. The interaction between acid and oxidized nitrogen atom differs from that between acid and neutral nitrogen. As was shown, PANI polarons are stabilized rather by acid anion than by proton. At the same time, proton remains relatively free (Colomban & Tomkinson, 1997), thus providing high proton conductivity of PANI (which is absent in ammonia salts and protonated amines). However, in this paper we will use the common term "protonation" to describe interaction between acids and nitrogen-containing polymer structures.

Among conducting polymers, PANI possesses the greatest number of revealed and characterized forms with different properties (Skotheim & Reynolds, 2007). Due to interaction with acids, each of three protonated states has a corresponding deprotonated form with low conductivity. Thus, polymer can exist in a minimum of six forms differing in both degree of oxidation and protonation state (Scheme 1). PANI is a unique polymer, since its conducting forms (emeraldine and pernigraniline) can be transformed into non-conducting states in two different ways. The first approach consists in introducing electrons into PANI and reduction of nitrogen atoms; the second method is removal of polaron-stabilizing acid which leads to transformation of polymer chain structure and disappearance of polyconjugation. In other words, we can destroy charge carriers or their "paths". In both cases, polymer conductivity decreases by eight or ten orders of magnitude.

Conductivity, optical, magnetic and other properties of PANI may change depending on its oxidation state and degree of protonation. The reduced polymer which does not contain charge carriers possesses a conductivity of 10^{-8} – 10^{-10} S·cm^{-1}; it is weakly colored and diamagnetic. As oxidation of protonated PANI proceeds within the potential window of 0 to +1 V (*versus* the reference hydrogen electrode (RHE)), concentration of polarons in the polymer increases. At the same time, electrical and ionic conductivity, intensity of its absorption bands in the visible and near IR regions also increase, and the polymer becomes paramagnetic. The highest electric conductivity is inherent to the most stable PANI form – emeraldine. Under usual conditions in solid state, emeraldine has an electronic conductivity of 10^{0} to 10^{1} S·cm^{-1}. However, samples with conductivity of the order of 10^{2} S·cm^{-1} were obtained. The charge carrier mobility in the highly conductive emeraldine ranges from 10^{-3} to 10^{-1} cm^{2}·V^{-1}·s^{-1} (Harima et al., 2001). Protonated emeraldine also possesses the highest proton conductivity (up to 10^{-2} S·cm^{-1}). Thus, oxidized and protonated PANI demonstrates a mixed electron and proton conductivity (Colomban & Tomkinson, 1997). When the degree of oxidation exceeds a certain optimum value, PANI conductivity lowers. Conductivity of protonated pernigraniline is less than 10^{-2} S·cm^{-1}. The absorption band of polaron in pernigraniline form is shifted to shorter wavelength as compared with that of emeraldine, indicating considerably higher localization of polaron.

Deprotonation of PANI is performed by neutralization of acid with the aid of base. Acid molecules are bound with PANI main chain at the expense of reversible ionic interactions. While protonation-deprotonation cycles continue, the acid is alternately absorbed and expelled from the polymer matrix until the new equilibrium between changed medium and modified polymer is established. Deprotonation may lower emeraldine conductivity by 10 orders of magnitude. Polymer becomes dielectric, its paramagnetism and density decrease, polaron absorption band becomes less intense and shifts to the short-wavelength region. Since these changes are reversible, reprotonation (i.e., PANI-acid interaction following deprotonation) leads to recovery of conductivity and attendant properties of the polymer (Skotheim & Reynolds, 2007).

3. Oxidative polymerization

In a vast majority of cases, PANI is synthesized by oxidative polymerization of aniline. Other methods, e.g., polycondensation of aniline derivatives with reactive functional groups (Y-An-X = $(\text{-An-})_n$ + XY), are used very seldom. Oxidative polymerization is used for the synthesis of polymeric (oligomeric) products from various classes of monomers (aromatic amines, phenols, thiophenols, aromatic hydrocarbons and heterocycles (Higashimura & Kobayashi, 2004). The monomers used in oxidative polymerization are characterized by pronounced electron donor properties and high oxidation tendency. These properties, in particular, are inherent to aromatic amines, phenols and thiophenols or sulphur- and nitrogen-containing heterocycles due to the presence of electron donor substituent in benzene or heterocyclic ring. Oxidation of monomer takes place under the action of inorganic (or organic) oxidizing agent or the applied potential. During this process, cation or cation radical sites are generated in monomer (polymer) molecule, thus initiating polymer growth.

Technically, oxidative polymerization may be considered as formation of covalent bond between monomer molecules at the expense of abstracting two protons. There are many types of linkage between monomer units. For example, in the case of aniline, "head-to-head", "tail-to-tail" and "head-to-tail" configurations are possible. In addition, in the last two cases chain assembly may occur due to substitution in phenyl ring with the formation of *ortho-*, *para-* and *meta-* monomer units. Thus, oxidative polymerization yields chains with a wide variety of monomer unit structures.

3.1. The chain mechanism of aniline polymerization

Oxidative polymerization is often considered as a kind of polycondensation, since chain growth is accompanied by the formation of low-molecular products. However, it is not always true. The formation of chain may proceed in two ways. The first one is recombination of cation radical oxidation sites. In this case, polymer growth process is classed as polycondensation, since fragments of any length may recombine. The second way of chain growth belongs to electrophilic substitution; in the case of aniline, oxidized nitrogen-containing structure attacks phenyl ring of another aniline molecule and substitutes one proton of the ring.

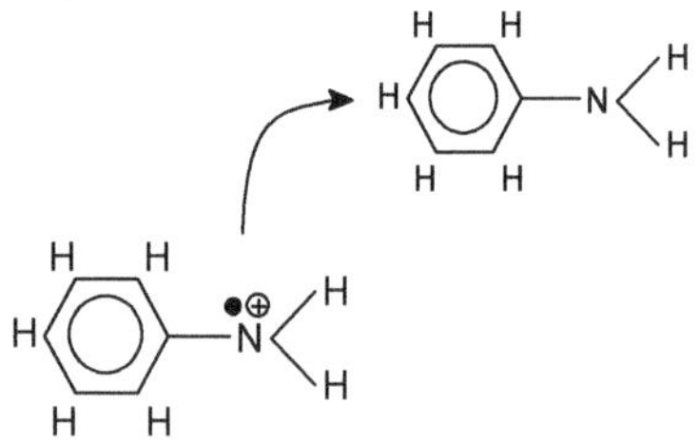

Scheme 2. Electrophilic substitution reaction

Both the ring and nitrogen-containing structure lose one proton; after that, monomer units bind with each other, and the chain becomes longer.

In the case of electrophilic substitution, both variants (polycondensation and chain-growth processes) are possible. To determine the type of addition, we should measure molecular mass of the polymer throughout the reaction. In the course of polycondensation, fragments with any molecular mass can react with each other; therefore, average molecular mass of the product increases slowly, monomer is rapidly transformed into low molecular weight oligomers coexisting with the high molecular weight fraction up to high conversion. Different molecular weight distribution is observed during chain-growth processes. In the course of chain-growth polymerization, monomer units are sequentially added to the polymer chain bearing active end group. Here, during oxidation monomer coexists with high molecular weight fraction of the product, and the content of low molecular weight oligomers is minimal. Oxidative polymerization of several monomers containing heteroatoms (including aniline) was found to proceed according to the chain-growth mechanism.

Y.Wei (Wei, 2001) singled out the polymerization of aniline into a special type of reactivation chain process. The chain growth involves repeated acts of activation/deactivation of the polymeric structure:

$$An_m \xrightarrow[\text{Activation}]{Ox} An_m^* \xrightarrow[\text{Deactivation}]{An} An_{m+1}$$

The dormant polymer chain (An_m) is activated by an oxidant (to form An_m^*), then it adds aniline molecule to yield a new dormant chain of higher molecular mass (An_{m+1}). In contrast with other types of chain-growth polymerization (e.g., radical process, where oxidant takes part only in the formation of initial radical), oxidative polymerization requires large amounts of oxidizing agent. Here, oxidant is spent in every step of monomer addition, and molar concentration of oxidant should be comparable with monomer concentration. Oxidant "works" throughout the polymerization and takes part in redox interactions until the last monomer molecule is added to polymer chain.

3.2. Chain growth according to electrophilic substitution mechanism

The process occurring during the formation of polyconjugated PANI chains belongs to electrophilic substitution reactions, since the attacking species is oxidized and acts as an electrophilic agent. Let us consider the most probable structure of monomer unit during electrophilic substitution process. Due to the presence of electron donor (heteroatom) in the monomer, this heteroatom will most likely be oxidized, and monomer units will be linked in the "head-to-tail" fashion. However, even when this type of linkage is repeated, PANI structure may be heterogeneous. Monomer units with *ortho-*, *para-* and *meta-*structure may be formed as a result of attacks of the electrophilic agent directed to different atoms of phenyl ring. Contents of different types of units in the polymer should vary. The probability of the formation of *meta*-structures is low, since aniline possesses electron donor substituent in its phenyl ring. Donor substituent creates excess negative charge on *ortho-* and *para-* carbon atoms of phenyl ring, and electrophilic attack is thus directed to *ortho-* and *para*-atoms. Therefore, according to the existing theory of organic reactions, in the case of aniline oxidation, the most probable monomer structures are *ortho-* and *para*-units linked in the "head-to-tail" fashion. The *ortho*-units should prevail, since phenyl ring contains two vacant *ortho*-positions (Cram & Hammond, 1964). The probability of the formation of other structures also exists, and we should expect the formation of irregular PANI chains.

Thus, from general theoretical considerations, the formation of regular PANI structure containing only *para*-substituted monomer units linked in the "head-to-tail" mode cannot be explained.

3.3. The active site of polymerization

The question of the nature of the terminal active group (active site) responsible for the polymerization of aniline was disputed for a long time (Percec V and Hill, 1996). This was assumed to be nitrenium cation or radical cation. Studies on the aniline oxidation in the presence of "traps" for both types of structures showed that alkyl-substituted phenols (2,6-di-*tert*-butyl-4-methylphenol) and electron-enriched alkenes (2,3-dimethoxybuta-1,3-diene) inhibit the polymerization by acting as scavengers of radical cations. At the same time, PANI is readily formed in the presence of electron-enriched arenes (1,3- and 1,4-dimethoxybenzene) well known as "traps" of nitrenium cations. This fact proves that oxidation of aniline proceeds via a radical cation site (Ding at al., 1999).

The activated terminal group of the chain is generated by an oxidant. In the initial step, it is the monomer that is oxidized; however, upon formation of oligomers, these are terminal amino groups of the oligomer (polymer) that are oxidized because they have lower oxidation potential. In the course of polymerization, the polymer chain performs the function of a mediator in the oxidation of monomer. It oscillates between the highest and intermediate oxidation states, i.e., pernigraniline and emeraldine. The active site formed at the end of the chain attacks the monomer molecule. This attack is directed at a hydrogen atom in the aromatic ring and results in replacing hydrogen by a polymeric fragment. The chain propagates due to the addition of new monomer units.

3.4. Oxidizing agents

Polymerization of aniline and synthesis of its conducting polymer may be performed by electrochemical or chemical methods. During electrochemical synthesis of PANI, electrode potential is increased (if only once) so that it exceeds +1.05 V (RHE) (Yang et al. 2007). A wide range of oxidants is used in the chemical synthesis of PANI. As a rule, compounds with high oxidation potentials exceeding +1.0 V (persulfates, dichromates, cerium (IV) salts, aurates etc.) are employed. This is associated with the fact that the onset of the propagation of the polymer chains in acidic media requires overcoming energy barrier corresponding to an oxidation potential of +1.05 V. Once the propagation starts, the oxidation potential of the reaction decreases. In the chemical synthesis of polyaniline, persulfates (having an oxidation potential of +2.01 V) are used most widely. However, experiments showed that aniline is also oxidized under the action of weak oxidants with a potential close to +1 V or even lower (Sapurina & Stejskal, 2012). Such agents can oxidize aniline not only in basic and neutral media, but also, strange as it may seem, in acidic media where potential barrier of oxidation is high. However, the use of weak oxidants causes some problems because it does not necessary lead to the formation of conducting products.

3.5. Kinetic features of polymerization

Studies of aniline polymerization in acidic aqueous media (where conducting PANI is formed) have shown that the oxidation of monomer proceeds non-monotonically. It starts with a slow process ("induction period"); during this period, aniline oligomers are formed. The induction period is followed by the rapid exothermic step of polymer chain propagation. The following kinetic dependences were suggested for description of this two-step process (Tzou & Gregory, 1992):

$$-d[An]/dt = k_1[An][Ox] + k_2\,\sigma[An]\,P$$

where

[An] is the molar concentration of aniline;

[Ox] is the molar concentration of the oxidant;

P is the surface of the reaction medium interface;

σ is the surface factor;

k_1 is the rate constant for the initial step (induction period) of polymerization;

k_2 is the rate constant for the polymer chain propagation.

Both the first and, presumably, the second steps are first-order reactions with respect to monomer. For the initial step, the reaction order with respect to the oxidant is also the first one. In the propagation step, monomer is oxidized with pernigraniline form of PANI rather than with the original oxidant. Polyaniline is insoluble, and the reaction is heterogeneous; hence, the P parameter and the σ factor are introduced into the equation in order to characterize the interface on which the reaction proceeds. The rate constant for the chain propagation (k_2) is three orders of magnitude higher than the rate constant for the initial step of polymerization (k_1). In other words, the formation of the polymer leads to impressive acceleration of oxidative polymerization. Similar acceleration was also observed upon the introduction of PANI "seeds" into the reaction medium. This process is called "autoacceleration" (Tzou & Gregory, 1992).

Polymerization rate is sensitive to the presence of different inert and insoluble materials in the reaction medium, for example, alumina oxide, silica gel, carbon, cellulose, synthetic polymers, etc. (Stejskal & Sapurina, 2005). This effect is stronger for materials with greater surface areas (P). The strong effect of the medium acidity exerted on the polymerization rate was revealed. This dependence is complex in nature and varies with the pH range. Under conditions of high acidity (for $pH<1$), direct proportionality between molar concentration of the strong acid and k_2 was observed. At the same time, the increased acidity has a little effect on the rate constant of the initial step; however, the induction period shortens substantially with decreasing pH.

3.6. Modern concepts of the mechanism of aniline polymerization

Presently, electrochemical and chemical synthesis of conducting PANI by oxidative polymerization is usually described by the following scheme (Scheme 3) (Wei at al., 1989). It involves: 1) oxidation of nitrogen atom of monomer followed by oxidation of end nitrogen atom of oligomer and polymer; 2) addition of monomer in the "nitrogen-carbon" fashion as a result of chain reaction (electrophilic substitution of proton in aromatic ring of monomer by oxidized polymer fragment).

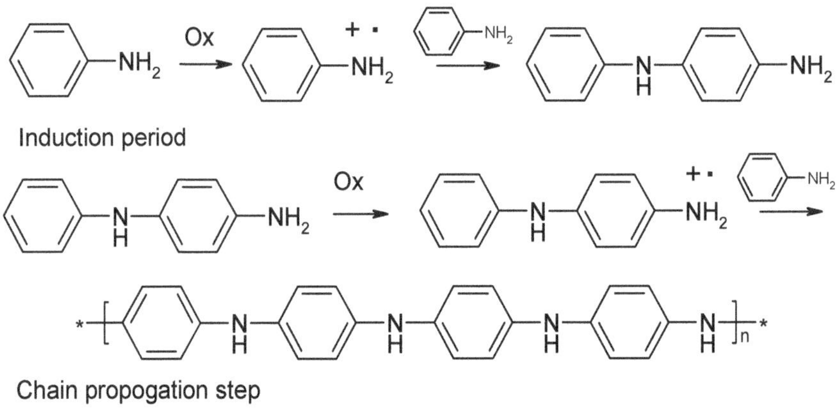

Scheme 3. Aniline oxidative polymerization

Presumably, in the induction period, aniline radical cations are formed; this process is followed by their recombination (according to the electrophilic substitution mechanism) to afford a dimer, namely, N-phenylphenylene-1,4-diamine (*p*-semidine). The subsequent propagation step is assumed to be similar to the electrophilic substitution process where the oxidized terminal amino group of oligomer (polymer) attacks *para*-position in monomer. It is assumed that the reaction involves monomer in its most reactive deprotonated form. The oxidation rate-limiting step is considered to be either the formation of aniline radical cations or their dimerization. The experimental observation that the introduction of small amounts of the aniline dimer (*p*-semidine) sharply accelerates polymerization can be considered as a substantiation of the proposed mechanism. The rationale is that the introduction of *p*-semidine (intermediate formed in the initial limiting step) eliminates this limiting process so that the polymerization begins with the rapid step of chain propagation.

The presented scheme of PANI synthesis is practically similar to the general mechanism of oxidative polymerization suggested in the 1950s for phenol polymerization and, subsequently, for a vast majority of monomers (Higashimura & Kobayashi, 2004). However, PANI synthesis seems anomalous compared with reactions of synthesis of dielectric polymers. Polyphenyleneoxides obtained by oxidation of phenols in aqueous basic or organic media are comparatively low molecular weight products with molecular masses of 20 000 - 30 000. The structure of polyphenyleneoxide monomer units is heterogeneous. The polymer contains units linked in a "tail-to-tail", "head-to-head", "head-to-tail" fashion; both *ortho*- and *para*- substituted aromatic rings are present. Polyphenyleneoxides with "head-to-tail" chains and high content of *para*-substituted monomer units possess better mechanical properties, higher degree of crystallinity and chemical stability than irregular polymers. To increase degree of homogeneity of polymer structure, oxidative polymerization of phenol should be carried out under special conditions: 1) use of "soft" organic oxidants (peroxidases or laccase in combination with peroxides or oxygen); 2) use of monomers with blocked *ortho*-positions (2,6-dialkylsubstituted phenols); 3) low concentrations of reactants and low temperatures. Only in this case it is possible to obtain more regular polyphenylene oxide containing a large amount of *para*-substituted units. However, polymers with high molecular weight still cannot be synthesized (Higashimura & Kobayashi, 2004).

By contrast, oxidative polymerization of aniline in acidic medium yields polymers with strictly regular structure. The studies of molecular structure of PANI chains by Raman, ESR, electron and X-ray photoelectron spectroscopy, chromatography and chemical analysis have shown that the polymer contains 95-98% of *para*-substituted monomer units linked in the "head-to-tail" fashion (Trivedi, 1997). Polyaniline contains only a small amount of "foreign" units (phenazine rings). The appearance of these rings is explained by chain branching and intermolecular crosslinking occurring during synthesis or after termination (do Nascimento, 2006). It is surprising that the synthesis of this highly regular polymer can be initiated by a wide range of oxidizing agents (from persulfates possessing high oxidation potential of +2.01 V to iron (III) with a potential of + 0.7 V). There is no need to use aniline with blocked *ortho*-positions. Moreover, polymerization occurs at a high rate; and neither high concentrations of reactants nor high temperatures lead to the formation of irregular chains. The high selectivity of oxidative polymerization of two other heterocyclic monomers (thiophene and pyrrole) should be noted; their polymers (polythiophene and polypyrrole) are also conductive. The peculiar features of synthesis of these polymers have much in common with PANI synthesis.

However, within the framework of the existing concepts of oxidative polymerization of aniline (Scheme 3), we cannot explain several important features of PANI synthesis.

1. It is unclear why oxidative polymerization of aniline is highly selective, and only *para*-structure of monomer unit is repeated during chain propagation.
2. No explanation is given for the non-monotonic course of oxidation and the great difference between rate constants of the first and second steps, although the scheme represents them as reactions of the same type leading to the formation of similar structures.
3. The scheme fails to explain the following changes in the oxidation potential in the course of polymerization: its increase up to +1.05 V followed by its drop down to +0.7 V - 0.8 V in the chain propagation step.
4. The pH of reaction medium having a great influence on polymerization is not taken into account.
5. The scheme does not explain "autoacceleration", i.e., heterogeneous process of chain growth, and influence of different types of surfaces on polymerization kinetics.

As a rule, in discussion of oxidative polymerization of aniline, it is stated that the existing concepts are controversial, and molecular mechanism of the process is still unclear. In what follows the authors will attempt to explain anomalies of aniline polymerization and, in the first place, the reasons for high selectivity of this process in acidic medium.

4. Influence of pH on aniline polymerization, properties of products and their supramolecular structure

Oxidative polymerization of aniline can occur within a very wide interval of acidity (from 1M base solution to 2M sulfuric acid solution), i.e. in the pH range 14 to -1. However, it is known from the experiment that oxidation products formed in different pH intervals possess radically different properties. Polymerization in basic, neutral and weakly acidic media gives brown powder with low conductivity. The synthesis in highly acidic medium yields highly conductive PANI, since only under these conditions conducting dark-green emeraldine can be obtained. Depending on pH of reaction media, the conductivity of protonated products of aniline oxidation may vary from 10^{-10} to 10^{2} $S{\cdot}cm^{-1}$. Cyclic voltammograms may contain from one to four pairs of waves; optical and magnetic properties of material may also vary. Oxidation products either dissolve in different organic solvents or are insoluble at all. However, all these materials obtained by aniline oxidation and demonstrating various properties are called "polyaniline". This general name creates considerable difficulties during comparison of experimental results, particularly, in discussion of PANI applications (Stejskal et al. 2010).

The composition and structure of non-conducting materials obtained at high pH are little studied. However, they have many potential applications in development of membranes for gas separation and pervaporation, as antioxidants, in catalysis, sensor devices and treatment of heavy metal-contaminated runoff water. In addition, some products synthesized in neutral and weakly acidic media demonstrate very complex and interesting supramolecular structure, and are of great scientific interest (Stejskal & Trchova 2012).

Let us consider the effect of pH on aniline polymerization, properties of products and their supramolecular structure.

4.1. Acid-base properties of aniline and poly (oligo) aniline

Aniline, its oligomers and polymers are organic bases taking part in acid-base equilibria, and their properties depend on pH. They include different types of nitrogen-containing structures which are protonated at different pH values. Aniline is the strongest base. At low ionic strength, i.e., at low concentrations of aniline in distilled water, the protonation constant of aniline takes on a value of ~ 4.6 (Lide & Frederikse, 1995). Increasing ionic strength in the presence of electrolytes leads to equilibrium shifts towards lower pK values. Thus, in solutions of salts with concentrations of 0.1 - 0.9 $mol{\cdot}L^{-1}$, protonation constant of aniline takes on a value of ~ 3.5. According to these data, at pH>3.5 aniline exists mainly as a neutral molecule, and at pH<3.5 it is protonated and acquires positive charge.

Polymeric (oligomeric) PANI chains include two types of nitrogen-containing structures: disubstituted amino groups (where nitrogen atom is not oxidized) and imino groups with oxidized nitrogen. Disubstituted amino groups can be protonated only at very low pH which are not used in the synthesis. As a rule, leucoemeraldine consisting of amino groups is not protonated. PANI in higher oxidation states contains imino groups which are protonated at higher pH values. However, protonation constant of imino group depends on the content of these groups in polymer. With increasing concentration of ionogenic groups, polymer basicity decreases. While imino groups of emeraldine can be protonated by acids of medium strength, during oxidation and increasing number of imino groups, multiple interactions between positive charges within the chain arise, and the protonation ability of imino groups decreases. In contrast with emeraldine (pK=3÷4), pernigraniline is protonated at pH<2.5 (Trivedi, 1997). In the course of protonation of imino groups, PANI is transferred into protonated emeraldine or protonated pernigraniline form. As a result, the whole polymer chain becomes covered with a net of positive charges and turns into polycation.

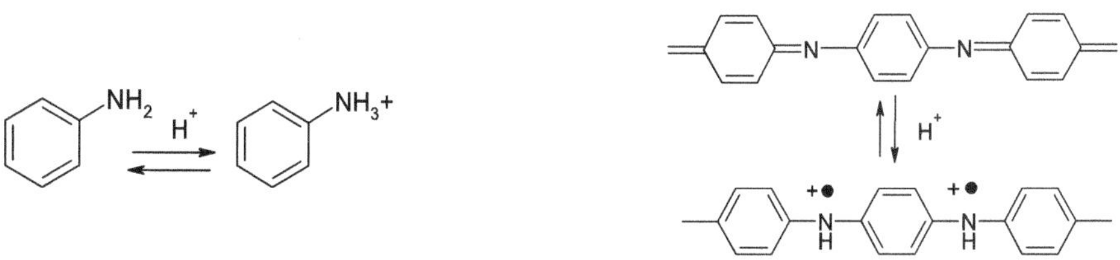

(a) (b)

Scheme 4. Protonation of aniline a) and polyaniline chain b)

It should be taken into account that oxidation potentials and, therefore, reactivity of neutral and protonated molecules differ considerably. Deprotonated nitrogen-containing structures have low oxidation potential and oxidize readily. Neutral, uncharged aniline molecule easily gives away electron and can be oxidized even at +0.3 V. Protonation results in the appearance of excess positive charge on nitrogen atom that hinders oxidation. Consequently, oxidation potential of nitrogen-containing structure increases up to more than +1.0 V (Li et al., 2002).

It should be also kept in mind that in the course of oxidation protons are released, and pH is being constantly lowered. The rate of pH change depends on many factors: the presence of alkali, acids or buffering agents in the reaction medium, intensity of oxidation processes, conversion degree etc. For example, in water for the initial aniline and persulfate concentrations of 0.4 and 0.8 mol·L^{-1}, respectively, and upon the complete monomer-to-polymer conversion, pH of reaction medium varies from 9 to 1, i.e., the reaction that has started in an alkaline medium ends under the conditions of acid-promoted polymerization (Stejskal et al. 2010).

Thus, the oxidative polymerization of aniline is a complex dynamic process accompanied by permanent changes in synthesis conditions. What is important is the pH value at which the process begins, and the nitrogen-containing groups (either protonated or non-protonated) that are involved. However, of no less importance is how quickly the pH drifts and at which pH values the polymerization comes to an end. The reaction often covers the pH ranges where nitrogen-containing fragments in different degrees of protonation dominate. As will be shown later, the transition from one pH region to another can be accompanied by some changes in kinetic parameters of the process. This can result in changing molecular structure of chains and, hence, varying properties of products, and also lead to transformations in PANI supramolecular structure. Now we will illustrate these points by aniline polymerization processes occurring in media with different pH (Konyushenko et al. 2010; Sapurina & Stejskal, 2010)

4.2. The influence of pH on aniline polymerization carried out under the action of strong oxidizing agents

In this section, two series of experiments are discussed demonstrating kinetics of aniline oxidation within various pH intervals. Acidity of reaction medium and temperature of polymerization mixture were monitored in the course of the synthesis. Every elementary reaction of monomer unit addition in the process of electrophilic substitution is accompanied by release of two protons; therefore, change of pH throughout the synthesis can serve as an objective characteristic of the process. Oxidation of aniline is an exothermic process and can easily be followed by temperature changes. The temperature dependence provides information on the intensity of oxidation processes in individual steps of the synthesis, thus supplementing the results of pH measurements. The interpretation of results involved the use of additional data obtained in the *in situ* analysis of oxidation with the use of spectral and potentiometric methods, as well as the data on the formation of paramagnetic sites.

Experiments were carried out in order to elucidate the kinetics of oxidation of aniline in different pH regions. We used a strong oxidant (ammonium persulfate), which has an oxidation potential of 2.01 V and is capable of oxidizing all intermediates of PANI synthesis. The reaction shown by Scheme 5 proceeded under ambient conditions. It was initiated by instantaneous mixing of reactant solutions and proceeded without stirring. The kinetic data were supplemented by information on the properties and the morphology of the reaction products. These were isolated by filtration, dried in air at room temperature and then analyzed by conductometry, spectroscopy, gel permeation chromatography and electron microscopy.

(1)

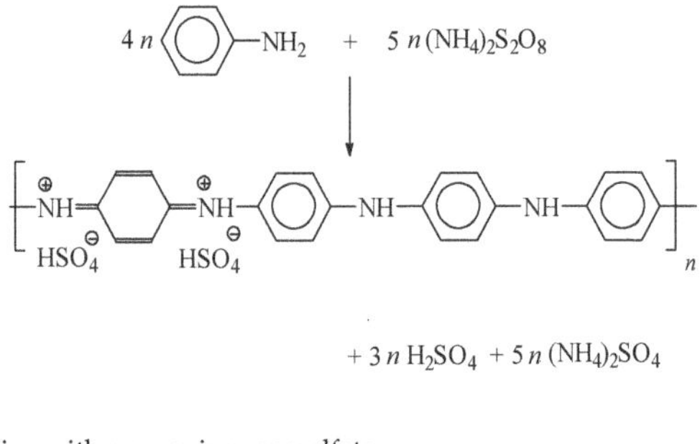

Scheme 5. Oxidation of aniline with ammonium persulfate.

4.2.1. *Oxidation of aniline at different initial pH values*

In the first series of experiments, oxidative polymerization of aniline (0.2 mol·L^{-1}) was carried out using the stoichiometric amount of ammonium persulfate (0.25 mol·L^{-1}), which provided the complete monomer-to-polymer conversion. Polymerization was started in the media of different acidity, namely, 1) in 0.2 M ammonia solution (pH=10); 2) in water (pH=6); 3) in 0.4 M acetic acid (pH=4.5), 4) in 0.2 M sulfuric acid (pH=2). The results of monitoring of temperature and pH are shown in Fig. 1 a,b.

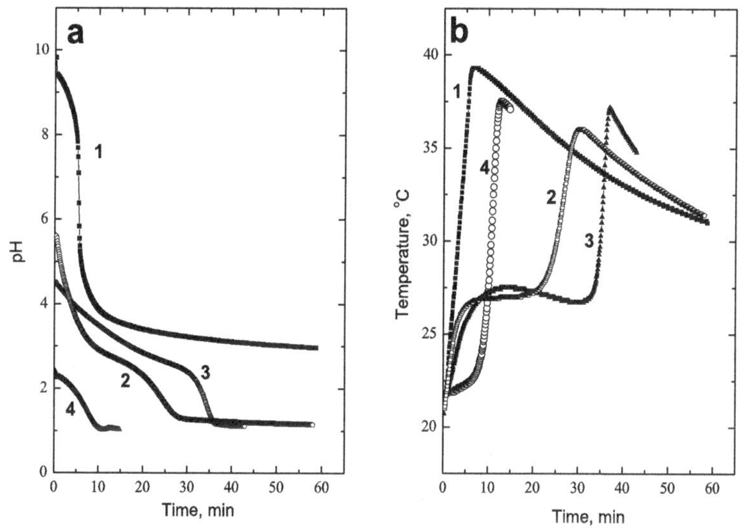

Figure 1. Changes in the temperature (a) and pH (b) of the medium in the course of aniline (0.2 mol·L^{-1}) oxidation initiated by ammonium persulfate (0.25 mol·L^{-1}) at different initial acidities; *(1)* in 0.2 M $NH_3 \cdot H_2O$; *(2)* in water; *(3)* in 0.4 M acetic acid; *(4)* in 0.1 M H_2SO_4.

It is evident that both parameters change synchronously. The decrease in pH corresponds to the exothermic steps: the sharper increase in temperature corresponds to the more pronounced decrease in pH. This also suggests that thermal effects are due to the processes of oxidation and proton liberation

The general features of aniline oxidation processes can be formulated as follows.

1. The initial rate of aniline oxidation increases with increase in the initial pH of the medium. It can be seen from slopes of temperature and pH dependences in the initial stage. With increasing acidity, initial oxidation rate decreases. In sulfuric acid (pH=2), it is close to zero (the so-called "oxidation induction period").
2. The oxidation processes in which the pH dependence does not pass through pH~2.5 are characterized by a single temperature wave of oxidation. These processes include the oxidation in ammonia, which begins at pH=10 and ends at pH>3, and the reaction in sulfuric acid, which proceeds at pH<2.2.

3. The oxidation processes that occur in the pH region including pH~2.5 have two temperature waves. These processes include the reactions in water and acetic acid, which start at pH=6 and pH=4.5, respectively, and end at pH=1.
4. The first oxidation wave falls in the range with pH>3 and resembles the monotonic process of oxidation of aniline in ammonia. The range 2.5<pH<3.5 is characterized by the abrupt deceleration of the oxidation processes. However, with the attainment of pH 2.5, the exothermic reaction regains its vigor, and oxidative polymerization is concluded by the second temperature wave. The character of pH variations in the second wave is identical to that typical of polymerization in sulfuric acid.
5. The pH decrease in water proceeds much faster than in acetic acid because this weak acid serves as a buffer and binds protons released in the course of polymerization.

4.2.2. Oxidation of aniline at different monomer: oxidant ratios

In the second series of experiments, oxidation of aniline (0.2 mol·L^{-1}) was carried out in 0.4 M acetic acid at different [Ox] : [An] molar ratios. The ammonium persulfate concentration was varied from 0.25 (stoichiometric amount) to 0.025 mol·L^{-1} (Fig.2). At low concentrations of oxidant, monomer is not completely oxidized. In all the experiments, polymerization started at pH=4.5 and continued up to the complete consumption of the oxidant. Figure 2 shows the temperature and pH dependences obtained in this series of experiments. It is evident that the higher the oxidant concentration, the lower the final pH; this fact is associated with the more complete monomer-to-polymer conversion and, correspondingly, the presence of a larger amount of liberated protons.

The second experimental series produced the relationships similar to those of the first series. For the oxidation at low [Ox] : [An] ratios (0.025, 0.5, 0.25), no acidity drift below pH=3 was observed. In this pH region, as well as in the case of oxidation of aniline in ammonia, the reaction proceeds monotonically and the single temperature wave is observed. For high [Ox] : [An] ratios (0.75, 1.0, 1.25), the final pH of the medium is less than 2.5. Moreover, the processes demonstrate two temperature oxidation waves. As well as in the first series, in the range 2.5<pH<3, the reaction proceeds extremely slowly. The second exothermic wave begins once the reaction medium pH reaches 2.5.

As follows from the data presented, irrespective of which reaction parameter was varied (initial pH of the medium or the [Ox] : [An] ratio), the processes have similar character if proceed in the same pH region. Judging from the sharp change in the oxidation character in a point with pH=2.5, different molecular mechanisms of the synthesis are realized in the pH ranges above and below this point.

4.3. Properties and morphology of products of aniline oxidation

In the first experimental series, the synthesis proceeded up to the 100% conversion of the monomer to the polymer (Table 1). For polymerization in ammonia (3<pH<10), the oxidation products are aniline oligomers with a weight-average molecular mass (M_w) of 4000 and polydispersity (M_w/M_n) of 1.3. Conductivity of the material in the completely protonated state does not exceed 10^{-10} S·cm^{-1}. The electronic spectrum of oligomers (Fig. 3) contains no polaron charge transfer band typical of the emeraldine polymer form with a continuous conjugation system. Studies on the morphology of the oxidation products showed that the microsphere structure dominated, i.e., the oligomers formed spherical particles with diameters ranging from 500 nm to 2 μm (Fig. 4a).

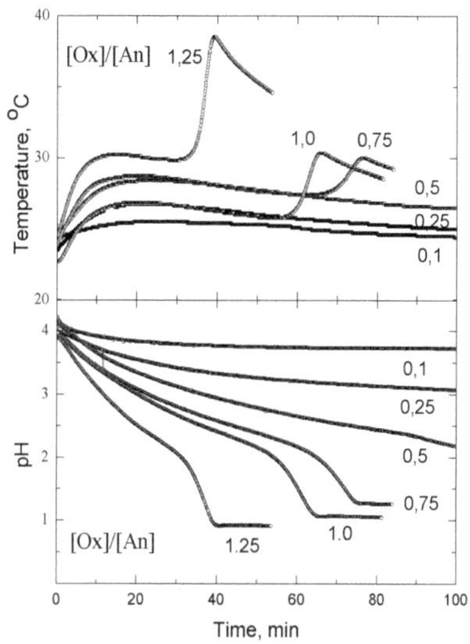

Figure 2. Changes in the temperature (a) and pH (b) of the medium in the course of aniline (0.2 mol·L^{-1}) polymerization in 0.4 M acetic acid under the action of different amounts of ammonium persulfate. The ratios of molar concentrations of oxidant and aniline are denoted by numbers. [Ox] : [An]=1.25 *(1)*, 1.0 *(2)*, 0.75 *(3)*, 0.5 *(4)*, 0.25 *(5)*, 0.1 *(6)*.

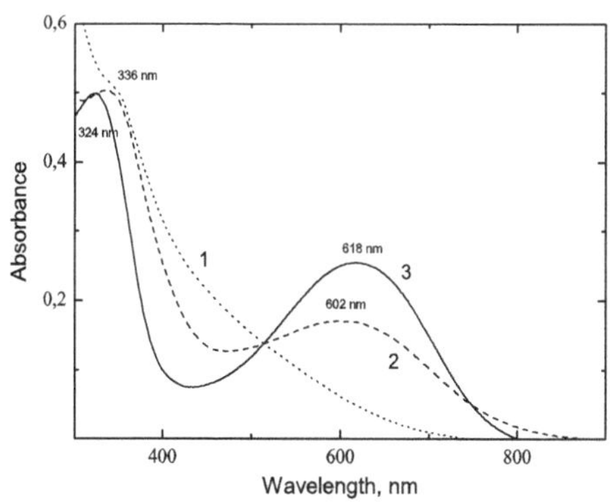

Figure 3. Electronic absorption spectra of the products of aniline (0.2 mol·L^{-1}) oxidation initiated by ammonium persulfate (0.25 mol·L^{-1}) taken in media with different acidities; *(1)* in 0.2 M $NH_3 \cdot H_2O$; *(2)* in 0.4 M acetic acid; *(3)* in 0.1 M H_2SO_4. Oxidation products are deprotonated and dissolved in N-methylpyrrolidone. Wavelengths of absorption peaks are given in nanometers.

Properties of products	$NH_3 \cdot H_2O$ (0.2 M)	Water	Acetic acid (0.4 M)	H_2SO_4 (0.1 M)
Initial pH	9.5	5.7	4.5	2.4
End pH	3.0	1.1	1.1	1.0
Conductivity of protonated form ($S \cdot cm^{-1}$)	$<10^{-10}$	0.055	0.036	3.7
Conductivity of deprotonated form ($S \cdot cm^{-1}$)	—	6.0×10^{-8}	7.9×10^{-9}	1.1×10^{-9}
Density of protonated form ($g \cdot cm^{-3}$)	—	1.35	1.338	1.402
Molecular mass M_w	4.090	-	32200	39400
Polydispersity, M_w/M_n	1.3	-	19.0	13.1
Morphology	Microspheres	Plates, nanotubes	Nanotubes	Granules

Table 1. Properties of products of aniline oxidation (0.2 $mol \cdot L^{-1}$) with ammonium persulfate (0.25 $mol \cdot L^{-1}$) at various pH values.

The aniline oxidation product synthesized in the presence of sulfuric acid ($1<pH<2.2$) is the polymer with $M_w \sim 40\ 000$ and $M_w/M_n=13.1$. The material demonstrates an absorption band at 618 nm typical of deprotonated emeraldine (Fig. 3) and the conductivity of the protonated emeraldine form (3.7 $S \cdot cm^{-1}$). The polymer has a globular structure typical of PANI with globules measuring from 100 to 200 nm (Fig. 4d).

The oxidation products obtained in acetic acid and water exhibited characteristics intermediate between those of aniline oligomers and high molecular weight PANI. As compared with PANI synthesized in sulfuric acid, molecular masses of the products were lower and their polydispersities were higher; their conductivities were two orders of magnitude lower and the band corresponding to polaron charge transfer was less intense and shifted to the short-wavelength range. These characteristics, as well as broadening of MWD of polymerization products suggested that PANI synthesized in the intermediate pH range ($1<pH<6$) is a mixture of oligomers and polymer.

In the second series of experiments (polymerization of aniline in acetic acid at various [Ox] : [An] ratios, Table 2), the relationships typical of the first series are reproduced. If the polymerization is completed at $pH>2.5$ ([Ox] : [An]=0.025, 0.5, 0.25), it produces non-conducting aniline oligomers without continuous conjugation systems. At the same time, if the synthesis is completed at $pH<2.5$, the products, in addition to oligomers, contain high molecular weight PANI. The extent of conjugation and, hence, the conductivity are higher.

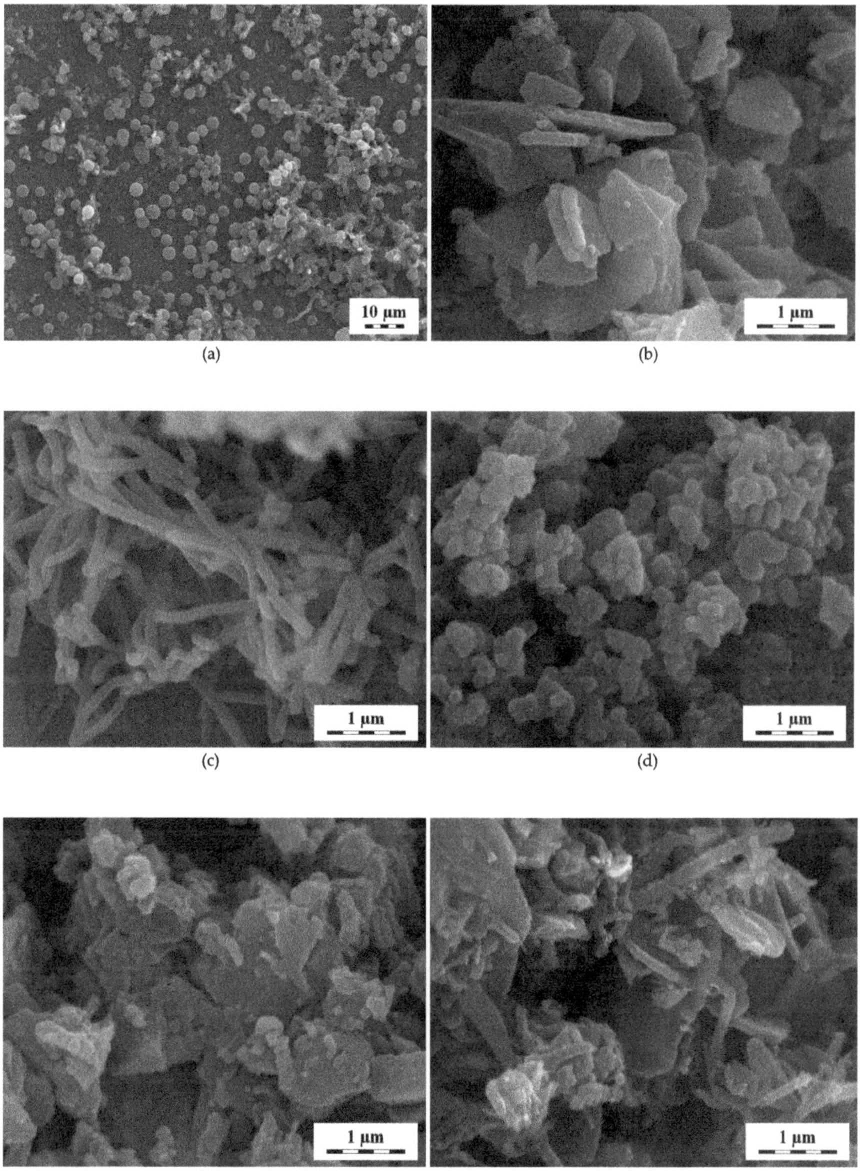

Figure 4. Scanning electron microscopy. Products of aniline (0.2 mol·L^{-1}) oxidation with ammonium persulfate in different experimental conditions: a) in aqueous ammonia (microspheres); b) in water (two-dimensional structures – "plates"); c) in 0.4 M acetic acid (nanotubes); d) in sulfuric acid (globular PANI morphology. e) Amorphous agglomerates of aniline oligomers synthesized upon oxidation of

aniline (0.2 mol·L^{-1}) under deficiency of ammonium persulfate (0.05 mol·L^{-1}) at pH>3.5. f) Products of aniline oxidation (0.2 mol L-1) in 0.4 M acetic acid with ammonium persulfate (0.25 mol·L^{-1}) isolated in the intermediate step of polymerization at pH ~ 2.5-3.

The morphology of PANI synthesized in the intermediate pH range (1<pH<4.5) that includes two temperature waves is characterized by transition from amorphous to organized morphologies (Fig 4e and 4f) and the presence of one-dimensional structures, namely, nanotubes and nanofibers. The one-dimensional structures begin to appear in products at pH~2.5. The appearance of nanofibers and nanotubes accompanies the transition from oligomeric products to polymeric ones. As the pH drifts to acidic values, the length of tubular fragments increases together with the fraction of one-dimensional structures. For PANI synthesized in acetic acid in the pH range 1<pH<4.5, the one-dimensional structures prevail (Fig.4c). However, further decrease in pH increases the granular precipitate fraction. Polyaniline synthesized in water also contains fibers and tubes with an outer diameter of 150 ± 200 nm; in addition to tubes, PANI precipitate contains the so-called "plates" (Fig.4b) and the granules measuring up to several hundred nanometers.

[Ox]/[An]	0.25	0.5	0.75	1.0	1.25
End pH	4.0	3.3	2.5	1.2	1.1
Conductivity of protonated form (S·cm^{-1})	$<10^{-10}$	2.4×10^{-10}	-	0.036	0.095
Density of protonated form (g·cm^{-3})	–	1.307	–	1,338	1.465
Molecular mass M_w	3600	2100	23600	17600	44600
Polydispersity, M_w/M_n	3.8	16.4	11.1	9.4	10.6
Morphology	Amorphous	Two-dimensional	Inclusions of nanotubes	Nanotubes	

Table 2. Properties of products of aniline oxidation (0.2 mol·L^{-1}) with different amounts of ammonium persulfate in 0.4 M acetic acid. The initial pH is 4.5.

Transition from one pH range into another leads to changes in molecular structures of chains and properties of products accompanied by transformation of supramolecular structure of PANI. During oxidation at pH<2.5, high molecular weight polyaniline with developed polyconjugation system and high conductivity is formed. By contrast, at higher pH, PANI oligomers are obtained. Oligomers have low conductivity and, judging by spectral characteristics, there is no polyconjugation. If the reaction proceeds within both pH ranges, the mixture of oligomeric and polymeric products is obtained. In various pH intervals morphology of products also varies. However, there is no direct correlation between conductivity, molecular mass and type of supramolecular structure. Depending on synthesis conditions, oligomers may exist in the form of microspheres or two-dimensional formations, and high molecular weight PANI may have one-dimensional (nanofibers or nanotubes) or three-dimensional structure (granules).

4.4. pH of reaction medium as a critical factor in aniline oxidation

The literature contains a vast amount of information concerning influence of various parameters of synthesis on properties and morphology of products obtained in aniline oxidation. There are many publications (Wan, 2009; Liu & Zhang, 2009; Venancio et al., 2006) demonstrating that changes in concentrations of monomer, oxidizing agent and their ratio, nature and concentration of acids which protonate PANI, or other additives have a significant effect on polymerization process, properties and morphology of products. However, authors do not take into account that this effect is not always obvious; usually it is indirect and carried out via changes in pH of the reaction medium. There are many examples of indirect influence of pH masked by other synthesis parameters. Thus, aniline and its polymers, being organic bases, increase pH. The increase in monomer concentration is equivalent to alkalization of the solution. Addition of acids and acidic salts decreases pH. The [monomer] : [acid] ratio and the nature of the acid used (its strength and buffering properties) are also important parameters. Weak acids exhibit the buffer effect suppressing the pH change during the synthesis.

The pattern of pH changes occurring during the reaction may be altered by certain factors one would think independent of pH. Thus, changes in [oxidant] : [monomer] ratio have an effect on pH, determine conversion degree and a number of released protons. The dilution of reaction medium also influences this pattern, because the range of pH changes in concentrated solutions is much wider than in diluted ones, and, as a result, the reaction may pass to another pH interval. The presence of organic phase immiscible with aqueous medium is also a striking example of such factor. This is the so-called aniline polymerization at the interface between two immiscible media; monomer is dissolved in organic phase, and oxidizing agent is present in water. The mechanism of "interface" polymerization and the factors leading to the formation of one-dimensional structures in this system are actively discussed in the literature. It was shown (Li et al., 2011) that characteristics of aniline oxidation, as well as properties and morphology of products, depend on pH and correspond to the profile of change of water phase pH.

Frequently, effects of various factors in oxidative polymerization of aniline mask the one factor which has a critical influence on the mechanism of synthesis. In polymerization with strong oxidant, it is the pH factor that is masked by a large number of various parameters of the reaction (Konyushenko et al. 2006a; Stejskal et al. 2006; Sapurina & Stejskal, 2010). Considerable changes in polymerization process, properties and morphology of products occur when the acidity falls within pH intervals coinciding with protonation constants of reactants (monomer and growing polymer chain). At the same time, if the experiment is carried out within the same acidity interval, changes in these parameters (concentrations of reactants and/or nature of additives) do not lead to dramatic changes in properties of products. Thus, at pH ≤ 2.5, polymerization always yields emeraldine with characteristic electronic spectrum (absorption band at wavelengths higher than 800 nm) and high conductivity (≥ 10^{-1} $S{\cdot}cm^{-1}$). By contrast, products of aniline oxidation at pH>2.5 definitely have low conductivity which can be raised neither by protonation nor by redox reactions. All these facts indicate that products obtained at pH<2.5 and pH>2.5 have fundamentally different molecular structure.

5. The influence of oxidation potential on the process of aniline oxidation, properties and morphology of products

Oxidation potential of nitrogen-containing structures, including aniline, increases with increasing acidity of the medium, due to protonation of nitrogen atom. The potential increases non-monotonically. In the pH range corresponding to protonation constant, oxidation potential changes discontinuously, in accordance with the pattern of change in concentration of protonated and deprotonated forms of reactants. Deprotonated form of aniline existing in alkaline and neutral media is oxidized at low potential (+0.3 V). In acidic media, start of polymer chain propagation requires overcoming the energy barrier corresponding to an oxidation potential of +1.05 V (Li et al 2002). Therefore, oxidants with potentials lower than 1.05 V capable of oxidizing aniline in alkaline and neutral media should not initiate polymerization at low pH. However, experiments have shown that oxidation of aniline in acidic media can be performed and leads to the formation of polyconjugated PANI. Such oxidants as $FeCl_3$ (+0.77 V), V_2O_5, (+0.98 V), $AgNO_3$ (+0.8 V), MnO_2 (+1.2 V), KIO_3 (+1.1 V), Pd^{+2} (+0.89 V), Au^{+3} (+1V) (Sapurina & Stejskal, 2012), possessing potential close to 1 or even lower, may oxidize aniline not only in alkaline or neutral media, but, strange as it may seem, in acidic media with high potential barrier of oxidation.

5.1. Oxidation of aniline under the action of a weak oxidant (silver nitrate)

In this section, we will discuss oxidation of aniline under the action of a weak oxidizing agent – silver nitrate $AgNO_3$ with a potential of +0.80 V (which is much lower than potential barrier of aniline oxidation in acidic media) (Sapurina & Stejskal, 2012). Oxidation was carried out at different pH values; the initial pH values of reaction medium were 11, 6 and 2.5. Thus, oxidation proceeded within three pH ranges where both the monomer and growing chains are deprotonated and have low oxidation potential or are present in protonated state with high oxidation potential. To aid the visualization, oxidation of aniline with $AgNO_3$ is compared with oxidation under the action of persulfate under similar conditions. In the course of oxidation, silver cation is reduced to give metallic silver. Silver forms composites with polymeric products; this composite was analyzed by electron microscopy, spectroscopy, conductometry, gel-permeating chromatography and thermogravimetry. The content of reduced silver in products was determined by thermogravimetry; this allowed us to control the participation of $AgNO_3$ in the reaction of aniline oxidation.

(2)

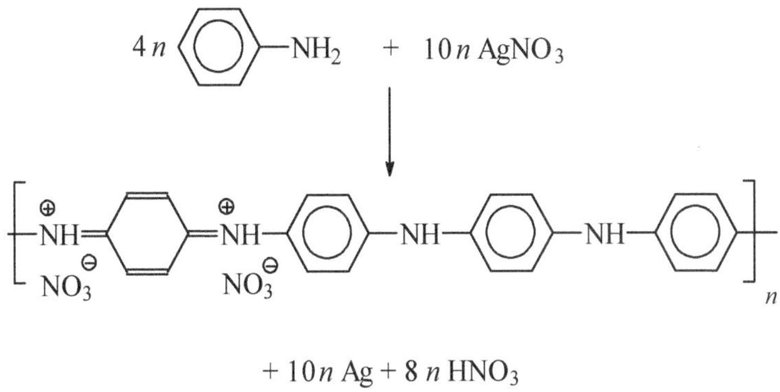

5.2. The oxidation process and properties of products

5.2.1. Oxidation at pH>3.5

At pH>3.5 both monomer and polymer chain are deprotonated. Only deprotonated nitrogen-containing structures with low oxidation potential take part in the reaction. As well as in the case of persulfate and $AgNO_3$, oxidation starts at once and proceeds monotonically with high rate. Fig. 5a shows aniline oxidation profiles for both oxidants in 0.5 M ammonia solution. At an initial aniline concentration of 0.2 mol·L^{-1} and $[Ox]_e/[An] = 2.5$, reactions are completed within several minutes. Silver mirror is formed on the walls of the reaction vessel in the course of oxidation of aniline with $AgNO_3$. In the case of persulfate, the initial pH of 10.0 lowers down to 8.9 by the end of the reaction. When $AgNO_3$ was used, the interval of pH change was 10.8 – 9.3. This fact indicates that both monomer and polymer chain remain deprotonated throughout the process.

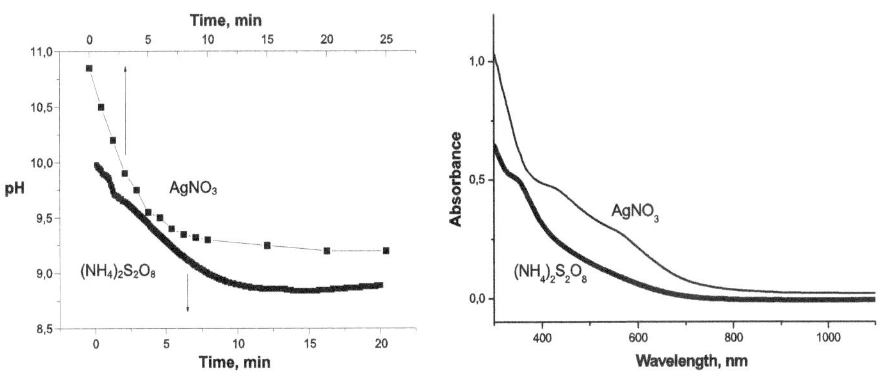

Figure 5. (a) Change of pH in the course of aniline oxidation under the action of ammonium persulfate and silver nitrate in 0.5 M aqueous ammonia. (b) Electronic spectra of products is lost of aniline oxidation with ammonium persulfate and silver nitrate; samples are dissolved in N-methylpyrrolidone.

Yields of oxidation products and their properties are given in Tables 1,3,4 . In the spectra of products obtained during oxidation of aniline with both oxidants, absorption bands lie in the wavelength region below 600 nm (Fig. 5b); this fact indicates the absence of polyconjugation in these samples. Conductivity of the products synthesized using persulfate does not exceed 10^{-10} S·cm^{-1}. The material obtained in the reaction with $AgNO_3$ is a composite containing 75 wt% of metallic silver. Its density is almost four times higher than that of PANI. Electric conductivity of this composite is 3300 – 5030 S·cm^{-1} (Table 4). Oxidation of aniline with both persulfate and $AgNO_3$ yields mainly amorphous products (Fig.6 a, b). Silver is present as particles with dimensions ranging 50-200 nm.

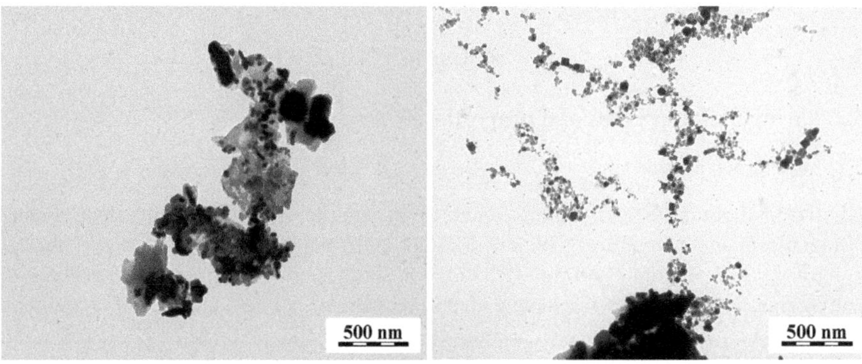

Figure 6. Products of aniline (0.2 mol·L^{-1}) oxidation by $AgNO_3$ in 0.5 M aqueous ammonia. Scanning electron microscopy

5.2.2. Oxidation at 3.5>pH>2.5

Oxidation of aniline at higher pH was studied in solutions of carboxylic acids. For example, in the solution of acetic acid (0.4 M) containing 0.2 mol·L^{-1} of aniline and 0.25 mol·L^{-1} of persulfate (or 0.5 mol·L^{-1} of silver nitrate), the initial pH values are 4.5 and 5.7, respectively. The considerable part of aniline is protonated, but imino groups of the chain are still in deprotonated state. Fig. 7a shows pH profiles of oxidation of aniline with both oxidizing agents. It can be seen that within the 3.5 > pH >2.5 region, oxidation is dramatically decelerated. This pH range lies between two stages of more intense oxidation with more sharp decrease in pH.

When persulfate is used, the stage of slow oxidation drift within the 3.5 > pH >2.5 range is completed after 30 min and culminates in vigorous exothermic reaction accompanied by intensive release of protons and decrease in pH. In the case of $AgNO_3$, at the same initial concentrations of reactants, oxidation drags on for hundreds of hours. In fourteen days, conversion of aniline oxidation in acetic acid reaches 30%, and pH decreases down to 3.5 (due to abstraction of protons). Then oxidation continues and somewhat accelerates.

The yield of aniline oxidation in the 2.5-3.5 pH range is very low. As was previously shown, in the case of persulfate, the products are non-conducting oligomers with low molecular masses (Table 2). In the case of $AgNO_3$, homogeneous dark brown powder was isolated; its density is 3.28 $g \cdot cm^{-3}$, and conductivity is 4350-7160 $S \cdot cm^{-1}$ (Table 3, 4). As well as after oxidation in 0.5 M ammonia solution, the obtained product contains metallic silver. The electronic spectrum of organic fraction of the product extracted with N-methylpyrrolidone is similar to the spectrum shown in Fig. 5b. This fact evidences that oxidation of aniline with $AgNO_3$ to 30% conversion gives products with low degree of polyconjugation.

Supramolecular structure of products of aniline oxidation in the pH range 2.5 to 3.5 differs from the structure of products obtained in alkaline medium. In the case of oxidation with persulfate, these are planar formations with dimensions of the order of microns and thickness of up to 100 nm and "sticks" with a diameter of up to 200 nm (Fig. 4f). In the case of oxidation of aniline with silver nitrate, two-dimensional morphology also prevails. The significant part of the material consists of planar disks with a diameter of several microns; sticks and thread-like formations are also observed. Transmission microscopy allows us to reveal localization of metallic silver. Silver in the form of dark stripes covers planar and one-dimensional objects or forms spherical nanoparticles with a diameter of the order of 10 nm which then turn into larger agglomerates (Fig. 8).

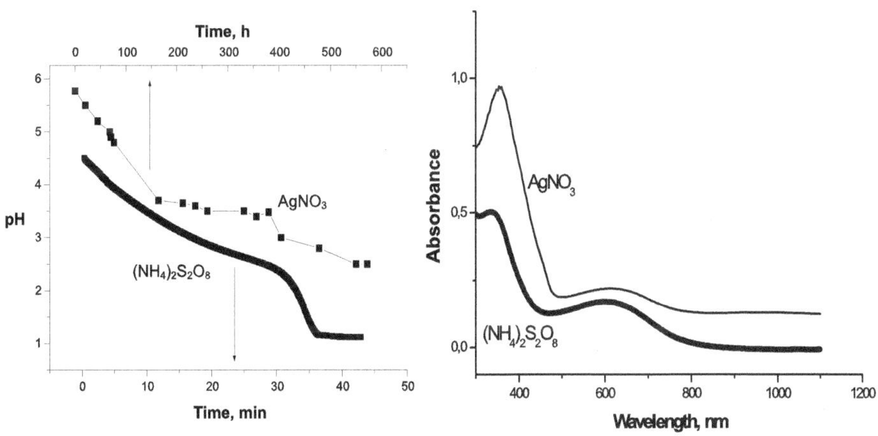

Figure 7. (a) Change of pH in the course of oxidation of aniline with ammonium persulfate and silver nitrate in 0.4 M acetic acid. (b) Electronic spectra of products of aniline oxidation with ammonium persulfate (100% conversion) and silver nitrate (70% conversion, pH < 2.5); samples are dissolved in N-methylpyrrolidone.

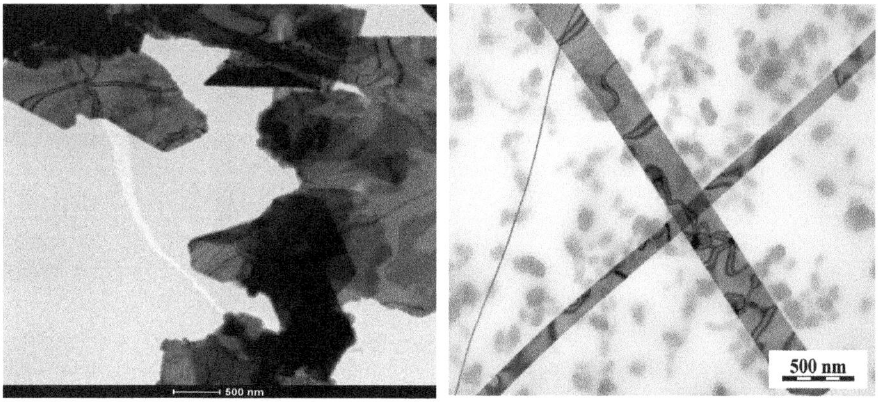

(a) (b)

Figure 8. Two types of supramolecular structures of products of oxidation of aniline (0.2 mol·L^{-1}) with $AgNO_3$ (0.5 mol·L^{-1}) within the 2.5 - 4.5 pH range (TEM images).

Parameter	0.5 M NH_4OH, 100% conversion	0.4 M CH_3COOH, 30% conversion	0.4 M CH_3COOH, 70% conversion	0.1 M HNO_3, 30% conversion
Initial pH	10.8	5,7	5,7	2.5
End pH	9.3	3.5	2.5	1.8
Oxidation time, (h)	0.24	168	540	540
Wavelength of polaron absorption band, (nm)	-	-	618	626
Molecular mass M_w	-	-	-	41400
Morphology of organic fraction	Amorphous	Two-dimensional One-dimensional	Fibers Tubes	Fibers Tubes
Morphology of silver	Spheres	Layer on oligomer	Spheres + sticks	Spheres

Table 3. Morphology and properties of composites formed in the course of oxidation of aniline (0.2 mol·L^{-1}) with silver nitrate (0.5 mol·L^{-1}) at various pH values.

5.2.3. Oxidation at pH<2.5

Oxidation process at pH<2.5 is illustrated by the reaction between 0.2 mol·L^{-1} solution of aniline and 0.5 mol·L^{-1} solution of $AgNO_3$ in 1.0 M nitric acid (Fig. 9a,b), as well as in acetic and formic acids (0.4 M) at high conversions of aniline; release of protons leads to decrease in pH down to 2.5 (Fig. 7a,b). The reference reaction was interaction between aniline (0.2 mol·L^{-1}) and ammonia persulfate (0.25 mol·L^{-1}) in 0.2 M sulfuric acid.

Two-step reaction between aniline and ammonia persulfate was completed in 10 minutes. The yield of emeraldine with a molecular weight of 40 000 and conductivity of 1 S·cm^{-1} is close to 100% (Table 1). Polymer consists of granules (which is characteristic of PANI obtained in acidic medium under the action of strong oxidants, Fig. 4d).

Oxidation in the presence of $AgNO_3$ in 1 M nitric acid proceeds very slowly. During the first 100 hours, no visible signs of oxidation appear. The 16% yield is achieved in two weeks. The density of the product is 3.6 g·cm^{-3}, it has a conductivity of 2200 S·cm^{-1} which indicates the presence of a large amount of metallic silver in the composite. In three weeks and a half, the yield is 30%, and conductivity of the product decreases down to 1000 S·cm^{-1}(Table 3, 4). Electronic spectra of organic fraction of the composite include intensive polaron absorption band with a maximum at 626 nm (Fig. 9b). IR spectra of deprotonated samples are also similar to those of emeraldine (Blinova et al. 2009). Thus, organic fraction of the composite is a polymer with high degree of polyconjugation.

The studies of products of aniline oxidation obtained in organic acids at high conversion (when pH decreases down to 2.5) also showed that the polymer with long conjugation system is formed. For example, organic fraction of composites obtained in acetic acid at a conversion of 75% (Fig. 7b) and in formic acid at a conversion of 80% demonstrates wide absorption bands in the spectra at 618 nm and 600 nm characteristic of emeraldine. However, intensities of polaron absorption bands are much lower than these in the case of reaction in a stronger (nitric) acid, this fact indicating that the content of polymer fraction is low. The densities of samples obtained in acetic and formic acids are 3.42 and 3.66 g·cm^{-3}, and conductivities are 326 and 0.85 S·cm^{-1}, respectively (Table 4).

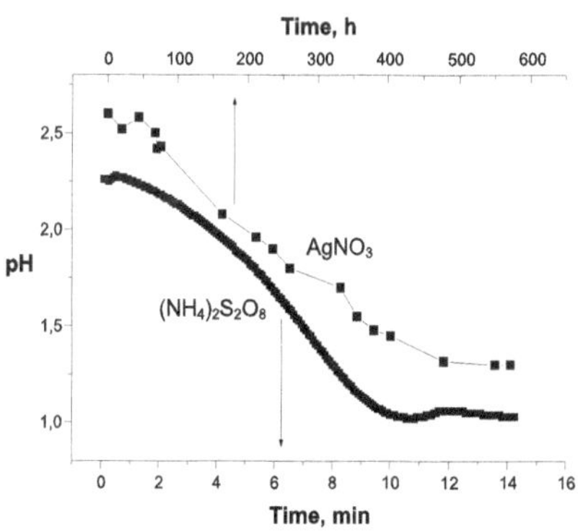

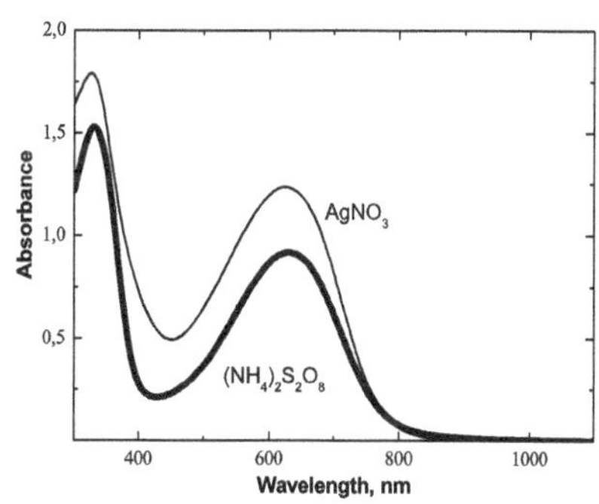

Figure 9. (a) Change of pH in the course of aniline oxidation under the action of ammonium persulfate in 0.1 M sulfuric acid and silver nitrate in 1 M nitric acid. (b) Electronic spectra of products of aniline oxidation with ammonium persulfate and silver nitrate; samples are dissolved in N-methylpyrrolidone.

Although at pH < 2.5 both the strong and weak oxidants oxidize aniline to the emeraldine state, supramolecular structures of products are fundamentally different. In the case of $AgNO_3$, instead of spherical particles typical for the products of persulfate oxidation (Fig. 4d), one-dimensional particles prevail. PANI in the form of nanotubes and hierarchical structures consisting of fibers with a diameter of 10 nm includes spherical particles of metallic silver with a diameter of 50 nm (Fig. 10 a, b). At high conversions, in acetic and formic acids silver "sticks" several μm in length and with a diameter of up to 300 nm appear. These sticks are covered with loose layer of polymer which starts to grow from silver surface as fibers (Fig. 10 c, d). In the product obtained in formic acid, this morphology prevails.

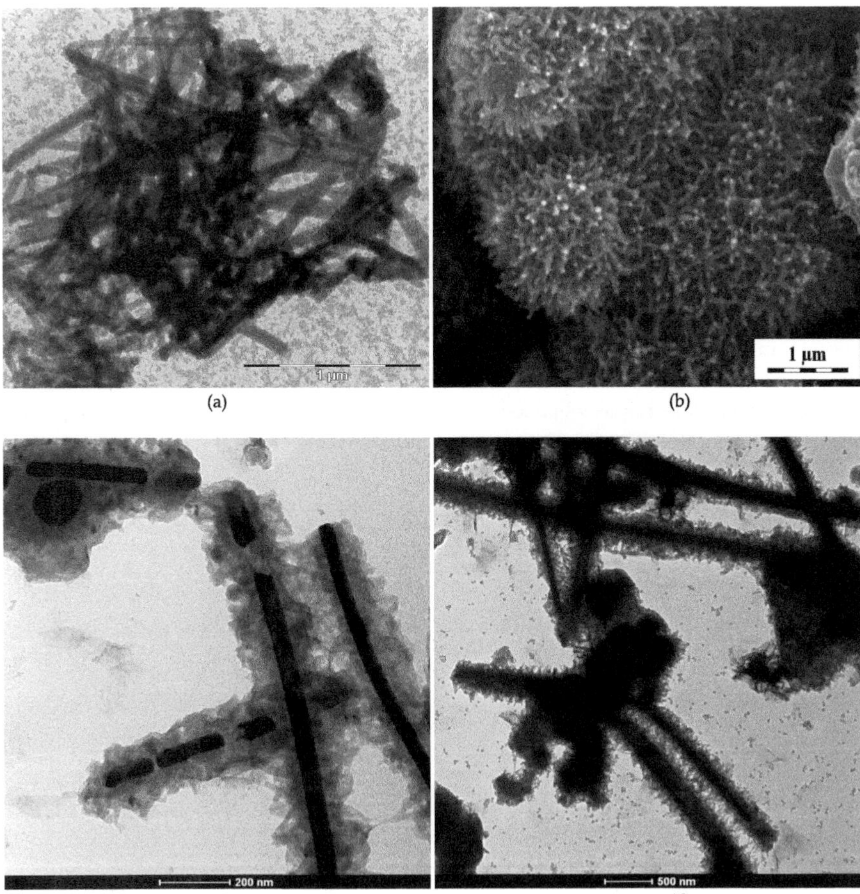

(a) (b)

(c) (d)

Figure 10. (a,b) Morphology of products of oxidation of aniline (0.2 mol·L^{-1}) with AgNO3 (0.5 mol·L^{-1}) in 1 M nitric acid (SEM image). (c, d) Products of oxidation of aniline (0.2 mol·L^{-1}) with $AgNO_3$ (0.5 mol·L^{-1}) in acetic and formic acids at high conversions (pH < 2.5) (TEM image).

5.3. The composition of PANI/Ag materials

Oxidation of one gram of aniline with a stoichiometric amount of $AgNO_3$ theoretically yields 1.31 g of PANI and 2.9 g of metallic silver. Therefore, according to stoichiometry of the reaction (2), the content of reduced silver in metal/polymer composite should be 68.9 wt%.

Oxidation to nonconducting aniline oligomers gives the composite with 74 wt% of silver. Thermogravimetric analysis allows to establish composition of silver-containing products. Mass fraction of metal is determined as a mass of incombustible residue left after heating of sample in air flow up to more than 600°C. According to TGA data, silver content in all the composites corresponds to the stoichiometry of redox process: in alkaline media it is 75 wt%, and in acidic media – about 70 wt% (Table 4). These data prove that it is $AgNO_3$ that acts as the main oxidizing agent.

Reaction medium	pH		Yield of composite, %	Content of Ag, wt%	Conductivity, $S \cdot cm^{-1}$	Density, $g \cdot cm^{-3}$
	initial	end				
0.5 M NH_4OH	10.8	9.3	60.5	75.0	3300-5030	4.22
0.4 M CH_3COOH	5.7	3.5	32.7	70.8	4350-7160	3.26-3.28
0.4 M CH_3COOH	5.7	2.5	75.7	69.0	326	3.42
0.4 M HCOOH	-	-	79.8	72.6	0.85	3.66
1.0 M HNO_3	2,5	1.8	30.0	70.7	709-1080	3.54

Table 4. Composition and properties of products of oxidation of aniline (0.2 $mol \cdot L^{-1}$) with $AgNO_3$ (0.5 $mol \cdot L^{-1}$) at various pH values.

5.4. Oxidation potential as an important factor in aniline oxidation

Comparison of results of oxidation of aniline with strong and weak oxidants shows that both the strong oxidant (persulfate) and weak one ($AgNO_3$) at pH < 2.5 oxidize aniline to give PANI in the emeraldine oxidation state, while at higher pH in both cases aniline oligomers without developed conjugation system are formed. The courses of these two reactions occurring in acidic media are different. While in alkaline medium both persulfate and $AgNO_3$ oxidize aniline in several minutes, and kinetic curves of oxidation are comparable. In going to acidic media, rate of oxidation drops considerably. The reaction proceeds for hundreds of hours; the lower the initial pH, the lower the conversion at any given instant of time. At the same time, the general views of time "profile" of pH for reactions with persulfate and $AgNO_3$ are similar. In the media with pH > 3.5 and pH < 2.5, faster pH change is observed, while in the pH range 2.5 to 3.5, oxidation becomes slower.

Another interesting aspect of using oxidant with low potential is its influence on PANI morphology. At pH ≤ 2.5 and high concentrations of reactants, oxidation by persulfate leads to the formation of a polymer with granular structure. Replacing strong oxidant by a weaker one leads to the formation of fibers, tubes and hierarchical structures. These transformations were observed not only in the case of $AgNO_3$ (Blinova et al. 2009), but also in the reactions with another oxidants having low potential: $FeCl_3$ (Ding et al. 2010), V_2O_5, (Li et al. 2010;), MnO_2 (Pan et al. 2007), KIO_3 (Gizdavic-Nikolaidic et al. 2010), CeO_2 (Chuang et al 2008). The authors of (Ding et al. 2007) revealed interrelation between the value of oxidation potential of oxidizing agents and diameter of one-dimensional PANI particles: lowering the potential leads to decrease in fiber diameter. Similar regularities are observed during electrochemical oxidation of aniline (Molina et al. 2007). In addition to one-dimensional objects, hierarchical structures (urchins, flowers, corals) are formed. Thus, transformation of supramolecular structure of PANI when employing weak oxidant is of a universal character.

The important feature of process with the use of $AgNO_3$ is the formation of metallic silver. In the course of the reaction, at any initial values of pH silver is reduced in stoichiometric amounts with respect to oxidized aniline. This fact proves that silver cation serves as an oxidizing agent. Depending on synthesis conditions, reduced silver precipitates in the form of various nano-dimensional structures (spheres, sticks or bands covering oligomer particles). The presence of silver provides high conductivity of the obtained materials, and these composites have many possible applications in polymer electronics. It is interesting that depending on synthesis conditions, in spite of equal silver content, conductivity of PANI/Ag composites changes over wide range (from 10^{-2} to 7 000 $S \cdot cm^{-1}$). Morphology and properties of PANI/Ag composites, as well as reasons for this difference in conductivity, are discussed in (Sapurina & Stejskal, 2012).

6. Molecular mechanism of aniline oxidation

In this section, we will discuss molecular mechanism of oxidative polymerization of aniline with the key factors responsible for chain formation being pH of reaction medium and oxidation potential of reactants (related to pH) (Sapurina & Stejskal 2008; Stejskal et al. 2010). The concepts of mechanism were formed not only on the basis of the given results. A vast amount of experimental data available in the literature give us additional information.

Fig.11 shows a typical kinetic curve of aniline oxidation (for the case of reaction proceeding in the wide pH interval, 6>pH>1). Temperature and pH profiles change synchronously with time and demonstrate three regions with different oxidation rates. Transitions between these regions are observed at pH values of 3.5 and 2.5. These pH values coincide with protonation constants of aniline (pK = 3.5) and imino groups of polymer chain (pK = 2.5). At pH > 2.5, oligomers without polyconjugation chain are formed; these materials possess low conductivity. At pH < 2.5, high molecular weight PANI can be synthesized; it demonstrates high degree of polyconjugation and high conductivity, its chains contain 95-98% of *para*-substituted monomer units. Dramatic change in the properties of products takes place at a pH value of 2.5 corresponding to the protonation constant of growing polymer chain.

Therefore, at pH 2.5, mechanism of polymer formation changes, and mixed irregular addition gives way to regular polymerization. These general features are in effect for different oxidants and protonating acids over a wide concentration range.

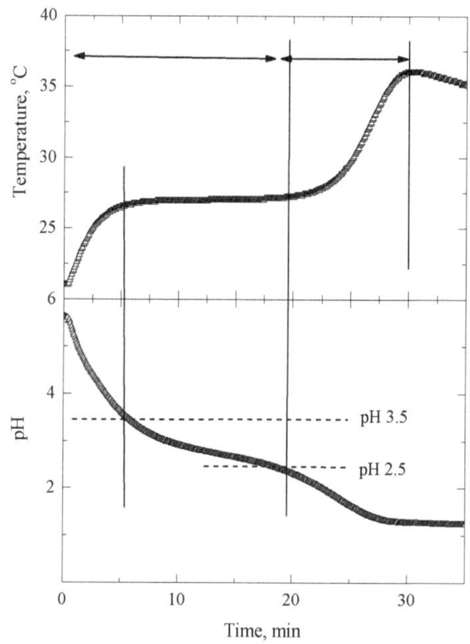

Figure 11. Temperature and pH profiles of oxidation of aniline (0.2 mol·L^{-1}) with ammonium persulfate (0.25 mol·L^{-1}) in water solution.

6.1. Oxidation at pH > 3.5

At pH >3.5 both monomer and growing chain are deprotonated, and oxidation of aniline (chain reaction of electrophilic substitution) proceeds in full accordance with concepts of theoretical organic chemistry. The most probable process is oxidation of nitrogen-containing structure of monomer and oligomer and linking monomer units in the "head-to-tail" fashion. However, other substitution paths are also possible, e.g., the process leading to "head-to-head" structures. Due to the presence of amino group (electron donor substituent) in aniline, the attack of oxidized polymer fragment is directed at *ortho*- and *para*-positions of monomer phenyl ring. Since phenyl ring contains two vacant *ortho*-positions, monomer units with *ortho*-structure should prevail. Chain growth is accompanied by secondary reactions. Monomer units of *ortho*-structure have low oxidation potential and undergo further oxidation leading to the formation of phenazine rings (Scheme 5). These reactions may be both intramolecular (similar to these in *o*-phenylenediamine oxidation) and intermolecular (with the participation of monomer). Thus, in addition to linear *ortho*- and *para*-monomer units, polymer chains should contain phenazine fragments. Scheme 4 demonstrates the most probable paths of electrophilic substitution leading to the formation of *ortho*- and *para*- monomer units; intramolecular oxidative cyclization of *ortho*-units and their transformation into phenazine rings are also shown.

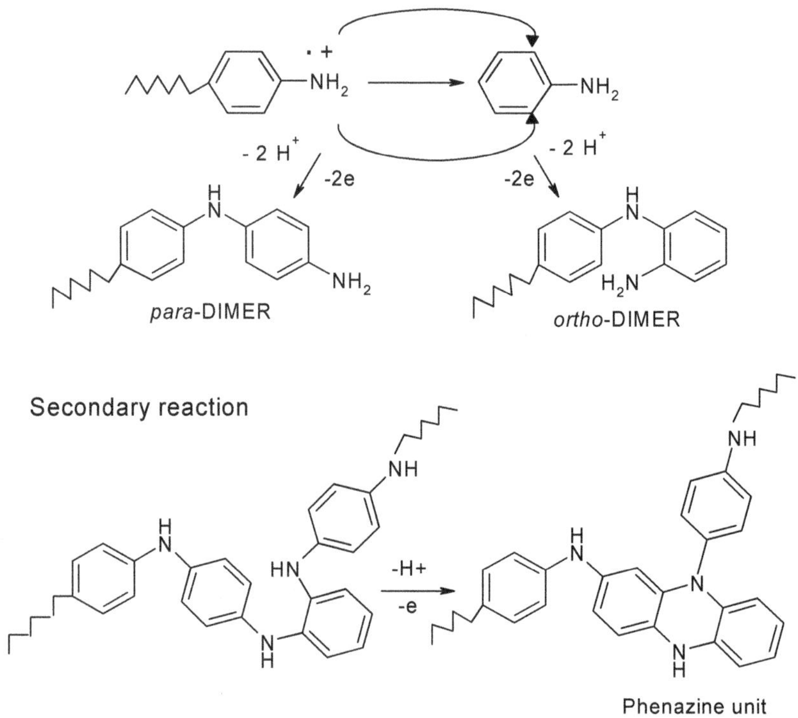

Scheme 6. Formation of monomer units of *para*- and *ortho*-structure after attack of aniline cation radical directed at *para*- and *ortho*-positions of aniline phenyl ring. Secondary reactions intramolecular oxidative cyclization of *ortho*-units and their transformation into phenazine rings are shown.

However, the probability of formation of other structures is also more than zero. Besides, hydrolysis and formation of quinones may take place in products in the presence of an oxidant. Thus, at medium and high pH values, oxidation of aniline should give chains with heterogeneous structure and prevailing linear or cyclized *ortho*-units (the presence of the latter was proved by spectroscopic methods (Stejskal & Trchova, 2012). Oxidation of deprotonated forms of reactants proceeds at low potentials (+0.3 - +0.5 V) and may take place under the action of weak oxidants; reaction rates in the cases of strong and weak oxidants are comparable. Monotonous decrease in open circuit potential taking place in the course of the reaction (Surwade et al 2009) is due to the fact that monomer oxidation gives way to oxidation of end amino group of di- and trimer (which possess lower oxidation potentials). However, in the case of longer oligomers, the difference is smoothed, this fact accounting for low molecular masses of oxidation products. In alkaline media, linear chains prevail, since chain propagation proceeds faster than cyclization processes (Losito et al. 2003). In oligomers with heterogeneous chain structure, no polyconjugation is observed, and we should expect low conductivity (see Tables 1 and 2). The process of aniline oxidation with the participation of deprotonated reactants (monomer and growing polymer chain) which yields aniline oligomers of heterogeneous structure lacking in polyconjugation will be termed "oligomerization".

6.2. Oxidation at 3.5 > pH > 2.5

At pH values ranging from 2.5 to 3.5, the major portion of aniline is protonated, and its oxidation potential increases. At the same time, under these conditions, imino groups in polymer chains are still not protonated, this resulting in imbalance of redox interaction between monomer and growing chain. The chain possessing low oxidation potential cannot oxidize protonated monomer with increased oxidation potential. Within the pH range 2.5 to 3.5, oxidation is drastically decelerated. Redox process proceeds only at the expense of oxidation of traces of neutral monomer (which has low oxidation potential). Neutral monomer is fed through shift of deprotonation equilibrium $AnH^{+}A^{-} = An + H^{+}A^{-}$ (Baizer & Lund, 1983).

How does pH change affect the structure of polymerization products? Oligomers obtained in alkaline media are mainly linear, and their degree of polymerization does several tens (Tables 1, 2). As oxidation slows down (i.e., pH decreases, and neutral aniline is spent), length and composition of oligomers change. The rate of monomer addition (i.e., chain growth) decreases, and intramolecular cyclization prevails. This change leads to the formation of short cycle-containing molecules. At pH < 3.5, the main product of oxidation is aniline dimer with *ortho*-structure (Stejskal et al., 2010). Virtually all these dimers undergo intramolecular cyclization to produce phenazine structures, since cyclization rate in acid media is higher than that in alkaline media (Losito et al. 2003).

Aniline dimerization and cyclization of dimers occur at low potentials and may start under the action of weak oxidants. As a result, aromatic phenazine rings with high oxidation potential are formed. This fact is corroborated by the slow increase in open circuit potential taking place in the course of oxidation in the pH range 2.5 to 3.5 (Ding et al. 2009; Ding et al. 2010). It is well-known that phenazines can take part in copolymerization and even homopolymerization, but phenazine starts adding monomer molecules ("germinates") at high oxidation potential (+1.05 V) (Inzelt & Puskas, 2004; Puskas & Inzelt, 2005). The value of "germination" potential of phenazine is similar to the potential of PANI growth start (Yang et al, 2007), and, consequently, we suppose that phenazine ring serves as a nucleate of future polymer chain. Thus, in the pH range 2.5 to 3.5, according to Scheme 6, slow accumulation of aromatic phenazine rings takes place; these rings have high oxidation potential and are intermediates of polymerization and nucleates of future polymer chains.

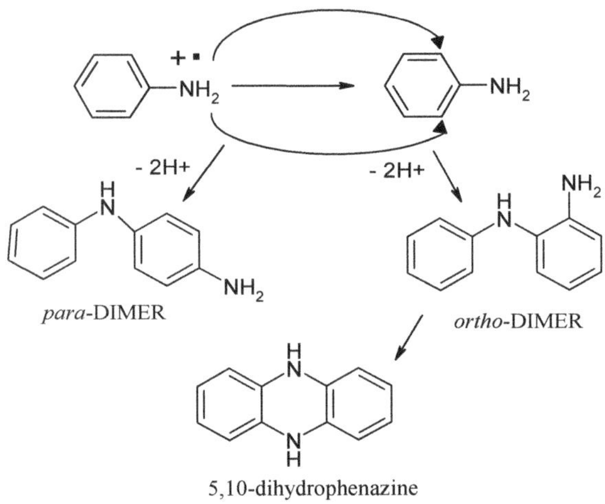

Scheme 7. Accumulation of aromatic phenazine rings.

6.3. Oxidation at pH < 2.5

The growth of polymer chains with regular structure becomes possible after the next protonation step. At pH < 2.5, the most important pH transition in aniline polymerization takes place. The protonation of monomer amino groups is followed by protonation of imino groups in oligomers, and their oxidation potential increases. This process restores balance of redox interaction between reactants. The chain is once again able to oxidize and add monomer, although in the protonated form which dominates in this pH range. The chain growth taking place at the expense of interaction between protonated chain and protonated monomer and leading to the formation of polyconjugated conducting polymers will be called "polymerization" (Scheme 8).

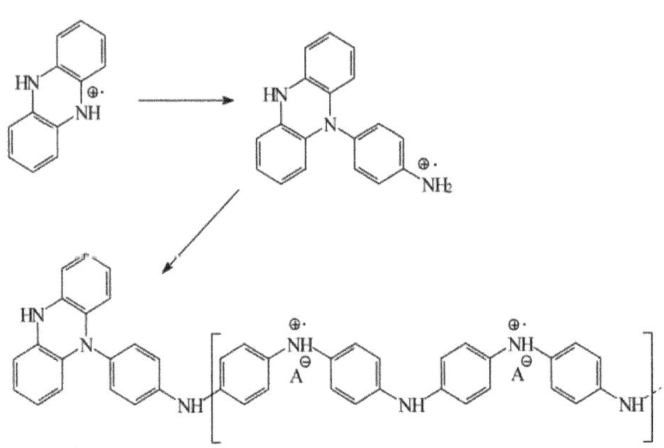

Scheme 8. The "germination" of phenazine nucleate and propagation of PANI chain.

6.3.1. The limiting step of oxidation

However, in acidic medium (where the majority of monomer molecules are protonated), oxidative polymerization of aniline cannot start with the reaction between cation radical of monomer and protonated aniline molecule, since (according to quantum-chemical calculations) interaction of two small particles bearing excess positive charge is not energy-favored (Ćirić-Marjanović, 2006). To start growth of regular chains, larger polyconjugated fragment is needed; only this fragment is able to lower the energy of interaction between oxidizing site and monomer. These particles are formed in the initial stage of oxidation ("induction period"). We suppose that they are phenazine rings. The role of phenazine as a bulk polyconjugated particle involves delocalization of positive charge of cation radical. At pH < 2.5, this induction period is similar to aniline oxidation in 2.5 – 3.5 pH range and proceeds according to the "oligomerization" mechanism. Aniline cation radical and deprotonated monomer (having low oxidation potential) take part in electrophilic substitution with preferential formation of *ortho*-dimers and their subsequent cyclization (Scheme 7).

Chain growth begins with addition of monomer unit to phenazine nucleate with cyclic structure and high oxidation potential. This is a limiting step of the whole process; its energy barrier corresponds to a potential of +1.05 V. After addition of monomer unit, the nature of growing end changes. Instead of tertiary nitrogen atom embedded into aromatic ring, new active site (end amino group) is formed; its structure is repeated in the process of polymer growth. Oxidation potential of amino group is lower than that of phenazine, and this fact leads to considerable acceleration of polymerization (Li et al., 2002).

The presented mechanism agrees with the behavior of open circuit potential. The character of variation of this potential in acid medium strongly differs from that observed in alkaline medium (Surwade et al 2009; Geng et al., 1998). In acid medium, the initial stage is accompanied by increase in the oxidation potential, which corresponds to accumulation of phenazine nucleates. As exothermic chain propagation starts, the oxidation potential decreases down to a value of +0.7 to +0.8 V which is characteristic of growing pernigraniline chains. Thus, the oxidation potential of aniline at pH < 2.5 passes through a maximum corresponding to the transformation of phenazine nucleate into active polymerization site. In order to overcome that maximum, a reagent with an oxidation potential higher than +1.05 V is necessary. On the other hand, oxidants with a potential lower than +1 V ensure both initial and final stages of oxidative polymerization.

Ammonium persulfate (having an oxidation potential of +2.01 V readily overcomes the barrier related to "germination" of phenazine nucleates, and oxidative polymerization is completed in several minutes. Weak oxidants, such as silver nitrate, cannot oxidize phenazine structure. Therefore, the induction period in the reaction with $AgNO_3$ takes hundreds of hours. The oxidation stops at the stage of generation and accumulation of phenazine nucleates. Nevertheless, if pH < 2.5, polymerization with the formation of polyconjugated chains does occur and is characterized by a fairly high yield. Obviously, in the absence of other strong oxidants phenazine ring is oxidized with atmospheric oxygen whose oxidation potential is equal to +1.23 V; in addition, oxygen is soluble in water and therefore accessible for the reactants. The formation of traces of polyaniline under the action of oxygen was reported in (Liao & Gu. 2002).

After "germination" of phenazine nucleates, i.e., transformation of these nucleates into active sites of polymerization, the oxidation potential decreases to +0.7 - +0.8 V (Geng et al. 1998) and, therefore, oxidants with low potential are again able to participate in oxidation process. Thus, weak oxidants, including silver nitrate, are capable of keeping both the initial aniline oxidation step (generation of phenazine units) and polymer chain propagation until the reaction is completed. The final stage involves oxidation of most part of monomer, and it consumes an equivalent amount of $AgNO_3$. The presence of a strong oxidant with a potential higher than +1.05 V is necessary only at the stage corresponding to the transformation of phenazine units into active polymerization sites. Therefore, the concentration of a strong oxidant, as well as its contribution to the overall oxidation process, may be insignificant.

6.3.2. The overcoming the "germination" stage

We believe that the oxidation of aniline to the emeraldine form in the absence of strong oxidants can held under the influence of a weak oxidant with oxygen. Phenazine units can be converted into active sites of polymerization under the action of atmospheric oxygen, but in this case the reaction is slow. An exception is likely to be a combination of atmospheric oxygen with Fe^{3+} which is known to catalyze oxidation processes (Davied et al. 1995). There are other examples of fast passing the energy barrier of phenazine "germination" and acceleration of polymerization. Oxidation of aniline can be performed with the use of a mixture of oxidants (Bober et al., 2011). Addition of a small amount of strong oxidant (when its concentration is lower than several percent of the amount required for complete oxidation) initiates the polymerization process. This allows to overcome high energy barrier and obtain a sufficient amount of active sites of polymerization. Weak oxidant then reacts with the main amount of aniline to accomplish chain growth at relatively low oxidation potentials.

Long induction period can also be shortened by change of the chemical structure of the nucleate. It is well-known that catalytic amounts of aromatic *para*-diamines sharply shorten the induction period and hence the overall time of polymerization (Tran et al. 2008; Bober et al., 2010). We propose the following mechanism of the process. Aromatic diamine is involved in the formation of phenazine nucleate. Its oxidation potential is lower than that of aniline; therefore, oxidation of diamine is preferred. The reaction between diamine and aniline at the dimerization stage yields aminophenazine (Scheme 9).

This aminophenazine nucleate gives rise to regular polyconjugated chain originating from exocyclic amino group, and high energy barrier of oxidative polymerization of aniline is thus reduced in the nucleate "germination" stage. As a result, the polymerization process is strongly accelerated due to acceleration of its slowest step. It should be noted that addition of *o*-phenylenediamine does not accelerate the process and that addition of *m*-phenylenediamine, by contrast, sharply decelerates polymerization of aniline via extension of induction period. The latter phenomenon is related to the formation of "defect" initiation site which cannot promote the growth of polyconjugated chain. After complete consumption of *m*-phenylenediamine for the formation of "defect" oligomers, polymerization of aniline is resumed.

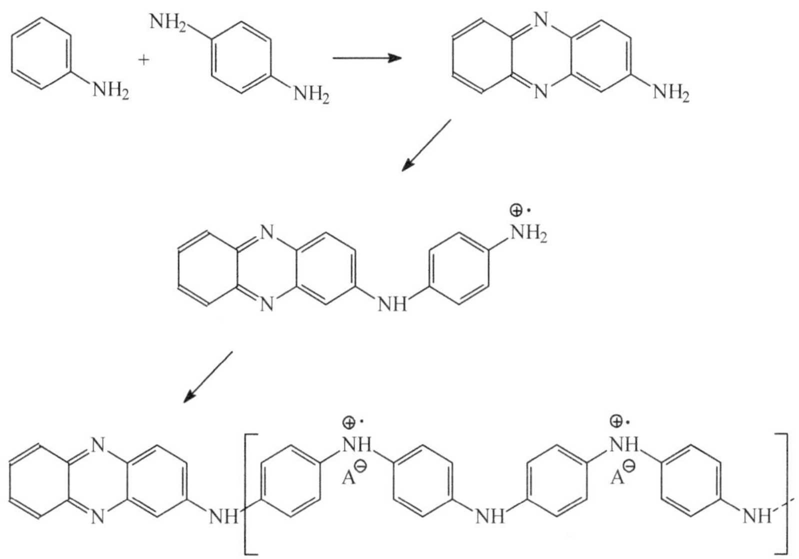

Scheme 9. Formation of aminophenazine nucleate giving rise PANI chain grows.

6.3.3. The growth of polyconjugated chains

Aniline polymerization, as well as oligomerization, proceeds according to electrophilic substitution mechanism. However, as opposed to oligomerization, the former yields polymers with strictly regular structure. The polymer containing only *para*-substituted units is formed. The reason for changes in reaction mechanism and transition from mixed *ortho*- and *para*-substitution into highly selective formation of *para*-structures is thus the following. At pH < 2.5 polymer grows in the protonated pernigraniline form where the chain bears excess charge (positive polarons). In this case, *para* substitution becomes energetically more favorable and is the only possible way of electrophilic substitution, since it reduces the energy of the chain. Only in the case of *para*-substitution, polyconjugation is formed which ensures delocalization of positive charges in pernigraniline and decreases repulsion energy. Thus, protonation of polymer chain forbids *ortho*-substitution (the most typical for aniline), as well as all other types of electrophilic substitution ("head-to-head", "tail-to-tail") and leads to the formation of regular polyconjugated *para*-structure and polymer chains possessing high conductivity.

6.4. Termination of chain growth

Polymerization stops after depletion of one or other reactant (monomer or oxidant). The state of PANI after termination depends on the strength of the used oxidant, as well as on molar ratio oxidant/monomer. When strong oxidant is taken in excess, the resulting polymer remains in pernigraniline form. If the reaction is carried out under stoichiometric conditions, or with aniline in excess, pernigraniline is reduced to emeraldine at the end of the reaction.

Pernigraniline is much less stable than emeraldine and takes part in irreversible reactions with electron donor agents. Apparently, during PANI synthesis and after its termination, end imino groups of polymer undergo hydrolysis. This supposition is corroborated by the presence of oxygen in deprotonated PANI samples, as well as by the presence of absorption bands of hydrolyzed structures in IR and NMR spectra of polymers (Kriz et al., 2011; Kellenberger et al., 2011). In the opinion of the authors (Gospodinova et al., 1994), hydrolysis of active sites proceeds continually both during oligomerization and after termination of polymer growth. The hydrolyzed terminal amino group effectively blocks any further growth reactions, because it cannot add monomer; therefore, aniline oxidation cannot be considered as a process proceeding without termination (i.e., "living" polymerization).

6.5. Concluding remarks

As a final matter, it should be noted that chain reaction of aniline oxidation in alkaline, weakly acidic and acidic media proceeds according to different molecular mechanisms, and, as a result, products with different molecular structures, morphologies and properties are formed. This is caused by the existence of monomer and growing chain in protonated and deprotonated forms which possess different oxidation potentials. Oxidation of aniline at pH > 2.5 occurs with the participation of deprotonated monomer and growing chain which have low oxidation potentials; therefore, it may be performed entirely under the action of weak oxidant. The products of this reaction are oligomers with low degree of polyconjugation and low conductivity.

At pH < 2.5, the reaction occurs in a stepwise mode, and variation of the oxidation potential during the process is described by a curve passing through a maximum. In the induction period, the oxidation follows low-energy path at a potential of +0.5 V and yields cyclic dimers having phenazine structure. The limiting stage is "germination" of phenazine rings (nucleates of future polymer chains), when the growth of linear polyconjugated chain starts. Phenazine has an oxidation potential of +1.05 V, and oxidative addition of monomer to this fragment is possible only in the presence of strong oxidants. If the oxidant has a potential lower than +1 V, the process is inhibited at the stage of formation of phenazine units. However, addition of 1 mol % of strong oxidant is sufficient to overcome this barrier. At the chain propagation stage, the oxidation potential again drops down to +0.7 V, and the reaction can proceed further under the action of weak oxidant.

Chain propagation involves protonated monomer and the chain possessing excess charge (positive polarons). Chain protonation constrains the reaction to produce regular polymer chains with *para*-coupled units. Regular polymer chain possessing extended conjugation system is characterized by the lowest energy due to delocalization of positive polarons over the conjugation system. Growth of other structures is terminated, since they become unfavorable from the viewpoint of energy.

According to protonation constants of nitrogen-containing groups of monomer and chain taking part in polymerization, the process can be divided into three ranges (Scheme 10):

- pH > 3.5, where all types of nitrogen-containing groups are deprotonated, and oxidation proceeds according to regularities of synthesis of typical dielectric polymers (e.g., phenols) with the formation of irregular chains having molecular masses of up to 5 000.
- 3.5 > pH > 2.5, where monomer is protonated, while imino groups of chains are not; imbalance of redox processes is observed, this leading to slow formation of cyclic dimmers with phenazine structure.
- pH < 2.5, where imino groups of chains, as well as monomer, are protonated, the balance of electron exchange between chain and monomer is restored, but all growth patterns except *para*-substitution are forbidden. This leads to the formation of regular polyconjugated chains with high molecular weight.

Thus, regular structure of PANI responsible for its unique properties is formed at pH < 2.5 due to protonation of imino groups in polymer chain.

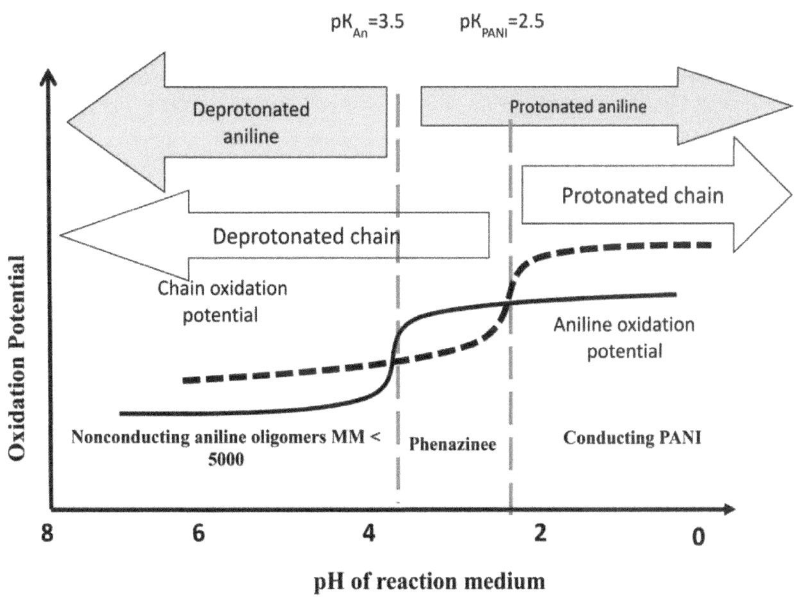

Scheme 10. Oxidation of aniline depending on protonation constant of monomer pK_{An} and that of chain imino groups pK_{PANI}.

7. The formation of polyaniline supramolecular structures

The important feature of PANI synthesis is heterophase nature of polymerization. As a rule, synthesis begins under homogeneous conditions, since both aniline and oxidant are well soluble in acidic aqueous media. However, already in the initial step, the reaction becomes heterophase. This process is called "precipitation polymerization", because it gives a precipitate (insoluble polymer). However, PANI precipitation cannot be considered as irregular agglomeration. Complex morphology of particles and layers formed by macromolecular chains in the course of polymerization gives evidence that self-organization of macromolecules occurs during synthesis. From theoretical and practical points of view, the mechanism of PANI self-organization and reasons for unusually high selectivity of oxidative polymerization of aniline are the main problems. It is the pattern of assembly of polymer chains in macroscopic sample and molecular structure of these chains that determine properties of material and application areas.

7.1. Types of supramolecular structures

Recently, numerous types of PANI supramolecular structures have been obtained. Their synthesis and potentialities are discussed in several reviews (Bhadra et al. 2009; Stejskal et al. 2010; Wang et al 2011). The basic structures are one-dimensional (nanofibers, nanotubes and nanorods) (Fig 4c, 8b, 10a, 12a,b); two-dimensional (ribbons, plates) (Fig 4b, 8a), and three-dimensional (granules, microspheres and nanospheres) (Fig 4d,f 12c). Their combination and more complicated organization give rise to a vast diversity of structures. Different hierarchical formations are built of fibers coming from the common centre or axis (Fig. 10 b, c, d,). One-dimensional structures can be packed in bundles or form a sort of network, flovers (Fig. 12 e) or dendrites. PANI fibers can also form two-dimensional structures called "micromats". These are rhomboid particles, "weaved" to produce mats of several layers of fibers coming from the vertex of the rhombus and crossing at right angle (Fig.12 d). Among three-dimensional structures, microspheres and hollow microspheres should be mentioned (Fig 4 a). Their outer diameters fall in the range from a half of micrometer to several micrometers, and the thickness of their walls varies from nano- to submicrometer sizes. The spheres can serve as centers of fiber growth or, conversely, the fibers can grow inside the spheres. Under certain synthesis conditions, one or another structural type dominates, thus virtually determining the morphology of the polymer as a whole.

PANI supramolecular structures may form not only in the volume of reaction medium, but also on the interface between two phases. In the course of aniline polymerization, thin films of polymer are generated on the surfaces contacting with polymerization medium (Malinauskas, 2001). These surfaces are reactor walls, macroscopic support immersed in polymerization medium, or nanoscopic carriers dissolved or dispersed in this medium. When the synthesis is completed, all these surfaces become covered with PANI layer with a thickness ranging from several nanometers to hundreds of nanometers. PANI layers, as a rule, have homogeneous thickness and are strongly attached to the surface. Adsorption onto macroscopic support leads to the formation of films, and combination of PANI with nanoscopic carrier may give colloidal dispersions of polymer (Sapurina & Stejskal, 2008).

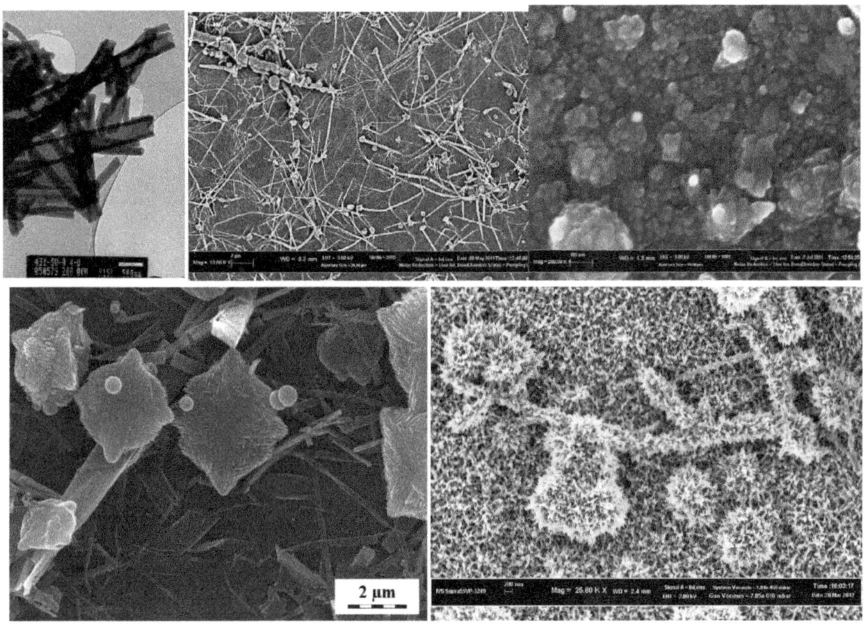

Figure 12. Supramolecular structures of PANI and aniline oligomers: a) PANI nanotubes, b) nanofibers, c) PANI granules. d) The structure of two-dimensional oligoaniline, "nanomats", in the presence of nanospheres and nanoribons. **e)** Hierarchical nanostructure of PANI flowers.

All the supramolecular structures are formed in the course of the synthesis, and the synthesis conditions exert the decisive effect on this process. Although the morphology is very sensitive to the polymerization conditions, it proved to be reproducible when the experimental conditions are strictly obeyed. Smooth transition from one structural type to another with the changes in the synthesis parameters is described in many publications (Steskal et al 1998; Stejskal et al 2008; Ding et al., 2008; Laslau et al 2009). In a number of publications, the stages of formation of supramolecular structures in the course of the synthesis are traced, the data on increase in globule diameter, lengthening of tubes and fibers are given. All these data evidence the processes of specific self-organization occurring during aniline polymerization and ceasing after its termination.

Recent studies have clearly shown that ability for self-organization of polymer chains is inherent to PANI (Huang & Kaner, 2006). The mechanism of formation of PANI supramolecular structures is common regardless of the results (thin films, stabilized colloids, precipitated particles of one-, two- or three-dimensional structure). Self-organization processes are the logical continuation of molecular reactions of synthesis allowing to explain the formation of a given type of morphology under different experimental conditions.

7.2. Mechanism of polyaniline self-organization during the synthesis

The suggested mechanism of PANI self-organization is based on the concepts of adsorption and agglomeration of nucleates (phenazine rings) (Sapurina & Stejskal 2008). According to molecular mechanism of synthesis, oxidation in pH range 2.5 – 3.5, as well as the initial step of oxidation at pH < 2.5, yields aniline oligomers having cyclic phenazine structure. It is known that phenazines are planar aromatic rings with high π-electron density. Phenazines are insoluble in aqueous reaction medium and tend to adsorb to surfaces and agglomerate. Phenazine adsorption occurs at the expense of π-electron interaction (one of the most energy-consuming types of physical bonds). Its energy is much higher than that of common van der Waals forces, it is comparable with hydrogen bond energy and estimated to be tens of $kcal{\cdot}mol^{-1}$ (Hoeben et al. 2005). Phenazines are planar aromatic structures belonging to the class of discotics. They are prone to crystallization with the formation of stacks and one-dimensional columns or two-dimensional planar templates. Intermolecular binding leading to the formation of columns occurs due to π-π interaction between aromatic rings. The processes of regular assembly of phenazine heterocycles are the most energetically favorable. However, they are not always realized during the synthesis and may give way to the formation of chaotic spherical agglomerates. We suppose that regular and chaotic types of phenazine ring agglomeration create different types of templates which then give rise to polymer chains. This hypothesis allows to explain the formation of various PANI supramolecular structures and gives insight into a number of important properties of the polymer.

7.3. The interrelation between processes of formation of PANI supramolecular structures and molecular reactions of PANI synthesis

The process of formation of polyaniline supramolecular structures corresponds to the paths of molecular reactions. In common with molecular reactions, it depends on pH of reaction medium and oxidation potential of oxidizing agent. Let us consider general rules of formation of supramolecular structures under different experimental conditions.

7.3.1. The role of acidity and oxidation potential

7.3.1.1. Oxidation at pH > 2.5

In alkaline media, aniline oxidation leads to the formation of oligomers with heterogeneous molecular structure and molecular masses of 3 000 – 5 000; these chains are little capable of self-organization. Oligomer chains are poorly soluble in alkaline medium, and, therefore, during oxidation, trivial aggregation and precipitation of amorphous structures occur. These heterogeneous oligomers form amorphous precipitate (Fig.4e; 6), inhomogeneous films possessing low adhesion to the surface and micro- and nanospheres (Fig 4a). The reason for the formation of spheres is a poor solubility of aniline at high pH and the conditions peculiar to emulsion polymerization (Sapurina & Stejskal 2010). In alkaline medium, aniline is present in the non-protonated form and hence poorly soluble. It exists as droplets in the reaction medium. These droplets may be stabilized by anions or surfactants present in the solution; consequently, microemulsions or nanoemulsions with sufficiently narrow particle size distribution appear. The oxidation proceeds mainly on the surfaces of droplets; they are covered with a layer of oligomer, but can contain the non-consumed monomer in their internal areas. In this case, hollow spheres are formed (see Fig.16). After addition of acid to the solution, the majority of products obtained in alkaline medium dissolve (as a result of protonation of nitrogen-containing structures). Most of these products are also soluble in organic solvents, such as chloroform, tetrahydrofuran, alcohols (Laska & Widlarz, 2005). Oligomers do not retain their initial morphology under heating; they melt and form monolith (Stejskal & Trchova 2012).

In going to neutral or weakly acidic media, molecular mass of oligomers decreases, and the content of phenazine rings rises. Oligomers become more homogeneous in composition and are prone to crystallization and self-organization. Products of oxidation in weakly acidic media demonstrate planar two-dimensional or one-dimensional morphology (Fig. 4b, f; 8a). These are particles like plates and bands or rhombic micromats and sticks (Fig 12d). We suppose that they are formed as a result of assembly of phenazine columns, as well as tri- and tetramers containing phenazine rings. Two-dimensional and one-dimensional structures differ in durability. Bands and plates with rather weak bonds between phenazines lying in one plane tend to dissolve at pH < 4 and rearrange to form fibers (Yu et al., 2011). At low pH, two-dimensional morphology is not formed; one-dimensional structures with interactions at the expense of more strong π,π-overlapping of adjacent planes prevail.

In the pH range 2.5 to 3.5, the majority of oligomers are phenazines insoluble in water. Compact phenazine molecules of the same type organize into particles of certain geometry. The materials obtained within this pH range possess the highest degree of crystallinity (Prathap & Srivastava. 2011). Here one-dimensional structures (needle-shaped crystallites with smooth surface or fibers which, as a rule, are assembled into more complex formations) prevail (Fig.8b). They consist of columns made of phenazine nucleates which have not yet started growing polymer chains. The formation of all types of oligomers proceeds at low oxidation potentials (starting from +0.4 V) according to oligomerization pattern, i.e., oxidation of reactive deprotonated form of aniline. Therefore, the reaction may proceed under the action of weak oxidants. Due to the absence of polyconjugation, these materials demonstrate low conductivity ($< 10^{-8}$ S·cm^{-1}).

7.3.1.2. Oxidation at pH < 2.5

At pH < 2.5, in the initial stage of the synthesis, as well as in the 2.5 – 3.5 range, phenazine nucleates of the same type are formed. There are no long oligomers of heterogeneous structure in this case. Aggregation of phenazines leads to the formation of template of future polymer particle, its shape being programmed in the first stage of synthesis. When the induction period is completed, nucleates start to grow polymer chains and reproduce shape of the template. In spite of the presence of only one type of nucleation centers, the structure of future PANI may vary. In this case, structure is determined by the conditions of self-organization of phenazine nucleates. At pH < 2.5, type of supramolecular structure depends on the duration of induction period (the stage of formation, accumulation and self-organization of nucleates). The long induction period favors organized assembly of phenazine into columns and is followed by the growth of one-dimensional structures, while short induction period resulting in fast "germination" of nucleates leads to chaotic agglomeration of nucleates and formation of spherical PANI particles (Sapurina & Stejskal 2012).

The use of a strong oxidant (e.g., persulfate) which initiates fast "germination" of nucleates and does not allow them to reach concentrations sufficient for assembly leads to the polymer assembled from granules (Fig.12c). In the presence of weak oxidants, when induction period is long, and phenazine cannot "germinate", mainly one-dimensional structures are formed (Fig.12a, b). The frames of hierarchical and one-dimensional structures are single phenazine columns or groups of columns assembled into more complex agglomerates. Phenazine agglomerates start growing polyconjugated chains under the action of a strong oxidant with a potential higher than + 1.05 V. In the absence of this oxidant, the transition occurs under the action of oxygen (+1.23 V).

Growth of polyconjugated chains leads to the formation of conducting material possessing properties of organic semiconductor. Supramolecular PANI structures were shown to be very stable; they are not destroyed during polymer carbonization and apparently soluble only in N-methylpyrrolidone. Carbonization of PANI does not lead to melting, as opposed to that of oligomers. The structure of one-dimensional and three-dimensional particles is retained, although their dimensions decrease (Zhang & Manohar 2006). We suppose that high energy of binding of initial phenazine fragments and interchain interactions cause high stability and low processability of PANI.

7.4. The condition of different types of supramolecular structures formation

All supramolecular structures were discovered accidentally when refining the methods of PANI synthesis. The combination of synthesis parameters leading to the prevalence of a certain type of PANI particles is called "method of the synthesis of a supramolecular structure". In the literature, special consideration is given to obtaining one-dimensional structures in the absence of any structuring template. Methods of synthesis of PANI fibers and tubes were developed by different researchers under different experimental conditions. Let us consider experimental conditions of different methods in terms of the suggested mechanism of formation of one-dimentional supramolecular structures. According to the mechanism, type of PANI supramolecular structure should be determined by the duration of induction period. Long induction period favors organized assembly of phenazines and formation of one-dimensional structures, while short induction period followed by fast "germination" of phenazines leads to chaotic agglomeration and formation of spherical particles. Analysis of synthesis conditions and polymerization process shows that this rule works for the following methods known from the literature.

- *The dilution method* (Zhang et al 2006). Polymerization of aniline is performed under the action of persulfate at pH < 2.5 and at low concentrations of reactants (< 0.001 mol·L^{-1}). Conducting PANI has the morphology of fibers several tens of nanometers in diameter and several micrometers in length. Here, long induction period (the stage of accumulation of phenazine nucleates) is observed, since at low concentrations of reactants, rate of nucleate "germination" decreases.
- *The method of soft templates* (Ding HJ et al 2008). Polymerization of aniline is performed in the presence of weak organic acids under the action of both strong and weak oxidants; concentrations of reactants are higher than 0.1 mol·L^{-1}. Nanofibers and nanotubes with external diameters of 50–500 nm and lengths of up to 10 μm are produced. In the presence of weak organic acids, the initial acidity of the reaction mixture is above pH 3. Thus, the reaction conditions enter the pH interval from 2.5 to 3.5, where the rate of aniline oxidation is dramatically reduced, and long induction period is observed. Under these conditions, reactions of chain growth are reduced and only slow accumulation of phenazine nucleates takes place.
- *The method of decreasing pH* (Stejskal et al 2010). Polymerization of aniline (> 0.1 mol·L^{-1}) proceeds under the action of different oxidants at low concentrations of strong acids or in the absence of any acids. Polymerization yields nanofibers and nanotubes with external diameters of 50-200 nm and lengths of up to micrometers. Equivalent to *soft templates method* at low concentrations of strong acids or in the absence of any acid oxidation falls within the acidity interval with pH > 2.5 characterized by slow accumulation of nucleates.
- *The method of polymerization at the interface* (Huang & Kaner, 2006). Polymerization proceeds at the interface of organic phase containing aniline (> 0.1 mol·L^{-1}) and aqueous phase containing an oxidant (persulfate, Fe^{+3}). The reaction is carried out in the presence of acids dissolved in one phase. PANI nanofibers grow at the liquid–liquid interface separating the reactants. There are two reasons for the formation of PANI nanofibers at the interface between immiscible liquids: (1) decrease in medium polarity at the organic solvent/water interface and suppression of protonation of nitrogen-containing structures (which is equivalent to increase in pH and going to the acidity interval pH > 2.5, where slow accumulation of nucleates occurs); (2) decrease in polymerization rate which is limited by diffusion of reactants at the interface (which is equivalent to dilution).
- *The method of mixed solvents* (Zhou et al, 2007). Aniline (> 0.1 mol·L^{-1}) is oxidized with persulfate in the mixed water/alcohol medium in the presence of acids. The reaction gives PANI fibers. Similarly to polymerization at the interface, dilution of water with organic solvent leads to decrease in medium polarity. This process suppresses protonation of reactants, and the reaction falls within the acidity interval with pH > 2.5 characterized by slow accumulation of nucleates.
- *Methods based on cycling of electrical potential or on galvanostatic technique* (Guo & Wang. 2008; Gao & Zhou 2007; Molina et al, 2010). Polymers are produced during the electro-oxidation of aniline at pH < 2.5 in the galvanostatic regime or during potential cycling when the initial potential of more than +1.05 V is reduced down up to negative values.

Nanofibers or a network of nanofibers are formed on an electrode. Decrease in oxidation potential suppresses "germination" of phenazine nucleates and promotes accumulation of these nucleates.

- *Methods of weak oxidants* (Sapurina & Stejskal 2012). The methods are equivalent to electrochemical techniques.
- *The method of catalytic additives of aromatic diamines with para-structure.*(Tran et al., 2008) Polymerization of aniline (> 0.1 mol·L^{-1}) proceeds under the action of persulfate at pH< 2.5 in the presence of catalytic amounts of *para*-semidine, *para*-phenylenediamine etc. Shortening of induction period leads to the presence of higher fraction of one-dimensional structures (fibers) in the products. This method is the only known exception to the general rule. Apparently, introduction of diamines leads to changes in nucleate structure and fast accumulation of high concentrations of nucleation centers (according to the Scheme 9) promoting the formation of one-dimensional templates.

In all cases, decrease in temperature favors processes of self-organization of phenazine rings into columns and decreases fractions of other morphology types, while increase in acidity results in transformation of one-dimensional structures into granules.

7.5. Stages of self-assembly occurring in the course of PANI synthesis

At pH<2.5, synthesis of PANI involves two steps: induction period and growth of polymer chains. The process of formation of supramolecular structures also may be divided into two stages; we term these stages "nucleation period" and "period of polymer formation". These phases fully coincide with the steps of molecular synthesis; let us examine the processes included in these steps.

7.5.1. Induction (nucleation) period

In acidic media (at pH < 2.5) conducting PANI (emeraldine) can be obtained as an insoluble precipitate or nanolayer on template surface or dispersion of polymer in aqueous (or organic) phase. Polymer particles included in these formations may have various morphologies (granular, one-dimensional or dendroid). How are all these structures formed? The induction period (formation of insoluble cyclic dimers) is at the same time nucleation stage (i.e., adsorption and association of phenazine nucleates and creation of all types of supramolecular structures). Adsorption of nucleates onto support leads to the formation of PANI films and adsorption onto stabilizer carrier to formation of colloidal particles (Stejskal & Sapurina, 2005). Association of phenazine oligomers results in the appearance of templates of presepitate (Scheme 11).

While phenazine aggregation in reaction medium depends only on synthesis conditions, formation of films and colloids is influenced by the nature of support and carrier. Analysis of literature shows that morphology of thin films and colloidal PANI particles is formed in accordance with the concept of selective interaction between phenazine and suface:

- PANI films grow more actively on negatively charged "anionic" surfaces than on positively charged "cationic" ones (Onoda et al 2006; Shinkai et al 2005).
- The rates of PANI layer growth on hydrophobic templates are higher than those of growth on hydrophilic materials (Onoda et al, 2006; Shinkai et al 2005).
- PANI morphology depends on hydrophilicity/hydrophobicity of the template. Films with homogeneous thickness are formed on hydrophobic surfaces, while on hydrophilic templates, island-like films or polymer layers with rough profiles are grown (Scheme 11) (MacDiarmid 1997).

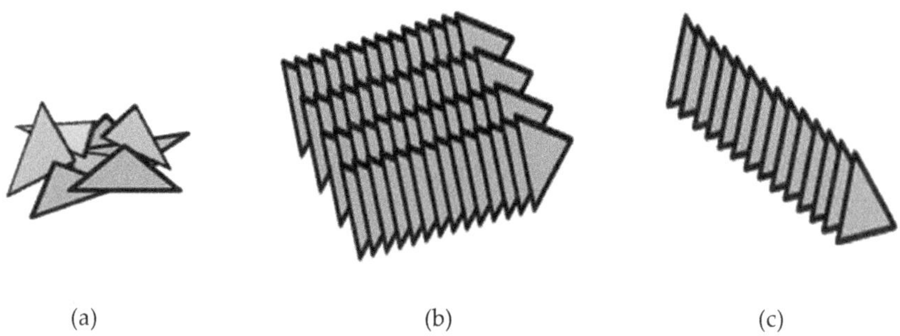

Scheme 11. Agregation of phenazine nucleates: (a) chaotic agglomeration b) formation of two-dimentional and c) one-dimentional templates.

Peculiarities of film growth and morphology of these films can be explained by the fact that phenazine nucleate is a hydrophobic organic base. As a base, it interacts strongly with anionic sites and is distributed more uniformly on hydrophobic surfaces. Specific influence of magnetic materials and magnetic field on the growth of PANI film was also revealed. Particularly, some authors observed preferential growth of polymer layer on the boundaries of ferrite domain walls (Fig.13) (Babayan et al. 2012), change of film morphology under the action of magnetic surface, influence of magnetic field on the rate of film formation and its structure (Cai et al. 1997; Ma et al. 2008). The above-listed phenomena may be related to magnetic cation-radical nature of phenazine nucleates showing up under the conditions of synthesis.

The influence of dispersed forms of carrier favoring the formation of PANI colloidal dispersions is also very specific. Depending on the chemical nature of the carrier, it can accelerate aniline polymerization; the more its surface area, the higher the polymerization rate (Stejskal et al. 2003). According to the molecular mechanism, nucleation stage proceeds at low oxidation potential, but culminates in the energy-consuming stage (addition of first monomer unit to phenazine). Certain types of carriers are able to activate phenazine rings and accelerate their "germination". Activating carriers include ferrites and carbon materials (soot, graphite, carbon nanotubes). In the presence of these additives, induction period can be considerably shortened, and polymerization rate increases many times (Konuychenko et al. 2006b; Babayan et al. 2012). It was also established that on electrodes made of carbon materials, aniline polymerizes at lower potentials than on platinum electrode (Matsushita et al. 2005). Apparently, in the case of carbon electrode, activation takes place due to the influence of π-electron structure of carbon on π-electrons of phenazine rings, and in the case of ferrites, activation is caused by magnetic interactions. After addition of first monomer unit to phenazine, chain growth proceeds with a relatively low oxidation potential. Thus, "germination" is an "energy barrier" between nucleation stage and the stage of polymer formation. After the start of chain growth, nucleation is virtually terminated. Polymer is formed on the basis of the grown nucleation centers.

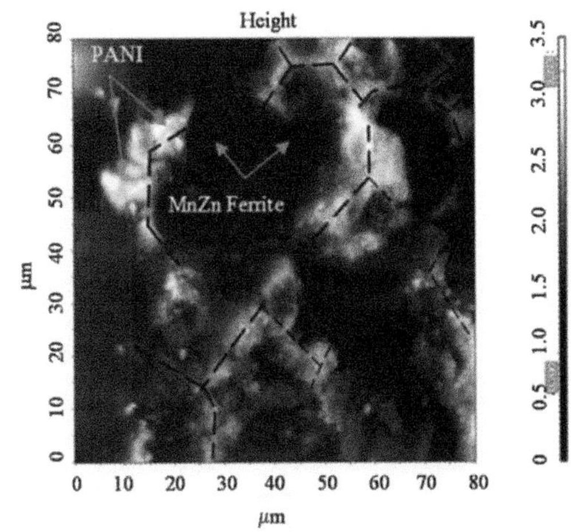

Figure 13. Thin PANI film with a «honeycomb» structure that repeats the microstructure of ferrite grain boundaries prepared on the surface of MnZn ferrite by *in-situ* polymerization

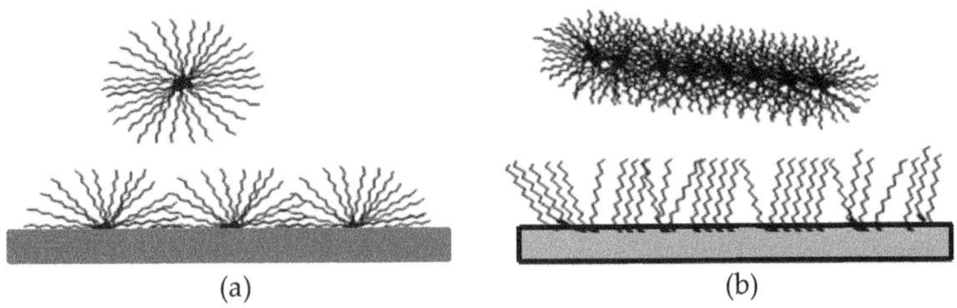

Scheme 12. The formation of PANI supramolecular particles of 3D and 1D morphology and films on (a) hydrophilic and (b) hydrophobic support.

7.5.2. Synchronous growth of PANI supramolecular structures

The existence of "energy barrier" between nucleation and growth stage has an influence on polymer morphology. Rather frequently PANI consists of particles of one type with practically similar dimensions. Fig. 14 shows spherical particles of PANI colloid and polymer nanotubes. In both cases, low size polydispersity is observed. This result is due to the fact that all growth processes proceed under similar conditions and during similar periods of time, i.e., growth starts and ceases simultaneously for all particles. With the start of growth reaction, nucleation is terminated, and new nucleation centers are not supplied any more. Only nucleates which have had sufficient time to start growth can now take part in propagation. Reaction is completed also synchronously after one of reactants is exhausted. All these processes are responsible for the fact that in the absence of secondary aggregation, PANI particles with a very narrow size distribution are formed. Polydispersity of spherical colloidal particles and globules comprising PANI precipitate is 1.05 – 1.1 (Stejskal et al.1996) Nanofiber diameters and nanotube wall thicknesses also have narrow size distributions.

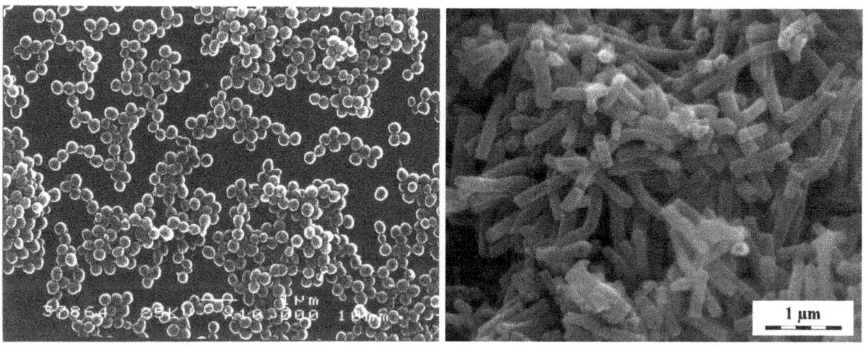

Figure 14. PANI colloidal particles a) and PANI nanotubes b) with low polydispersity of particle size. Scanning electron microscopy.

7.6. The growth of polymer chains possessing electronic and ionic conductivity

By the beginning of the growth stage, initial fragments of future polymer chains are already associated or attached to the surface, and the foundations of supramolecular "architecture" of PANI are laid. During exothermic growth reaction, templates (aggregated phenazine nucleates) which have added monomer units proceed to increase polymer mass at a high rate. The growth of PANI and other conducting polymers is dramatically different from the well-known propagation of dielectric polymers. Conducting polymer grows not as an individual chain, but as a conglomerate of chains possessing electronic and ionic conductivity (Blinova et al. 2007). During the growth process, both monomer and oxidant are soluble, i.e. dispersed at the molecular level. At the same time, growing chains are agglomerated; they are already bound at the expense of interactions between phenazine nucleates and/or between phenazine nucleates and the surface and therefore "feel" each other.

Molecular weight of the polymer increases in the following manner (Scheme 13). Oxidant oxidizes one terminal amino group in agglomerate of polymer chains. Positive polaron generated after oxidation is easily delocalized within the whole agglomerate and stabilized by mobile acid anion. Oxidation sites created by oxidant are redistributed within the whole conducting particle and minimize its energy. This redistribution determines the following additions of monomer units. Addition of monomer unit may occur not in the point where interaction with oxidant took place. PANI particle takes part in redox interaction as a single entity (Sapurina & Stejskal, 2008; Stejskal et al. 2010). Conducting polymer propagates as a self-organizing agglomerate of polymer chains; these chains response to oxidation/reduction processes as a single object and control shape of polymer particles. Cooperative interaction between oxidant, polymer particle and monomer increases probability of the process manifold and serves as a factor accelerating polymerization (the so-called "autoacceleration").

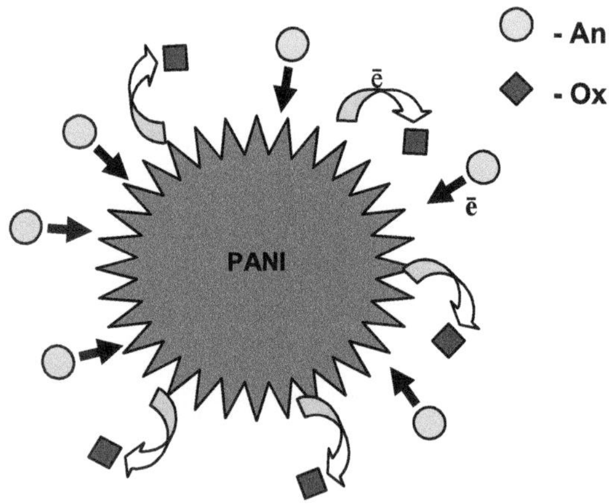

Scheme 13. Mechanism of heterophase PANI growth involving electronic and ionic transport processes (see text).

7.7. Peculiarities of formation of three-dimensional and one-dimensional structures

PANI granules, as well as spheres, belong to three-dimensional formations, but their structure and properties differ dramatically. Spherical shape is created by aniline droplet which turns into an agglomerate of non-conducting oligomers after oxidation. The basis of conducting granule is a chaotic cluster of phenazine nucleates which has initiated growth of regularly packed polyconjugated chains.

The formation of granules proceeds according to the diffusion-limited aggregation mechanism. Initial phenazines form a cluster, and this cluster starts growth of polyconjugated chains. Then, the second cycle of cluster-cluster diffusion-limited aggregation may take place, and a particle resembling blackberry is formed. The second cycle is observed in the beginning of the propagation step and inherent to particles which are not bound with surface. Apparently, this process is caused by growth of polymer chains and changes in surface properties of the initial cluster and, therefore, its interactions with medium. "Granule" is a common term for a particle assembled from clusters. In the course of the propagation step, granule grows at the expense of increase in dimensions of individual clusters, while the number of clusters in the particle remains constant. The cluster is a sphere; the granule has a quasi-spherical shape. In the final stage of the synthesis, insoluble granules aggregate to form micrometer-size particles of irregular shape (Spivak et al. 2010).

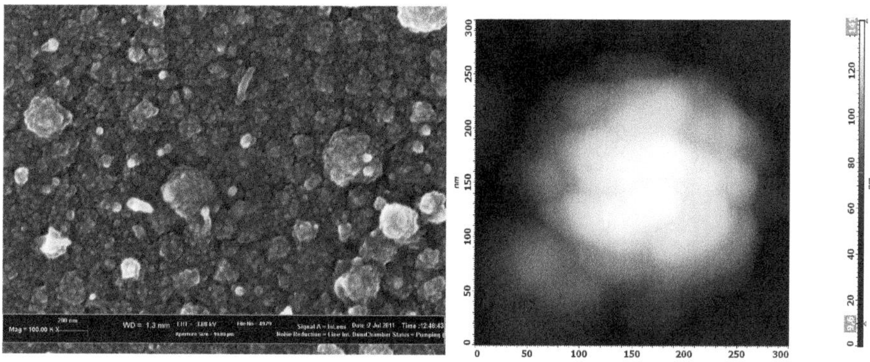

Figure 15. Hierarhical structure of PANI granules. a) SEM image, b) Atomic Forse microscopy.

Similarly to granules, one-dimensional structures (fibers and tubes) grow in the course of propagation. Individual column of phenazine nucleates may serve as a template for fiber growth. The presence of larger initial template is necessary for tube growth; this template should shape the inner cavity of polymer nanotube. Depending on synthesis conditions, templates may vary. The most frequently observed templates are crystalline oligomers of aniline forming tubular needle-shaped structures (Stejskal et al. 2006). However, PANI tubes with square and rectangular cross-sections were also obtained; these structures arise with the participation of inorganic crystals (Ding et al. 2009). Template should not necessary be one-dimensional and have the length corresponding to tube size. In the propagation step, lengthening of fibers occurs, and, by analogy with fibers, tube length also increases with retaining the initial shape of this tube. We suppose that during growth, central rod of one-dimensional structure lengthens at the expense of build-up of new cyclic oligomers. However, the formation of phenazine rings occurs not in the solution, but directly on the surface of butt end of growing fiber or tube. In essence, this process is analogous to growth of organic single crystal. It is less energy-consuming than formation of new phenazine sites in solution and occurs at lower oxidation potential (Sapurina & Stejskal 2008). Simultaneously, wall thickness also increases. Polymer chains arising in heterogeneous medium are regularly organized and interact with adjacent chains via hydrogen bonds. Apparently, the outer diameter of lengthening one-dimensional structure is controlled by the fact that the wall-forming chains "feel" their environment. The whole system, in common with crystal, controls its dimensions during growth.

Heterophase growth of polymer particles (involving processes of electronic and ionic transport of oxidation sites) allows to reach high level of self-organization. With that, a certain shape of particle (from spheres with clear boundaries to tubes with definite diameters) is formed. One is tempted to compare oxidative polymerization of aniline with the processes of nucleation and growth of crystals. Slow and energy-consuming stage of formation of initial crystallites occurs in solution. The subsequent crystal growth proceeds directly on the surface of a crystallite and develops very fast. Just as crystallite surface catalyzes processes of "assembly" of crystal structure, redox-active conducting polymeric agglomerate catalyzes and organizes assembly of monomer units.

To summarize the section devoted to stages of assembly of PANI supramolecular structures and influence of oxidation potential and pH of reaction medium on PANI morphology, we should note the following points. The formation of uniform phenazine nucleates always leads to high supramolecular order (be it organization of oligomers or polymer chains). In contrast with the case of oligomers of heterogeneous structure, there are no amorphous formations. The assembly proceeds according to reactions of molecular synthesis. Supramolecular architecture of conducting PANI laid down in the induction period (during organization of phenazine rings) is fixed and completed in the propagation step during the formation of hydrogen bonds between regularly packed polymer chains. These types of intermolecular interactions are the basis of a diversity of supramolecular PANI structures.

8. The principle of organization of PANI supramolecular structures

Comparison between different types of PANI supramolecular structures obtained under similar experimental conditions reveals many common features. Not only particles of one type (e.g., granules) have similar dimensions; linear dimensions of structures of different types also coincide. Thus, in one reaction vessel, PANI films and colloids may be obtained simultaneously (or films and granular precipitate). Radii of colloidal particles are equal to film thicknesses, dimensions of precipitated PANI granules are also in good agreement with the thicknesses of films (Riede at al. 2002; Sapurina et al 2002). Thickness of nanotube walls is always identical to radius of fibers and simultaneously equal to thickness of walls of nanospheres obtained in the same experiment as one-dimensional structures (Fig.16). These facts indicate that all the supramolecular PANI objects have common structure. The correspondence between linear dimensions of supramolecular objects (film thickness, thickness of polymer tube wall, fiber radius) and molecular mass of PANI forming a given structure was also revealed. Moreover, thickness of an object coincides with the length of stretched polymer chain (when calculation of chain length was performed with regard for zigzag structure (Sapurina et al 2002) and with the use of experimentally established number-average molecular weight of polymer).

Experimental data concerning correlation between linear dimensions of different types of supramolecular structures and dependence between dimensions and molecular mass of polymer can be explained by the hypothesis that PANI structures are organized on the principle of radial arrangement of polymer chains. In spherical particles, polymer chains radiate from the center. Fibers and walls of PANI tubes consist of polymer chains organized perpendicularly to the axis of one-dimensional structure, and PANI films attached to a surface have a structure of polymer brush . This type of organization may be promoted by diphilic nature of growing chain. PANI chain consists of cyclic nucleate and linear "tail". Under synthesis conditions, phenazine nucleates are deprotonated, and, being deprotonated nitrogen-containing structures, they are hydrophobic. At the same time, growing polymer chains are protonated and hydrophilic. Naturally, hydrophilic chains grow mainly towards aqueous phase, and away from hydrophobic nucleate. Diphilic "nucleate-chain" structure resembles micelle-forming surfactant molecule, with the difference that hydrophobic fragments of PANI chains are strongly attached to each other or a surface.

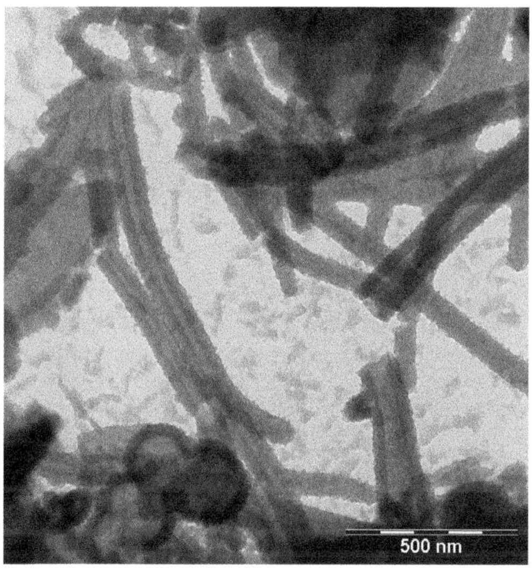

Figure 16. PANI nanotubes and nanowires together with hollow PANI spheres. (TEM image)

9. Supramolecular structure of polyaniline – The model of inhomogeneous metal

Since the end of the 1980s, mechanism of charge transport in conducting polymers has been studied with the use of four representatives of this class (polyacetylene, polyaniline, polypyrrole and polythiophene) as examples (Skotheim, 2007). All these conducting polymers demonstrate similar features. In the doped conducting form, they possess comparatively low conductivity (less than 10^2 S·cm^{-1} and 10^4 S·cm^{-1} for polyacetylene) characteristic more of semiconductors than metals. Activation character of temperature dependence of conductivity is also typical for a majority of samples. At the same time, the data obtained in the studies of magnetic, optical, dielectrical properties, as well as thermopower, indicate the presence of conduction electrons in doped polymers. Thus, polymers demonstrate dual character and reveal properties of both semiconductors and metals. Low charge carrier mobility (10^{-1} – 10^{-4} $cm^2 \cdot V^{-1} \cdot s^{-1}$ for PANI (Harima et al., 2001) and their high concentration ($3 \cdot 10^{20}$ $spin \cdot g^{-1}$) comparable with carrier concentration in metals provide evidence that there are mechanisms of localization of charge carriers in macroscopic samples.

To explain all these experimental data, the model of structurally inhomogeneous conducting polymer was suggested; in this model, the sample is considered as an inhomogeneous metal. The polymer is represented as "a system of conducting islands in dielectric matrix". According to the model, crystalline "islands" assembled of regularly packed polyconjugated chains exhibit metallic conductivity. Charge carrier mobility and polarizability inside this island are high. However, islands are distributed in a weakly conducting medium which creates a barrier for charge transport. The medium with low conductivity is a set of separate transport paths connecting these conducting islands into one common net. This model presents the best description of properties of conducting polymers, allows to explain localization of charge carriers and one-dimensional transport which is frequently observed and corresponds to the linear dependence $\sigma = f(T^{1/2})$ (Epstein et al, 1994; Alechin 2006).

The conductivity of a crystalline island is estimated to be about 10^6 $S \cdot cm^{-1}$, i.e., 4-6 orders of magnitude higher than that of a typical macroscopic sample. It was shown that the island possesses three-dimensional metallic conductivity. Crystalline areas demonstrate rather narrow size distribution and are uniformly dispersed in amorphous matrix. Depending on the used method of synthesis, doping and subsequent treatment, dimensions of crystalline areas may vary from 5 nm to 100 nm (Mazerolles et al., 1999). The level of conductivity of macroscopic sample is determined by degree of overlapping between boundaries of conducting "islands" and may vary over wide limits.

In deprotonated polymers, volume fraction of islands is low, but it increases with increasing degree of doping. Transition from non-doped polymer to doped one demonstrates percolation character. At the doping degree of 20%, percolation transition occurs, and conducting paths between metallic islands start to form; therefore, conductivity increases dramatically. The transport between islands requires activation. With increasing temperature, activation barriers between conducting elements may be more easily overcome, and conductivity grows. Changes in conductivity and inductive capacity depend on temperature and degree of protonation, and this dependence demonstrates fan-like pattern. Increase in the volume of crystalline island during swelling in organic solvent also may promote transition of macroscopic sample into metallic state. The suggested model of "disorganized metal" explains the presence of metallic state in conducting polymers, low level of macroscopic conductivity and activation-requiring transport in macroscopic sample (caused by inhomogeneity in polymer structure).

The model of structurally inhomogeneous PANI was proved directly in the experiment. Inhomogeneous morphology of the polymer, as well as inhomogeneity in the properties of macroscopic samples, was revealed by X-ray structural analysis, luminescence, ellipsometry, electron and atomic force microscopy (Leite et al 2008; Lux et al 2003). Crystalline islands were observed as dense inclusions in amorphous matrix with the aid of high-resolution transmission electron microscopy and atomic force microscopy (Mazerolles et al 1999). (Fig.17). The model of inhomogeneous metal describes properties of conducting polymers well, but it has a drawback: the reasons for appearance of island structure are not understood. While molecular structure of polyconjugated PANI chains demonstrates very high homogeneity, the mechanisms leading to the formation of this considerable inhomogeneity in structure and properties of macroscopic sample are unclear.

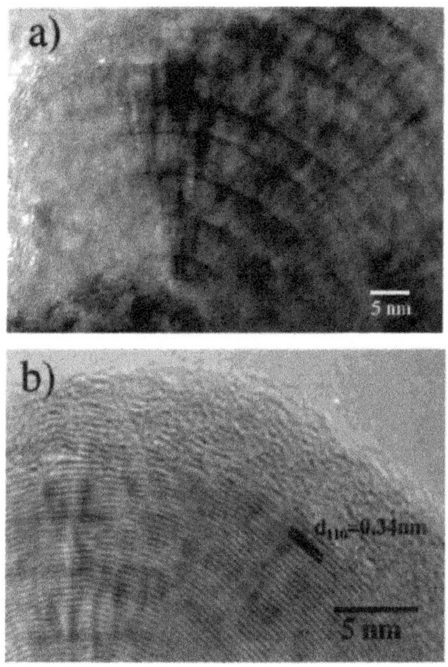

Figure 17. High resolution TEM image of PANI particles of granular structure.

From Mazerolles L, Folch S, Colomban P. Study of polyanilines by high-resolution electron microscopy. Macromolecules 1999;32:8504–8.

The mechanism of self-organization of polymer chains during doping process is an attempt to explain the formation of crystalline islands (Zuo et al, 1987; Zuppiroli et al 1994). According to this mechanism, appearance of polarons in solid phase of polymer is accompanied by processes of local self-ordering of chains and self-assembly to crystalline island structures, while disappearance of polarons leads to destruction of polymer islands. The suggested mechanism of self-organization in the course of doping is beneath criticism. Driving forces of this self-organization are unclear. The reasons for very irregular "gathering" of polymer chains in crystalline conglomerate (so that densities of neighbor areas of a sample differ by several times) are not understood either. According to this mechanism, we have to assume that self-organization into island structures occurs due to motions of macromolecules in solid polymer. Therefore, it is difficult to understand how high degree of chain ordering inside island leading to "metallic" transport is achieved. It is also impossible to explain the extremely high rate of self-organization processes. Changes in polymers proceed at the rates corresponding to mobility of low-molecular weight compounds in liquid phase (doping and dedoping of PANI may occur with a frequency of up to 30 Hz).

In our opinion, self-organization of PANI with the formation of "island" structures occurs in the stage of polymer synthesis. The formation of inhomogeneous structure of polymer proceeds according to the mechanism of heterophase growth of PANI as individual particle (Sapurina & stejskal 2008). We suppose that the PANI particles formed during polymerization are the inhomogeneous elements ("island" and its surrounding area).

Let us consider granular structure of PANI. Typical dimensions of spherical PANI clusters are tens of nanometers; these parameters are in agreement with sizes of conducting islands. The central part of a particle demonstrates high degree of organization of polymer chains. Similarly to the "island", this is a three-dimensional crystal, and it may support three-dimensional metallic conductivity (Joo et al 1994). On the periphery of a particle, organization is disturbed; every crystalline island is surrounded by amorphous mantle. This fact explains why at high degree of polymer crystallinity (60% for PANI) (Stejskal et al 1998) transport mechanism most often remains activation-dependent. Conductivity of macroscopic sample is limited by intervals between crystalline zones; here, transport is carried out via transport paths, this being in agreement with the frequently observed one-dimensional conductivity. Synchronous growth of particles of one type in the course of PANI synthesis guarantees uniform distribution of crystallites of the same size within polymer sample.

Behavior of PANI under external influence and during aging also agrees well with the model "particle – inhomogeneity element" (Travers et al, 1999). Oxidation/reduction, protonation/deprotonation and thermal aging of PANI result in changes on the periphery of a particle. These processes first cause response from the Curie spins present in amorphous zone. As a rule, they do not affect the central area of a particle. Reduction and deprotonation of emeraldine do not lead to complete disappearance of polarons. Non-conducting forms of PANI contain unpaired spins (10^{16} spin·g^{-1}), as well as acid which stabilizes positive polaron. It was shown that the content of the Pauli spins (populating crystalline areas) does not decrease after deprotonation. Loss of conductivity does not cause considerable reorganization of supramolecular structure (Pouget et al., 1994). The problem of structural inhomogeneity of PANI is very important, since it is the heterogeneity on supramolecular level that limits conductivity of polymers.

10. Concluding remarks

This review attempts to consider synthesis of PANI from a general viewpoint, starting with molecular reactions and assembly of supramolecular structures and ending with comparison between general concept of synthesis and the model of inhomogeneous metal explaining properties of conducting polymers. The information is presented succinctly, and we have no opportunity to give experimental proof of different points, but it can be found in the given references.

Apparently, PANI synthesis is not unique. Similar regularities are observed during synthesis of other conducting polymers by oxidative polymerization. During oxidation, pyrrole, thiophene and their derivatives demonstrate dependence between their properties, pH of reaction medium and oxidation potential; they are self-organized in the course of synthesis and form different types of supramolecular structures. There are reasons to believe that, similarly to PANI, high selectivity of synthesis and formation of polyconjugation in other polymers are related to the presence of excess positive charge on the growing chain. Evidently, the formation of aggregates between growing chains of conducting polymers involves electronic and ionic transport. Processes of molecular synthesis and assembly of supramolecular structures determine properties of conducting polymers. Understanding these processes would allow us to obtain materials with better characteristics.

Author details

I.Yu. Sapurina
Institute of Macromolecular Compounds RAS, Sankt-Petersburg, Russia

M.A. Shishov
St.-Petersburg State Electro-technical University "LETI" Saint-Petersburg, Russia

Acknowledgement

The book chapter was written with the support of grant of Russian Academy of Sciences and Goskontract N16.516.11.6034. The autors is also thankful to Dr. Jaroslav Stejskal for longstanding collaboration and support

11. References

Aleshin AN. (2006) Polymer nanofibers and nanotubes: Charge transport and device applications. *Adv Mater*;18:17–27.

Babayan, V. Kazantseva, N. E. Moučka, R. Sapurina, I. Spivak, Yu. M. Moshnikov V. A. (2012) Combined effect of demagnetizing field and induced magnetic anisotropy on the magnetic properties of manganese–zinc ferrite composites, *Journal of Magnetism and Magnetic Materials* 324, 161–172

Baizer M.M., Lund H., (1983) *Organic Electrochemistry*, Marcel Dekker, New York, 2nd ed, Part 1, P 52.

Bhadra S, Khastgir D, Singha NK, Lee JH. (2009) Pregress in preparation, processing and applications of polyaniline. *Prog Polym Sci*;34,783–810.

Blinova NV, Stejskal J, Trchová M, Ćirić-Marjanović G, Sapurina I. (2007) Polymerization of aniline on polyaniline membranes. *J Phys Chem B*;111:2440–8.

Blinova NV, Stejskal J, Trchová M, Sapurina I., Ćirić-Marjanović G. (2009) The oxidation of aniline with silver nitrate to polyaniline–silver composites. *Polymer*;50:50–6.

Bober P, Stejskal J., Trchova M., Prokes J., (2011) Polyaniline-silver composites prepared by the oxidantion of aniline with mixed oxidant, silver nitrate and ammonium peroxydisufate: The control of silver content, *Polymer* 52, 5947-52

Bober P., Stejskal J., Trchova M., Prokec J., Sapurina I., (2010) Oxidation of Aniline with Silver Nitrate Accelerated by p-Phenylenediamine: A New Route to Conducting Composites. *Macromolecules* 43, 10406- 13

Cai, L.T. Yao, S.B. Zhou, S.M. (1997) Effects of the magnetic field on the polyaniline film studied by in situ conductivity measurements and X-ray diffraction, *Journal of Electroanalytical Chemistry*, 421, 45-48.

Cai, L.T. Yao, S.B. Zhou, S.M. (1997) Improved conductivity and electrical properties of polyaniline in the presence of rare-earth cations and magnetic field, *Synthetic Met*, 88 205-208.

Chuang F.-Y., Yang S.-M. (2008) Cerium dioxide/polyaniline core-shell nanocomposites, *J. Colloidal and Interface Surface* 320. P. 194-9

Ćirić-Marjanović G., Trchová M., Stejskal J., (2006) MNDO-PM3 study of the early stages of the chemical oxidative polymerization of aniline *Collect. Czech. Chem. Commun* 71. № 7. P.1407-26.

Colomban Ph., & Tomkinson J.,(1997), Novel form of hydrogen in solids: the "ionic" proton and the "quasi-free" proton *Solid State Ionics* 97 123-34

Cram DJ and Hammond GS, (1964) *Organic Chemistry*, McGraw-Hill, London, 2nd ed, Chapter 11

Davied S., Nicolau YF., Melis F., Revillon A., (1995) Molecular weight of polyaniline synthesized by oxidation of aniline with ammonium persulfate and with ferric chloride, *Synth Met* 69 125-126.

Ding H., Long Y., Shen J., Wan, M., (2010) $Fe_2(SO_4)_3$ as a binary oxidant and dopant to thin polyaniline nanowires with high conductivity, *J. Phys. Chem. Bб* 114, 115-19

Ding HJ, Shen JY, Wan MX, Chen ZJ. (2008) Formation mechanism of polyaniline nanotubes by a simplified template-free method. *Macromol Chem Phys*;209:864–71.

Ding HJ, Wan MX, Wei Y. (2007) Controlling the diameter of polyaniline nanofibers by adjusting the oxidant redox potential. *Adv Mater*;19:465–9.

Ding Y, Padias AB, and Hall HK, (1999) Chemical Trapping Experiments Supports a Cation-Radical Mechanism for the Oxidative Polymerization of Aniline, *J. Polym. Sci., Part A: Polym. Chem.* 37:2569-2579

Ding ZF, Currier RP, Zhao YS, Yang D. (2009), Self-assembled polyaniline nanotubes with rectangular cross-sections. *Macromol Chem Phys*; 210:1600–6.

Ding Z, Sanchez T., Labouriau A., Iyer S., LarsonT., Currier RP., Zhao Y.,Yang D., (2010) Characterization of Reaction Intermediate Aggregates in Aniline Oxidative Polymerization at Low Proton Concentration *J.Phys Chem B*, 114, 10337-10346

Epstein AJ., Joo J., Kohlman RS., Du G., MacDiarmid AG., Oh EJ., Min Y., Tsukamoto J., Kaneko H., Pouget JP., (1994) Inhomogeneous disorder and the modified Drude metallic state o conductng polymers *Synth Met* 65, 149-57.

Gao Y, Zhou Y. (2007) Polyaniline nanofibers fabricated by electrochemical polymerization: A mechanistic study. *Eur Polym J*; 43:2292–7.

Geng Y, Li J, Sun Z, Jing X and Wang F, (1998) Polymerization of aniline in an aqueous system containing organic solvents *Synth. Met.* 96:1-6

Gizdavic-Nikolaidic M.R., Stanisavljev D.R., Easteal A.J., Zujovic Z.D. (2010) A Rapid and facile synthesis of nanofibrilar polyaniline using microwave radiation, *Macromol. Rapid Commun* 31, 657-61

Gospodinova N., Mokreva P., Terlemezyan L., (1994) Influence of hydrolysis on the chemical polymerization of aniline, *Polymer* 35, N14 , 3102-06

Guo SJ, Wang EK. (2008) Simple electrochemical route to nanofiber junctions and dendrites of conducting polymer. *Langmuir*;24:2128–32.

Hagiwara T, Demura T and Iwata K, (1987) Synthesis and properties of electrically conducting polymers from aromatic amines, *Synth Met* 18:317-21

Han J., Liu Y., Guo R., (2008) A novel templates method to nanofibers of polyaniline derivatives with size control, *Rapid Communication J Polym Sci PartA Polym Chem,* 46, 740-746

Harima Y., Patil R., Yamashita K., Yamamoto N., Ito S., Kitani A., (2001) Mobilities of charge carriers in polyaniline films, *Chem Phis Letters* 345, 239-244

Higashimura H., Kobayashi S. (2004) *Encyclopedia of Polymer Science and Technology, "Oxidative Polymerization"* Ed J.I. Kroschwitz, New York, Wiley,; 3 rol ed. Vol 10, P. 740.

Hoeben F.J.M. Jonkheijm P., Meijer E.W., Schenning A.P.H.J., (2005) About Supramolecular Assembling of π –Conjugated Systems, *Chem Rev* 105, 1491-1546,

Huang JX, Kaner RB. (2006) The intrinsic nanofibrillar morphology of polyaniline. *Chem Commun*;367-74

Inzelt G., Puskas Z.,(2004) Absorption and presipitation diring te redox transformation of phenazine, *Electrochem Acta* 49 , 1969-80

Joo J, Prigodin VN, Min YG, MacDiarmid AG, Epstein AJ, (1994) Phonon induced non-metal-metal transition of the doped polyaniline, *Phis Rev* 50, 12226-29

Kellenberger A., Dmitrieva E., Dunsch L., (2011) The stabilization of charge state at phenazine-like units in polyaniline under p-doping: an *in-situ* ATR-FTIR spectroelectrochemical study, *Phys. Chem. Chem.Phys.*, 13, 3411-3420

Konyushenko E.N., Trchova M., Stejskal J., Sapurina I., (2010), The Role of Acidity in the nanotubular growth of polyaniline, *Chemical Papers* 64 (1) 56-64

Konyushenko EN, Stejskal J, Šeděnková I, Trchová M, Sapurina I, Cieslar M, Prokeš J. (2006a) Polyaniline nanotubes: Conditions of formation. *Polym Int*;55:31–9.

Konyushenko EN, Stejskal J, Trchová M, Hradil J, Kovářová J, Prokeš J, Cieslar M, Hwang J-Y, Chen K-H, Sapurina I. (2006b) Multi-wall carbon nanotubes coated with polyaniline. *Polymer*;47:5715–23.

Kriz J., Konyushenko Trchova M., Stejskal J., (2011) NMR investigation of aniline oligomers produced in the oxidation of aniline in alkaline medium, *Polym Int* 60, 1296-1302.

Laska J., Widlarz J (2005) Spectroscopic and structuralcharacterization of low molecular weight fraction of polyaniline *Polymer* 46, 1485-95

Laslau C, Zujovic ZD, Zhang LJ, Bowmaker GA, Travas-Sejdic J. (2009) Morphological evolution of self-assembled polyaniline nanostructures obtained by pH-stat chemical oxidation. *Chem Mater*;21:954–62.

Leite FL, Alves WF, Mir M, Mascarenhas YP, Herrmann PSP, Mattoso LHC, Oliveira ON. (2008) TEM, XRD and AFM study of poly(*o*-ethoxyaniline) films: new evidence for the formation of conducting islands. *Appl Phys A – Mater Sci Process*;93:537–42.

Li X.-G., Huang M.-R., Duan W., Yang Y.-L., (2002) Novel multifunctional polymers fron aromatic diamins by oxidative polymerization, *Chem. Rev* 102. № 9. 2925-30.

Li Y., Wang Y.,Jing X., Zhu R., (2011) Early stage pH profile: the key factor controlling the construction of polyaniline micro/nanostructures *J Polym Res* 18: 2119-2131

Liao C., Gu M., (2002) Electroless deposition of polyaniline film via autocatalytic polymerization of aniline, *Thin Solid Films* 408 37-42

Lide DR, Frederikse HPR, (1995) *CRC Handbook of Chemistry and Physics* (Eds),, 76th Ed., CRC Press, New York; p. 8–49.

LiG., Jiang L., Peng H., (2010) One-dimentional polyaniline nanostructures with controllable surfaces and diameters using vanadic acid as an oxidant, *Macromolecules.* 40. P. 7890-94

Liu P, Zhang L. (2009) Hollow nanostructured polyaniline: Preparation, properties and applications. *Crit Rev Solid State Mater Sci*;34:75–87.

Losito I., Palmisano F., Zambonin P.G., (2003) o-phenylenediamine electropolymerization by cyclic voltammetry combined with electrospray ionization-ion trap mass spectrometry *Anal. Chem* 75. № 19. P. 4988-95.

Lux F, Hinrichsen G, Pohl MM. (2003) TEM evidence for the existence of conducting islands in highly conductive polyaniline. *J Polym Sci A: Polym Chem*;32:1957–9.

Ma, L.,Yan, . J. Gan, M.Y. Qiu, W. He, L. Li, J.F. (2008) Application of QCM technique in the kinetic study of polyaniline film formation in the presence of a constant (0.4 T) magnetic field, *Polymer Testing*, 27, 683-687.

MacDiarmid AG, (1997) *Synth Met* 84:27 . MacDiarmid AG, Polyaniline and polypyrrole: where are we headed? Synth Met;84:27–34.

Malinauskas, A. (2001) Chemical deposition of conducting polymers, *Polymer,* 42 3957-3972.

Matsushita M, Kuramitz H and Tanaka S, (2005) Electrochemical oxidation of aniline in neutral pH medium: Application to the removal o aniline based on the electrochemical polymerization on a carbon fiber. *Environ Sci Technol* 39:3805-10 .

Mazerolles L, Folch S, Colomban P. (1999) Study of polyanilines by high-resolution electron microscopy. *Macromolecules*;32:8504–8.

Molina J., del Rio A.I., Bonastre J., Cases F. (2010) Influence of the scan rate on the morphology of polyaniline grown on conducting fabric. Centipede-like morphology. *Synth. Met* 160. № 1-2. P. 99-107.

doNascimento GM, Silva CHB, Temperini SLA. (2006) Electronic structure and doping behaviour of PANI-NSA nanofibers investigated by resonance Raman spectroscopy. *Macromol Rapid Commun*;27:255–9.

Onoda M, Tada K and Shinkuma A, (2006) *Thin Solid Films* 499:61 . Onoda M, Tada K, Shinkuma A. In situ polymerization process of polypyrrole ultrathin films. *This Solid Films*;499:61–72.

Pan LJ, Pu L, Shi Y, Song SY, Xu Z, Zhang R, Zheng Y. (2007) Synthesis of polyaniline nanotubes with a reactive template of manganese oxide. *Adv Mater*;19:461–4.

Percec V and Hill DH, (1996) *Cationic Polymerization: Mechanism, Synthesis, and Applications;* Matyjashewski K, Pugh C (Eds); Marcel Dekker, New York; p. 646.

Pouget J.P., Oblakowski Z., Nogami Y., Albouy P.A., Laridjani M., Oh E.J., Min Y., MacDiarmid A.G., Tsukamoto J., Ishiguro T., Epstejn A.J., (1994) Resent structural investigation of metallic polumers, *Synth Met* 65 131-140.

Prathap M., Srivastava R., (2011) Morphological controlled synthesis of micro-/nano-polyaniline, *J Polym Res* 18, 2455-2467

Puskas Z., Inzelt G., (2005) Formation and redox transfrmatin o polyphenazine, *Electrochem Acta* 50 , 1481-90

Riede A., Halmstedt M., Sapurina I., Stejskal J. (2002) In-situ polymerized polyaniline films. 4. Film Formation in Dispersion Polymerization of Aniline J. of Colloid and Interface Science 248 413-418.

Sapurina I Yu., Stejskal J., (2012) Oxidation of Aniline with Strong and Weak Oxidants, *Russian J. General Chem* 82 N2 256-275,

Sapurina I, Osadchev AYu, Volchek BZ, Trchová M, Riede A, Stejskal J. (2002) In-situ polymerized polyaniline films 5. Brush-like chain ordering. *Synth Met;*129:29–37.

Sapurina I, Stejskal J. (2008) The mechanism of the oxidative polymerization of aniline and the formation of supramolecular polyaniline structures. *Polym Int;*57:1295–325.

Sapurina I.Yu., Stejskal J., (2010) The effect of pH on the oxidative polymerization of aniline and the morphology and properties of products , *Russ Chem Rew* 79 (12), 1123-1143

Shinkai S, Takeuchi M and Bae A-H, (2005) *Supramol Chem* 17:181. Shinkai S, Takeuchi M, Bae A-H. Rational design and creation of novel polymeric superstructures by oxidative polymerization utilizing anionic templates. *Supramol Chem;*17:181–6.

Skotheim TA and Reynolds JR, (2007) *Handbook of Conducting Polymers. Conjugated Polymers: Theory, Synthesis, Properties and Characterization,* CRC Press, Boca Raton,

Spivak, Y.M. Moshnikov, V.A. Sapurina, I. Kazantseva, N.E. (2010) Atomic force microscopy of polyaniline with globular structure, in: International scientific and applied conference opto-nano electronics and renewable energy sources, Technical University of Varna, Bulgaria,.

Stejskal J, Riede A, Hlavatá D, Prokeš J, Helmstedt M, Holler P. (1998) The effect of polymerization temperature on molecular weight, crystallinity, and electrical conductivity of polyaniline. *Synth Met;*96:55–61.

Stejskal J, Sapurina I, Trchová M, Konyushenko EN, Holler P. (2006) The genesis of polyaniline nanotubes. *Polymer;*47:8253–62.

Stejskal J, Sapurina I, Trchová M, Konyushenko EN. (2008) Oxidation of aniline: Polyaniline granules, nanotubes, and oligoaniline microspheres. *Macromolecules;* 41:3530–6.

Stejskal J, Sapurina I. (2005) Polyaniline: Thin films and colloidal dispersions. *Pure Appl Chem;*77:815–26.

Stejskal J, Trchova M, (2012) Aniline oligomers versus polyaniline *Polym Int* 61, 240-251,

Stejskal J, Trchová M, Fedorova S, Sapurina I, Zemek J. (2003) Surface polymerization of aniline on silica gel. *Langmuir;*19:3013–8.

Stejskal J., Kratochvil P., Helmsted M., (1996) Polyaniline dispersions. 5 Poly(vinyl alcohol) and Poly(N-vinilpyrrolidone) as Steric Stabilizers, *Langmuir,*12, 3389-92.

Stejskal J., Sapurina I., Trchová M. (2010) Polyaniline nanostructures and the role of aniline oligomers in their formation *Prog. Polym. Sci.*, Vol. 35. № 12. P. 1420.

Surwade S.P., Dua V., Manohar N., Manohar S.K., Beck E., Ferraris J.P., (2009) Oligoaniline intermediates in the aniline-peroxydisulfate system, *Synth. Met* 159. № 5-6. P. 445-55.

Tran HD, Li D, Kaner RB. (2009) One-dimensional conducting polymer nanostructures: Bulk synthesis and applications. *Adv Mater*;21:1487–99.

Tran HD, Wang Y, D'Arcy JM, Kaner RB. (2008) Toward an understanding of the formation of conducting polymer nanofibers. *ACS Nano*;2:1841–8.

Travers JP, Sixou B, Berner D, Wolter A, Rannou P, Beau B, Pepin-Donat B, (1999) Is granularity the determining feature for electron transport in conducting polymers? *Synth Met* 101 359-62.

Trivedi DC. (1997) Polyanilines. In: *Handbook of Organic Conductive Molecules and Polymers*, Vol. 2, Nalwa HS, ed.; Wiley, Chichester,; pp. 505–72.

Tzou K and Gregory RV, (1992) Kinetic study of the chemical polymerization of aniline in aqueous solutions, *Synth. Met.* 47:267-77

Venancio EC, Wang PC, MacDiarmid AG. (2006) The azanes: A class of material incorporating nano/micro self-assembled hollow spheres obtained by aqueous oxidative polymerization of aniline. *Synth Met*;156:357–69.

Wan MX. (2009) Some issues related to polyaniline micro-/nanostructures. *Macromol Rapid Commun*;30:963–75.

Wang Y., Tran HD., Kaner RB., (2011) Application of Oligomers for nanostructured conducting polymers *Macromol. Rapid Commun* Vol. 32. P. 35-49

Wei Y, (2001) Nonclassical or Reactivation Chain Polymerization: A General Scheme of Polymerization, *J Chem. Educ.* 78:551-53

Wei Y, Tang X, Sun Y and Focke W, (1989) A stady of the mechanism of aniline polymerization, *J. Polym. Sci., Part A: Polym. Chem.* 27:2385-96

Yang L.Y.O., Chang C.Z., Liu S.H., Wu C.G., Yau S.L. (2007) Direct visualization o an admolecule and its electropolymerization on Au(111) with in-situ scanning tunneling microscope *J. Amer. Chem. Soc.*129. P. 8067-77

Yu X., Fan H., Wang H., Zhao N., Zhang X., Xu J., (2011) Self-assembly hierarchical micro/nanostructures of leaf-like polyaniline with 1D nanorods on 2D foliage surface, *Material Letters* 65 2724-2727.

Zhang DH, Wang YY. (2006) Synthesis and applications of one-dimensional nano-structured polyaniline: An overview. *Mater Sci Eng B*;134:9–19.

Zhang X, Kolla HS, Wang X, Raja K and Manohar SK, (2006) Fibrillar growth in polyaniline. *Adv Funct Mater*;16:1145–52.

Zhou S, Wu T, Kan JQ. (2007) Effect of methanol on morphology of polyaniline. *Eur Polym J*;43:395–402.

Zuo F, Angelopoulos M, MacDiarmid AG, A Epstein, (1987) Transport stadies of protonated emeraldine polymer: A granular polymeric metal system *Phis Rev B* 36 3475-78

Zuppiroli L, BussacMN, Paschen S, Chauvet O, Forro L, (1994) Hopping in disordered conducting polymers, *Phis Rev B* 50 5196-203.

Chapter 9

Bulk Preparation of Radiation Crosslinking Poly (Urethane-Imide)

Chengfei Zhou

Additional information is available at the end of the chapter

1. Introduction

Ionizing radiation can be used for understanding mechanism of polymerization reaction as well as for initiation of the polymerization process. Some of the advantages of the radiation initiated polymerization over conventional methods are: (i) absence of foreign matter, like initiator, catalyst, etc., (ii) polymerization at low temperature or in solid state, (iii) rate of the initiation step can easily by controlled by varying dose rate and (iv) the initiating radicals can be produced uniformly by γ-irradiation. The gamma ray induced radiation polymerization, broadly speaking, in addition to the narrow sense "radiation polymerization (also known as radiation initiated polymerization. That is, using high energy ionizing ray to irradiate monomers, and generates ions or radicals, so that form the active center and the polymerization reaction.), also includes the radiation graft polymerization and radiation crosslinking polymerization, etc. [1-18].

Indeed, radiation crosslinking polymerization is one of the important research fields of radiation polymerization. Remarkably, the radiation crosslinking polymerization has been great development since Charles found polyethylene radiation effect and predict these radiation effects may lead to industrial applications. Radiation crosslinking polymerization can be carried out in solution, can also be carried out by bulk polymerization. Radiation crosslinking via bulk polymerization is relatively simple and has the advantages of high product purity. Therefore, as a powerful means for modification of polymerric materials to be widely applied in many fields, and, radiation crosslinking modification of polyurethanes is one of most important aspects [19-34].

In the study of the radiation crosslinking polyurethane, a lot of work was done in polyurethane synthesis, in its molecular backbone to import some unsaturated bond, application of γ-irradiation to realize the radiation crosslinking in order to obtain a better crosslinking effects, improved thermal stability and other properties of polyurethane. There are reports, in polyurethane molecular backbone structure into the diacetylene moieties, by ray irradiation, the diacetylene moieties can produce crosslinking polymerization (often called "cross polymerization") and the formation of covalently crosslinked conjugated polydiacetylene network [35-47]. And studies show that, if the polyurethane main chain structure is only hard segment containing unsaturated bond, then after irradiation, often produce crosslinking reaction only in hard segment microdomains, and the glass transition temperature (Tg) of the soft segment can remain unchanged. In addition, in recent years, the study of the radiation crosslinking polyurethane is also extended to the field of poly(urethane-imide)(PUI)[48].

2. Preparation foundation of radiation crosslinking poly(urethane – imide)

Poly (urethane –imide) is a new polymeric material, it has the excellent properties of both polyurethane and polyimide, such as good mechanical properties, thermal stability, high mechanical strength, electrical insulating properties, chemical resistance, hydrolysis resistance, radiation resistance, abrasion resistance and biological compatibility, etc. Poly (urethane-imide) is a block copolymer comprised of flexible chain segment and rigid chain segments. However, in the block structure of poly(urethane-imide), the rigid chain segment is polyimide [49-67]. Moreover, the bulk preparation of radiation crosslinking poly(urethane-imide) is also based on the bulk preparation of polyurethane. Therefore, here to discuss the bulk preparation of radiation crosslinking polyurethane before other description.

Radiation crosslinking polymerization of polyurethane is often to achieve via unsaturated bonds of molecular backbone structure, in addition, also to carry out by the introduction of unsaturated terminal using end capping methods, and, also to enhanced radiation crosslinking polymerization by adding crosslinking agent in the system. In our study, mainly through the preparation of double terminated polyurethane to realize its radiation crosslinking polymerization[68], specific method as shown in figure 1. The specific method are: (i) using 2-hydrooxyethyl methylacrylate (β-HEMA) as end capping agent, (ii) using bulk polymerization method to synthesize double terminated polyurethane prepolymer, (iii) adding crosslinking agent octavinyloctasilasesquioxane(OVS), (iv) by γ-irradiation, through free radicals react to form OVS-polyurethane crosslinked structure. The β-HEMA is a typical monomer, which are often used to radiation polymerization, here is to consider its hydroxyl groups can react with isocyanate groups, as end capping agent to use. Although the preparation of the OVS-polyurethane by radiation polymerization has been rarely reported, but there are many relevant reports [69,70].

As an example, according to the above method, the radiation crosslinking polyurethane was prepared by using polycarbonate diol and liquefied 4, 4`-diphenylmethane diisocyanate (Liquefied MDI). And, by γ-irradiation, radiation dose is 50kGy, irradiation dose rate for 10kGy / h. And, the sample 1 is unirradiated polyurethane prepolymer, the sample 2 is the radiation crosslinking polyurethane elastomer (without OVS), the sample 3 and sample 4 are the OVS-polyurethane radiation crosslinked samples (the adding quantity of OVS were 7% and 11%, respectively). The structure and properties of obtained sample was investigated with Fourier transform infrared spectroscopy(FTIR), X-ray diffraction(XRD), dynamic thermomechanical analysis(DMA), and thermal gravimetric analysis(TGA).

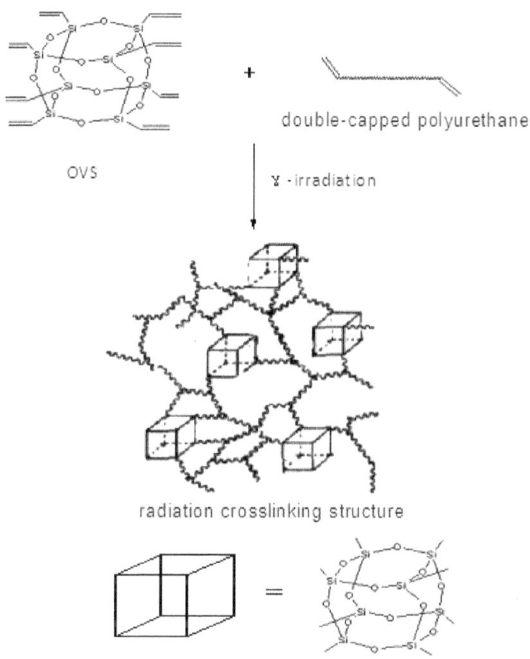

Figure 1. Schematic diagram of OVS-polyurethane radiation crosslinked system

Figure 2 is the FTIR spectra of the radiation crosslinking polyurethane. Unirradiated sample (sample 1), appears a weak absorption peak at 905cm^{-1}, which is the contribution of C=C double bond, but in the irradiated samples (2, 3 and 4), the peak disappear, it is obvious that the formation of cross-linked structure is mainly due to the crosslinking reaction of the unsaturated double bonds. On the other hand, Figure 3 is the XRD spectra of the radiation crosslinking polyurethane. whether linear (unirradiated sample, sample 1) or radiation crosslinking (sample 2), only the one passivated diffraction peak, only the presence of local regular structure, and no obvious crystal phenomenon. However, the OVS-polyurethanes radiation crosslinking samples(3 and 4), appears sharp diffraction peaks in the diffraction angle of 7.9, 8.8, 10.9, 11.7, 18.4, 19.9, 21.9, 24.3, 25.5, which belonged to the crystallization of OVS contained in OVS-polyurethane. In addition, Figure 4 is the TDA results of the radiation crosslinking polyurethane. Visible by the Figure 4, the thermal stability of radiation crosslinking polyurethane is greatly improved, the temperature of 5% weight loss for sample 2, 3 and 4 are 269.6°C, 242.6°C and 239.0°C, respectively, but the temperature of 5% weight loss of unirraded sample (sample 1) is only 133.3°C.

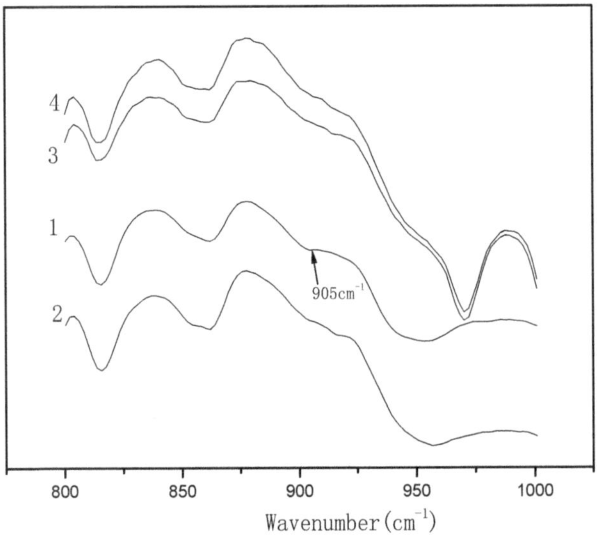

Figure 2. FTIR spectra of radiation crosslinking polyurethane

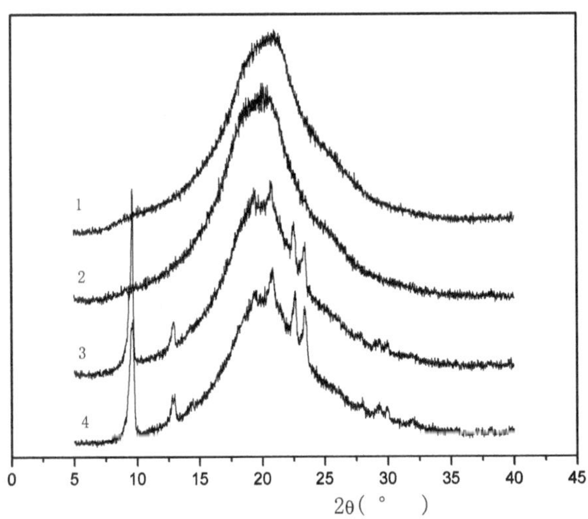

Figure 3. Figure 3 XRD spectra of radiation crosslinking polyurethane

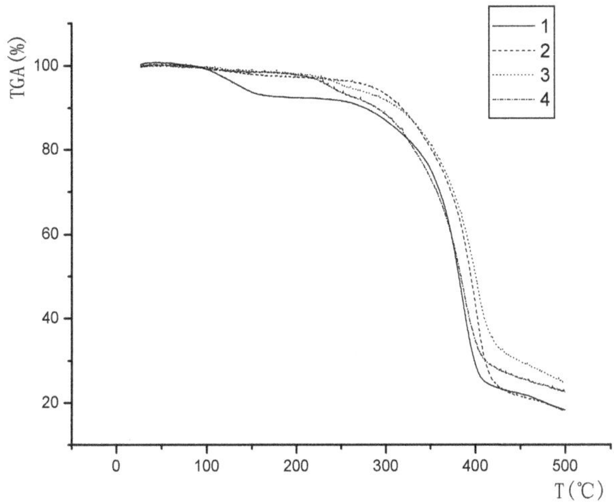

Figure 4. TGA curves of radiation crosslinking polyurethane

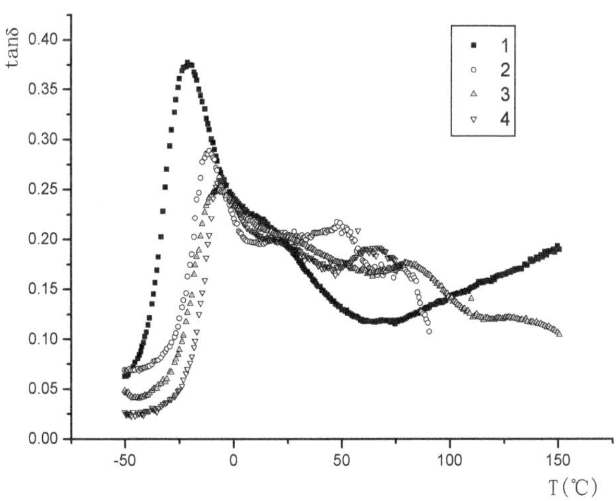

Figure 5. DMA charts of radiation crosslinking polyurethane

Figure 5 is DMA charts of radiation crosslinking polyurethane. The results show that, the tag δ-T curves of the radiation crosslinking polyurethane are two tag δ peak. From the molecular structure of polyurethane, the flexible chain segment (often called soft segment)

in macromolecules present random coil state, the soft segment and rigid chain segments (often called hard segment) of polyurethane are gathered in one block, forming the microphase separation structure. If the degree of phase separation is better, then the tag δ-T curve will appear the two tag δ peak, which belonged to soft segment and hard segment, respectively, but if between the two segment is compatibility, will become a tag δ peak. From that, the prepared radiation crosslinking polyurethane has a good degree of microphase separation. From Figure 5 it can be seen that, prior to irradiation (sample 1) ,only one tag δ peak from the soft segment, which is mainly the contribution of the long chain of polycarbonate diol. After irradiation (sample 2), because the β-HEMA polymerization and formation of polymeric chain segment, it appears a obvious tag δ peak due to poly(2-hydroxyethyl methacrylate) (PHEM) hard segment, and, the tag δ peak of soft segment shifts to higher temperature, which belonged to the confined effect of radiation crosslinking. After adding OVS, two tag δ peaks shift to more high temperature, and, the adding amount of OVS increased, this change more, This should be attributed to the contribution of OVS on radiation crosslinking polymerization.

3. Radiation crosslinking of double bond teminated poly (urethane-imide)

Already mentioned, the difference with polyurethane, the hard segment of poly (urethane-imide) is a polyimide. Then, during the preparation process of the radiation crosslinked poly (urethane-imide), to produce imide structure by the reaction of anhydride groups and isocyanate, thereby forming a polyimide hard segment, as shown in Figure 6. However, similar with polyurethane, if the molecular structure of poly(urethane-imide) does not contain unsaturated bond, it is difficult to achieve the ideal effect of radiation crosslinking polymerization. Therefore, the preparation of radiation crosslinking poly(urethane-imide), the same need to import some unsaturated group into its molecular backbone (soft segment or hard segment) or its chain end, to enhance adiation crosslinking effect. In fact, so far, although preparation research of poly(urethane-imide) made a lot of progress[49-67], but report closely related with radiation polymerization also not too much.

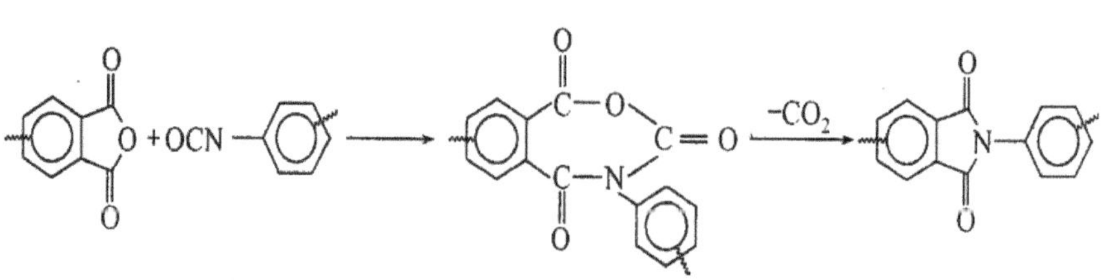

Figure 6. Schematic diagram of polyimide structure

Of course, we can prepare radiation crosslinking poly(urethane- imide) by using introduction of unsaturated bond into its molecular structure, especially hard segment structure, also through double end capping to prepare the radiation crosslinking poly(urethane-imide). This is because to analysis from the radiation crosslinking polymerization, the research ideas of radiation crosslinking poly(urethane-imide) should follow the basic preparation principle of polyurethane. However, in our study, mainly through the preparation of double bond terminated poly(urethane-imide) to realize radiation crosslinking polymerization, as shown in Figure 7. The specific method are: (i) using β-HEMA as end capping agent, (ii)using bulk polymerization method to synthesize double terminated poly (urethane-imide), (iii) by γ-irradiation, to produce radiation crosslinking poly(urethane-imide).

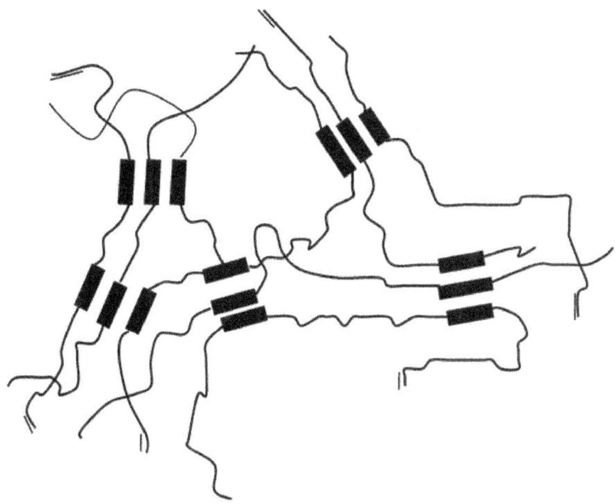

Figure 7. Schematic diagram of double capped poly(urethane-imide)

Here is to illustrate detailed preparation method via the specific discusstion. Radiation crosslinking poly(urethane-imide) was prepared using polycarbonate diol, pyromellitic dianhydride (PMDA) and hexamethylene diisocyanate (HDI). And, by using γ-irradiation, irradiation dose is 25kGy, 50kGy, 75kGy and 100kGy, respectively, the irradiation dose rate for 10kGy / h. The structure and properties of obtained sample was investigated with FTIR, XRD, DMA, TGA and static thermodynamic analysis (TMA). Also, focus on the effect of radiation dose to the structure and properties of. radiation crosslinking poly(urethane-imide).

Figure 8 is the FTIR spectrum of the radiation crosslinking poly(urethane-imide), marked with the 0kGy for unirradiated poly(urethane-imide) sample. Figure 8 shows that, absorption peaks appeared at 1840 cm^{-1} and 1768 cm^{-1},1371 cm^{-1}, 1116 cm^{-1},730 cm^{-1}, which may belong to the imine characteristic absorption of imine I, imine II, imine III and imine IV, respectively, the characteristic absorption peaks exist due to the poly(urethane-imide) sample does contain imide structure. In addition, the absorption peaks occurred at 3100cm^{-1} and 1634cm^{-1}, the absorption peaks can be attributed to the characteristic absorption of the terminal unsaturated structure (=C-H and C=C) in poly(urethane-imide) sample. In general, after irradiation, the most important reaction is radiation crosslinking and radiation degradation, and the unsaturated degree will also have some changes. From Figure 8 also shows, the absorption peaks of =C-H ($3100cm^{-1}$) and C=C ($1634cm^{-1}$) , although declining with the increase of irradiation dose, but not too obvious.

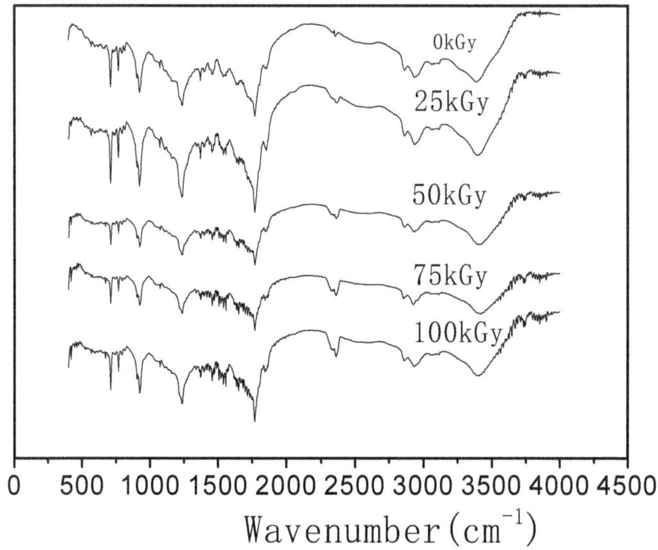

Figure 8. TIR spectra of poly(urethane-imide) with different irradiation dose

In addition, because of different kinds of isocyanate, long chain diols and anhydride, poly(urethane-imide) sometimes as amorphous but orderly structure, sometimes show some degree of crystallinity, this often is to analyze by using X ray diffraction(XRD). As shown in Figure 9, the unirradiated poly(urethane-imide) sample(0kGy), appear sharp diffraction peaks at the diffraction angle of 14.3, 16.4, 20, 21.9, 27.5, 28.8 and 30, and, is dispersive diffraction peaks at the diffraction angle of 20, indicating that both local regular structure exist in the poly(urethane-imide) sample, but also has certain crystalline. However, the effect of γ-irradiation on the crystallinity of polymer, but more complex. In general, γ-irradiation led to crosslinking, will limit the crystal growth, so that the crystallinity decreased. In contrast, γ-irradiation induced degradation, the molecular weight is reduced, thus, molecular chain length is shorter, is conducive to the orientation and crystallization, thereby, in favor of crystallinity increased. Figure 9 shows, to irradiate by using irradiation dose of 25kGy, showed a strong sharp diffraction peaks at the diffraction angle of 20.3, but the intensity of this peak gradually reduced with the increase of radiation dose; and the intensity of the sharp diffraction peak at 21.9 is increased with the increase of irradiation dose; also to irradiate by irradiation dose of 100kGy, this peak becomes very strong. In addition, when irradiation dose was 50kGy, a strong sharp diffraction peaks appeared at the diffraction angle of 18.3, but other samples are not this peak, which may be to irradiate by irradiation dose of 50kGy, is probably conducive to the formation of a specified level of crystalline structure.

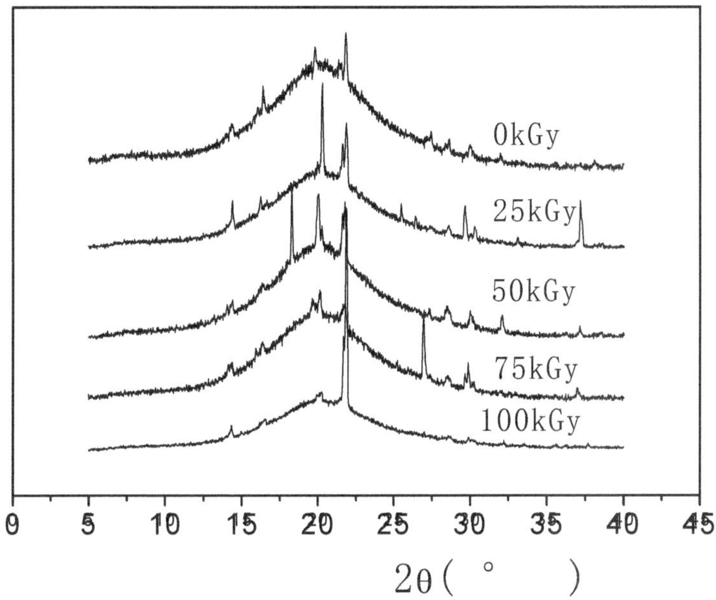

Figure 9. XRD spectra of poly(urethane-imide) with different irradiation dose

From the molecular structure of poly(urethane-imide), the carbamate bond, ether or ester bond all has the good flexibility, presents the random coil state, may be referred to as a flexible chain segment or soft segment; and imide chain segment is extended into a bar at ambient temperature, called the rigid chain segments or hard segment. And, the intermolecular cohesive energy and hydrogen bonding of the hard segment is enhanced by introduction of imide structure,. In this way, flexible chain segments and rigid chain segments are gathered in one block, forming the microphase separation structure. If the phase separation is better, then the two tag δ peak will appear in the DMA chart due to soft segment and hard segment, respectively, but if the phase separation of two segment is poor, will become a tag δ peak. As shown in Figure 10, a tag δ peak exist only in the tag δ-T chart of unirradiated poly(urethane-imide) sample(0kGy), indicating certain compatibility between hard segment and soft segment. After irradiation, the tag δ peak was to shift to high temperature, and increased with the increase of irradiation dose, when the irradiation dose was 100kGy, the tag δ peak shifts to approximately 150°C; in addition, after irradiation, the peak value of tag δ decreased significantly, and the shape of tag δ peak is changed, when the irradiation dose was more than 50kGy, the tag δ peak appeared in the form of obvious shoulder peak.

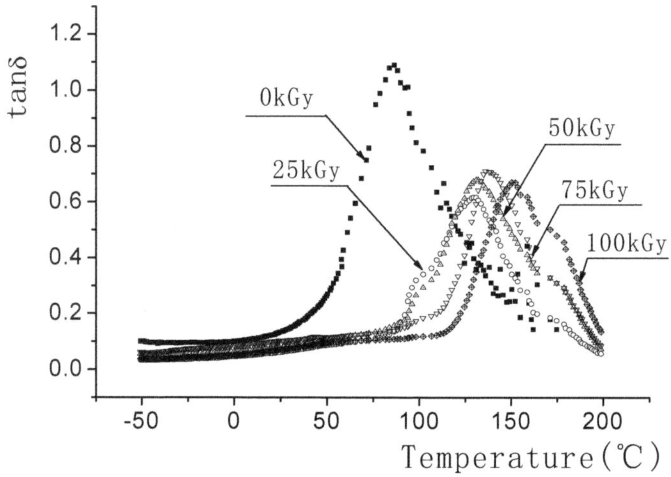

Figure 10. DMA charts of poly(urethane-imide) with different irradiation dose

Irradiation dose (kGy)	Temperature of 5% weight loss (°C)	Temperature of 15% weight loss (°C)	Temperature of 50% weight loss (°C)
0	204.0	245.5	489.9
25	210.2	283.8	494.3
50	220.0	299.2	497.5
75	220.3	306.8	496.4
100	211.6	293.2	489.2

Table 1. TGA results of poly(urethane-imide) with different irradiation dose

Also, from TGA test(table 1), to irradiate by using different radiation dose of 25kGy, 50kGy, 75kGy, the thermal stability of radiation crosslinking poly(urethane-imide) is improved gradually, and, to achieve the best when the radiation dose was between 50kGy and 75kGy, this can be attributed to the degree of crosslinking increases with the increase of radiation dose. However, when irradiation dose was 100kGy, the its thermal stability is worse than 75kGy, may be the irradiation dose if too high, the degradation reaction will increase strongly, led to the degree of crosslinking drop. The TDA test (Figure 11) also proved this point, that is to say, the radiation crosslinking poly(urethane-imide) obtained by the radiation dose of between 50kGy and 75kGy, as has the best degree of cross-linking, thus has the best heat resistance.

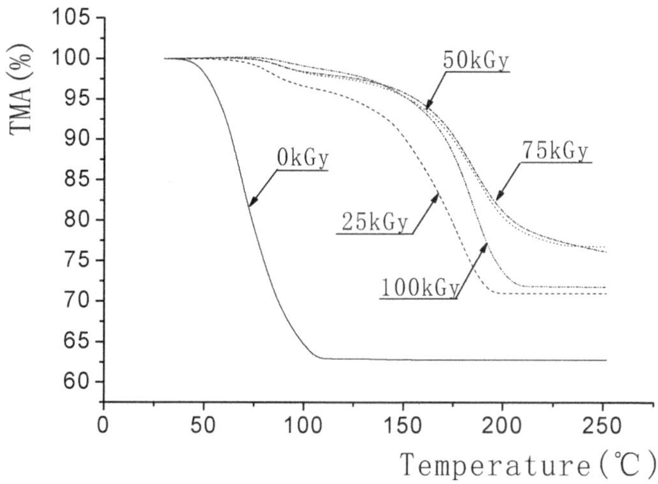

Figure 11. TMA charts of poly(urethane-imide) with different irradiation dose

4. Radiation crosslinking OVS-poly (urethane-imide) foam

In fact, radiation cross-linked poly (urethane imide) can be changed through the raw material and the conditions of the radiation crosslinking polymerization, to produce a variety of products with different properties. Can be prepared into high modulus specialty plastics, can also be prepared into high elastic rubber, but also can be made into films, fibers and foams, etc.. For the preparation of foam, can be made into rigid foam, can also be made into foamed elastomer. And, can also according to the specific requirement, made into a variety of foam products.

In the preparation of radiation crosslinked poly (urethane-imide) foams, can be achieved by the introduction of the unsaturated bond to the main chain molecular structure, can also be prepared via the double bond end capping. Moreover, can improve the effect of radiation crosslinking polymerization by adding a crosslinking agent. Also, the foaming problems of the poly(urethane-imide) foam can be solved using carbon dioxide gas (such as foaming agent H_2O react with isocyanate to produce CO_2, etc.). Here, only discuss the preparation of OVS-poly(urethane-imide) radiation crosslinking foam[71-73].

In the preparation of OVS-poly(urethane-imide) radiation crosslinking foam, specifically, was prepared using polyester polyol , polyphenyl polyisocyanate (PAPI), and 3,3',4,4'-Benzophenonetetracarboxylic dianhydride(BTDA), and β-HEMA as end capping agent, adding crosslinking agent OVS. And, by using γ-irradiation, irradiation dose is 25kGy, 50kGy, 75kGy and 100kGy, respectively, the irradiation dose rate for 10kGy / h. Also, focus on the effect of radiation dose to the structure and properties of. radiation crosslinking OVS-poly(urethane-imide) foam.

Irradiation dose (kGy)	Temperature of 5% weight loss (°C)	Temperature of 15% weight loss (°C)	Temperature of 50% weight los (°C)	Temperature of 80% weight loss (°C)
0	178.6	212.5	486.3	642.8
25	186.2	222.3	515.1	640.8
50	193.7	232.8	596.0	717.2
75	189.1	220.6	508.6	644.6
100	186.0	218.6	504.0	646.5

Table 2. TGA results of radiation crosslinking poly(urethane-imide) foam

First, use TGA to explore the irradiation dose on the effect of the thermal stability of radiation crosslinking OVS-poly(urethane-imide) foam, in order to facilitate comparison, select four weight loss points of 5%, 15%, 50% and 80% to determine TGA temperature, the results are shown in table 2. The results show that, when the weight loss was 5%, 15%, 50% and 80%, respectively, the effect of irradiation dose all showed the same trend, namely, the corresponding TGA temperature increases with the increase of radiation dose, when the irradiation dose reaches 50kGy, the TGA temperature reaches the maximum, then the TGA temperature drop with the increase of irradiation dose, and, the irradiation dose increased to higher, then more down low. Further analysis, the crosslinking structure between the vinyl group of OVS and terminal double bonds of poly(urethane-imide) will be formed by γ-irradiation, and the degree of crosslinking is closely related to radiation dose. If the irradiation dose is too small, the degree of crosslinking is insufficient, but if the irradiation dose is too large, the radiation degradation will increase, will cause the decrease of the degree of crosslinking, and this trend will increases with the increase of the irradiation dose.

Figure 11 is DMA charts of OVS- poly(urethane-imide) radiation crosslinking foam. First of all, from the relations of the irradiation dose and the elastic modulus (E'), the E' of the glassy state increased significantly with the increase of radiation dose, and when the irradiation dose was 50kGy, reached the maximum, and then, but decreased with the further increase of the irradiation dose, and, the irradiation dose increased, the E 'of the glass state is lower. This indicates that when the irradiation dose was 50kGy, the E' of the glassy stste was the highest. This can be attributed to the radiation dose of 50kGy can be the greatest degree of crosslinking, so that the foam has a high rigidity, also displays the maximum modulus. However, in the rubbery state, the effect of irradiation dose on the E' had no evident. Secondly, from the relations of the irradiation dose and the loss modulus (E"), the peak value of E" increases obviously with the increase of radiation dose, when irradiation dose was 75kGy, reached the maximum, and then, the irradiation dose increased again, appeared in rapid decline.

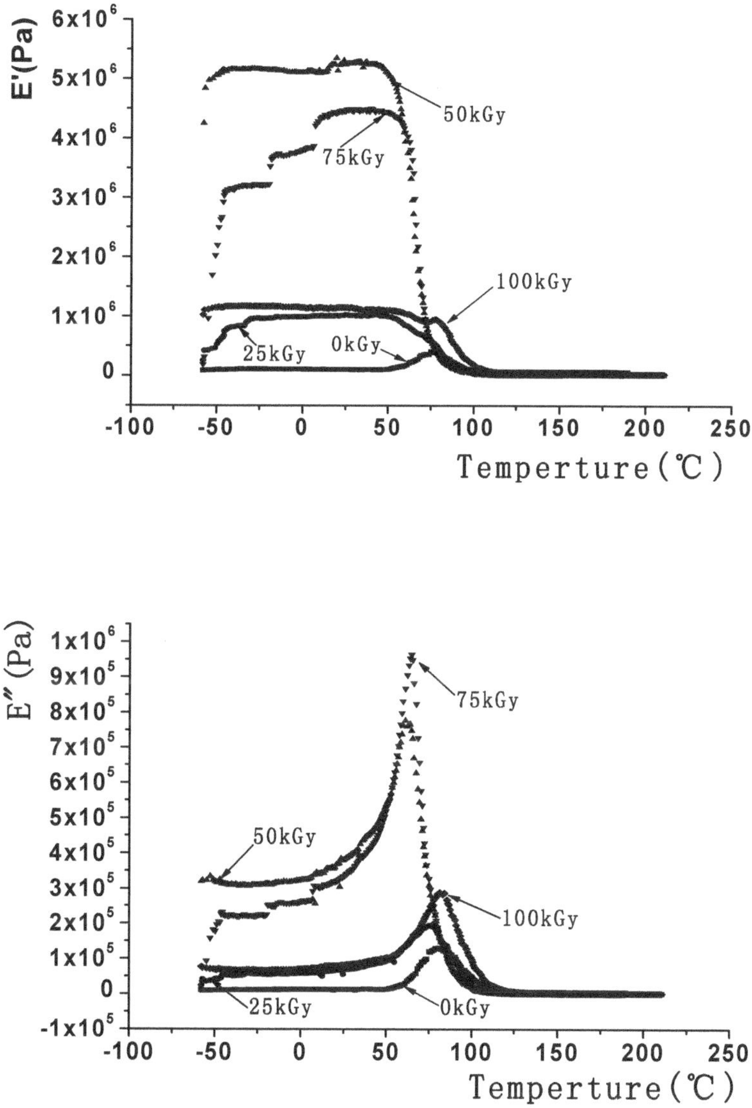

Figure 12. DMA charts of radiation crosslinking poly(urethane-imide) foam

5. Preparation of poly(urethane-imide) via other radiation polymerization

In addition to the above method, Can also use other radiation polymerization method to prepare radiation crosslinking poly (urethane-imide). Such as the use of radiation crosslinking of organic polyols oligomer to obtain. Among them, Can be the first to use γ-irradiation to produce radiation crosslinked organic polyol, as shown in Figure 13, then the radiation crosslinking poly (urethane-imide) prepared by radiation crosslinked organic polyol. Of course, this method, the application of the organic polyols oligomer containing unsaturated bonds would be more effective.

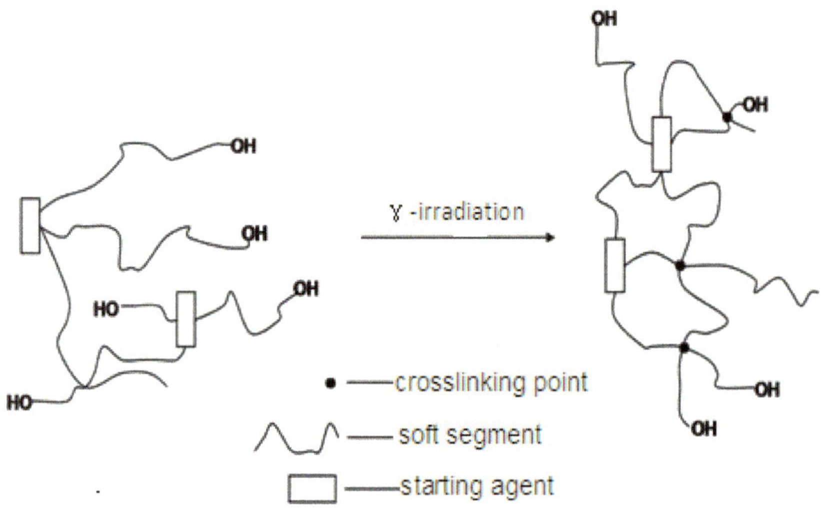

Figure 13. Schematic diagram of radiation crosslinking organic polyol

Sample	Viscosity (mPa s, 25°C)						
	0kGy	25kGy	50kGy	75kGy	100kGy	125kGy	150kGy
Polyether polyol	724	886	960	1210	1250	973	959
Rosin ester polyol	10170	10510	10640	12420	12540	12060	10890

Table 3. Effect of irradiation dose on viscosity of organic polyol

In which, we choose the rosin ester polyol (molecular backbone containing heterocyclic structure) and a polyether polyol, to execute radiation crosslinking modification by using γ-irradiation, irradiation dose is 25kGy, 50kGy, 75kGy, 100kGy, 125kGy and 150kGy, respectively, irradiation dose rate for 10kGy/h. And through the viscosity measurement to inspect the viscosity changes of organic polyols after irradiation, which explore the irradiation effect, experimental results as shown in Table 3. The results show that, polyether polyol and rosin ester polyol are all the same, that is, the viscosity increase with the increase of irradiation dose, and when the radiation dose was 100kGy, reached the maximum, but then the viscosity will gradually decline with the increase of irradiation dose. The increase of viscosity can be considered to be caused by radiation crosslinking, so that to achieve the best crosslinking effect when the radiation dose was 100kGy. In addition, the poly (urethane-imide) foams also has prepared using the obtained radiation crosslinking polyols(radiation dose for 100kGy), and, the properties of the obtained foams are shown in table 4.

Sample	Density (g/cm^3)	Open cell content (%)	Oxygen index (%)	Temperature of 5% weight loss (°C)
polyether polyol	0.030	91.9	34	184.4
rosin ester polyol	0.044	3.5	35	190.8

Table 4. Properties of radiation crosslinking poly (urethane-imide) foam

On the other hand, there is also an important method, which can be generally processed by γ-irradiation used for preparation of polymer polyols(POP) [74,75], and then to prepare poly (urethane-imide) by the POP. The γ-irradiation preparation of polymer polyol not only includes the radiation grafting and other radiation polymerization, moreover, these radiation reactions are carried out in special dispersion medium (organic polyol), which having a specific significance in radiation polymerization. Therefore, it is very necessary to elaborate.

Polymer polyol can be prepared using in situ polymerization method, that is to say, the hydrogen atoms of methine in polyether backbone can be shifted by γ-irradiation, to generate macromolecular free radicals, and cause graft polymerization with vinyl monomers. However, for this method, because the chain transfer constant of methylene is very small, it is difficult to improve the solid content of polymer polyol. It is best method to import containing double polyether polyol, often referred to as" macromer technique" [76]. This method is firstly to prepare the macromonomer (polyether polyol containing double bonds), and add it in polyether polyols(often referred to as polyether matrix), and then, produce graft copolymerization with vinyl monomers (styrene(St) and acrylonitrile(AN)) by γ-irradiation. In this case, the preparation method of macromonomer has many, and, the preparation method as shown in Figure 14 is one of the most commonly used methods.

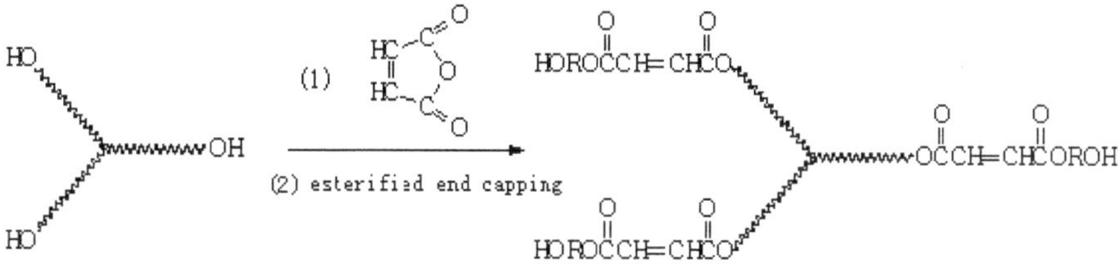

Figure 14. Schematic diagram of polyether macromonomer

The radiation polymerization reactions in this method is given in figure 15. In the obtained POP, the St-AN copolymer is present in the polyether matrix in the form of fine particles. The graft polyether due to radiation polymerization of macromonomer with St and AN, is to distribute on the surface of particle as steric stabilizer, as shown in Figure 16. As an example, first, to synthesize three kinds of macromonomer, Figure 17 is the FTIR spectrum of three kinds of macromonomer. And, styrene / acrylonitrile ratio is 55 / 45, the mixture of St/AN added based on the 55% of polyether matrix. And, by using γ-irradiation, irradiation dose is 5kGy and 25kGy, respectively, the irradiation dose rate for 10kGy / h.

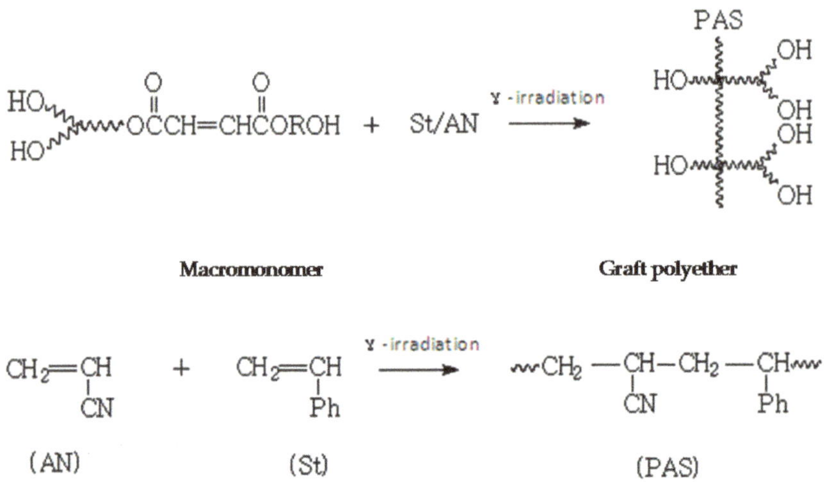

Figure 15. Schematic diagram of radiation polymerization reaction for preparing polymer polyol

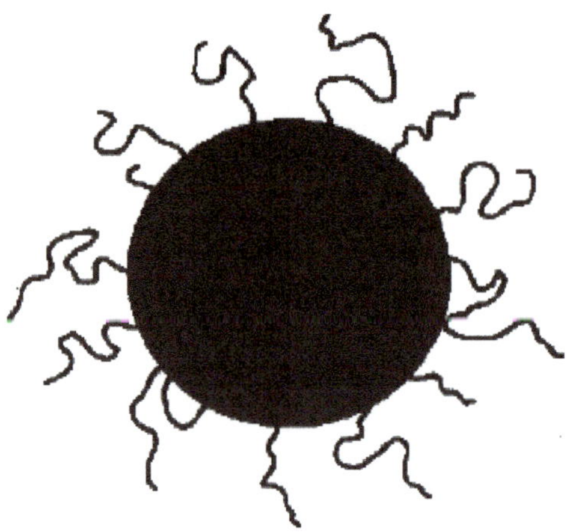

Figure 16. Schematic diagram of dispersion particle

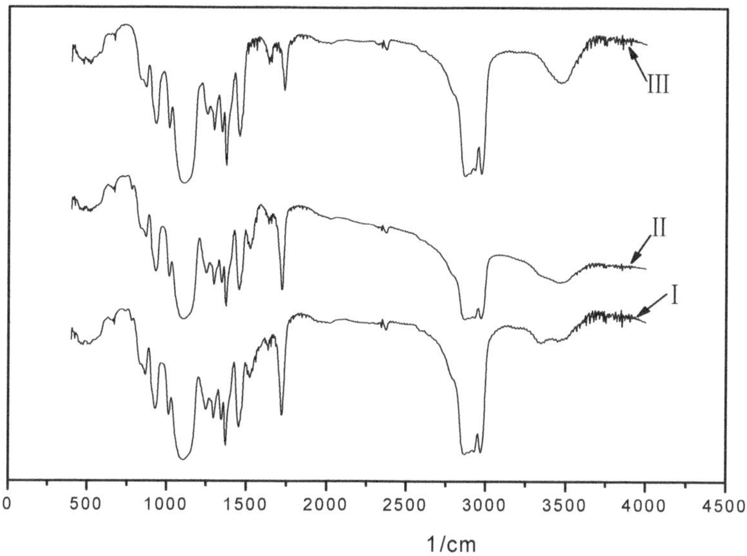

Figure 17. FTIR spectra of macromonomer

The dispersed phase particles of polymer polyols were isolated by high-speed centrifugation method, and, the morphological structure and properties of dispersed phase particles was investigated by Scanning electron microscope (SEM), XRD and DTA. Figure 18 is the SEM image for the dispersion particle, the results show that, the macromonomer and irradiation dose will influence the morphology of dispersed phase particles in polymer polyols. Among them, dispersion particle morphology has the more obvious difference for the polymer polyols using different macromonomer under the condition of 25kGy. And, from the XRD measurement results (Figure 19), the irradiation dose also may affect the ordered structure of dispersed phase particles, This is because the shoulder peaks of the diffraction peaks appear more obvious in the low radiation dose (5kGy). In addition, from the DTA results as shown in Table 5, the radiation dose also have certain effect on the thermal stability of dispersed phase particles. The dispersion particles preparing at the higher radiation dose (25kGy) showed relatively better thermal stability.

Also, poly (urethane-imide) foams was prepared using the obtained polymer polyols, the some properties of the obtained foams are listed in table 6. The obtained poly (urethane-imide) foams has distinct characteristics, these particles can effectively play a reinforcing effect, especially it can remarkably improve the bearing capacity of foam. And, the its principle is also to follow a general theory of polymer composites.

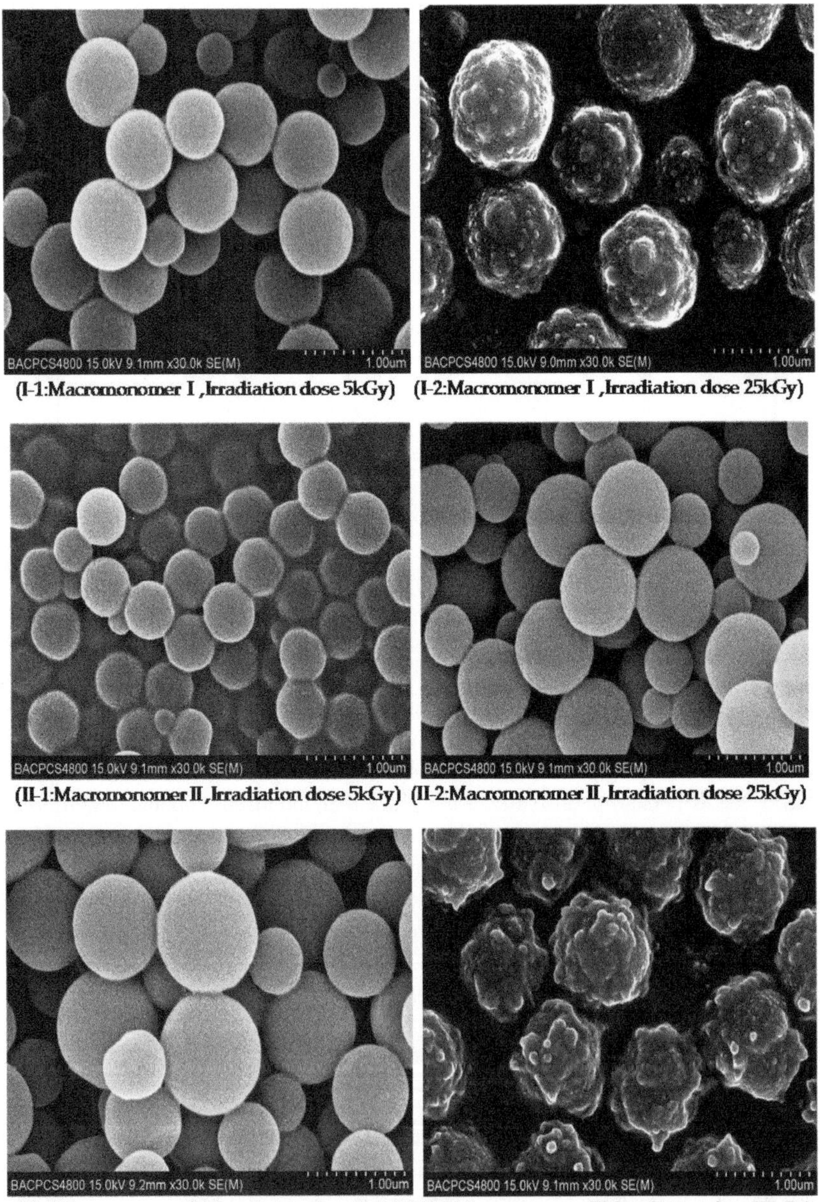

Figure 18. SEM image for dispersion particle

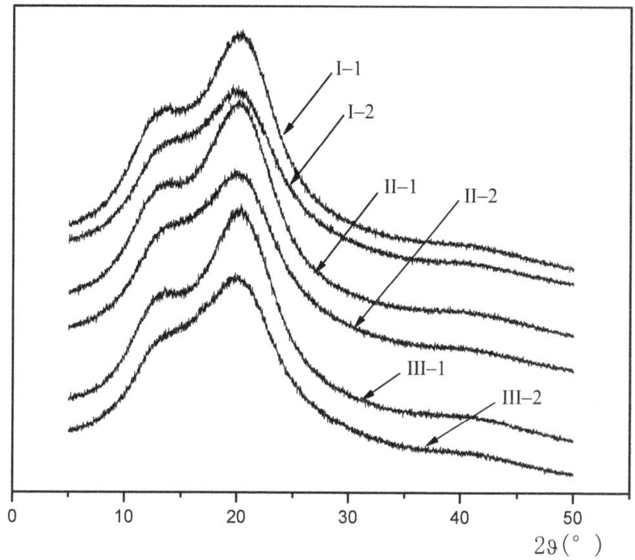

Figure 19. XRD spectra of dispersion particle

Sample	Temperature of 5% weight loss (°C)	Temperature of 15% weight loss (°C)	Temperature of 50% weight loss (°C)	Temperature of 80% weight loss (°C)
I-1	370.7	398.2	426.0	441.4
I-2	378.4	406.1	428.0	443.1
II-1	323.0	368.1	410.8	429.9
II-2	394.2	411.7	432.5	450.8
III-1	377.7	400.4	425.1	439.6
III-2	380.2	407.7	429.3	445.8

Table 5. TGA results of dispersions particle

Sample	Density (g/cm^3)	Open cell content (%)	Oxygen index (%)	Temperature of 5% weight loss (°C)
PUI foam I	0.039	93.1	34	189.6
PUI foam II	0.042	91.8	34	186.1
PUI foam III	0.044	91.3	34	185.6

Table 6. Properties of poly (urethane-imide) foam prepared by polymer polyols

6. Conclusions

In recent years, radiation polymerization obtained significant progress, while the radiation crosslinking polymerization also obtained the great development as an important aspect of radiation polymerization. Among them, the radiation crosslinking polymerization of polyurethanes has also made remarkable progress, and have expanded to the radiation crosslinking polymerization of poly(urethane-imide)s. Poly(urethane-imide) is a newly developed polymeric materials, also, the bulk polymerization of poly(urethane-imide) become the future direction due to the advantages of low pollution to the environment. Therefore, it has very far-reaching significance that the preparation of radiation crosslinking poly(urethane-imide) via bulk polymerization. And, it is based on this consideration that the research work should be carried out in this respect.

In the work described obove, it has elaborated emphatically that the preparation of radiation crosslinking poly(urethane-imide) can be conveniently carried out by using γ-irradiation and bulk polymerization. It is worth emphasizing that the bulk preparation of radiation crosslinking poly(urethane-imide) not only is a research direction, which developed recently in radiation polymerization field, also has the vast development prospects. Moreover, radiation crosslinking OVS–poly(urethane-imide) is a kind of novel nano-composites, therefore, the research in this area will achieve further development by mutual penetration of radiation polymerization with nanometer science.

Author details

Chengfei Zhou

Beijing Research Center for Radiation Application, Key Laboratory of Radiation Technology and Advanced Materials of Beijing Academy of Science and Technology, Beijing, P.R. China

7. References

[1] Myong-Goo Lee. Synthesis of conductive microspheres by radiation polymerization. Polymer, 2002, 43: 4307-4309

[2] Esmaiel Jabban, Samyra Nozari. Swelling of acrylic acid hydrogels prepared by γ-radiation crosslinking of polyacrylic acid in aqueous solution. European Polymer Journal, 2000, 36(12): 2685-2692

[3] Agnes Safrany, Barbara Beiler, Krisztina Laszio, et al., Control of pore formation in macroporous polymers synthesized by single-step γ-radiation-initiated polymerization and cross-linking. Polymer, 2005, 46(9): 2862-2871

[4] Baljit Singh, S Kumar. Synthesis and characterization of psyllium-NVP based drug delivery system through radiation crosslinking polymerization. Nuclear Instruents and Methods in Physics Research Section B: Beam Interactions with Materials and Atoms, 2008, 266(15): 3417-3430

[5] Abd H L, Mohdy E I, Agnes Safrony. Preparation of fast response superabsorbent hydrogels by radiation polymerization and crosslinking of N-isopropylacrylamide in solution. Radiation Physics and Chemistry, 2008, 77(3): 273-279

[6] Noriaki Seko, Masao Tamada, Fumio Yoshii. Current status of adsorbent for metal ions with radiation grafting and crosslinking techniques. Physics Research Section B:Beam Interaction with Materials and Atoms, 2005, 236(1-4): 21-29

[7] Naotsugu Nagasawa, Ayaka Kaneda, Shinichi Kanazawa, et al. Application of poly(lactic acid) modified by radiation crosslinking. Nuclear Instruments and Methods in Physics Research Section B: Beam Interaction with materials and Atoms, 2005, 236(1-4): 611-616

[8] Radoslaw A Wach, Hiroshi Mitomo, Naotsugu Nagasawa, et al. Radiation crosslinking of carboxymethylcellulose of varius degree of substitution at high concentration in aqueous solutions of natural pH. Radiation Physics and Chemistry, 2003, 68(5): 771-779

[9] Nursel Pekel, Fumio Yoshii, Tamikazu Kume, et al. Radiation crosslinking of biodegradable hydroxypropylmethylcellulose. Carbohydrate Polymers, 2004, 55(2): 139-147

[10] Hennink W E, Nostrum C F. Novel crosslinking methods to design hydrogels. Advanced Drug Delivery Reviews, 2002, 54(1): 13-36

[11] Salmi A, Benfarhi S, Donnet J B, Decker C. Synthesis of carbon-polyaorylate nanocomposite materials by crosslinking polymerization. European Polymer Journal, 2006, 42(9): 1966-1974

[12] Baljit Singh, Manu Vashishtha. Development of novel hydrogels by modification of sterculia gum through radiation cross-linking polymerization for use in drug delivery. Nuclear Instruments and Methods in Physics Research Section B: Beam Interactions with Materials and Atoms, 2008, 266(9): 2009-2020

[13] Acharya A, Mohan H, Sabharwai S. Radiation induced polymerization and crosslinking behavior of N-hydroxy methyl acrylamide inaqueous solutions. Radiation Physics and Chemistry, 2002, 65(3): 225-232

[14] Jinhua Chen, Masaharu Asano, Tetsuya Yamaki, et al. Chemical and radiation crosslinked polymer electrolyte membranes prepared from radiation-grafted ETFE films for DMFC applications. Journal of Power Sources, 2006, 158(1): 69-77

[15] Andrzei G Chmielcwski, Mohammad Haji-Saeid, Shamshad Ahmed. Progress in radiation processing of polymers. Interaction with Materials and Atoms,2005, 236(1-4): 44-54

[16] Erdener Karadag, Dursun Saragdin, Olgun Guven. Water absorbency studies of γ-radiation crosslinked poly(acrylamide-co-2,3-dihydroxybutanedioic acid) hydrogels. Nuclear Instruments and Methods in Physics Research Section B: Beam Interaction with Materials and Atoms, 2004, 225(4): 489-496

[17] Baljit Singh, Lok Pal. Radiation crosslinking polymerization of sterculia polysaccharide-PVA-PVP for making hydrogel wound dressing. Internationl Journal of Biological Macromolecules, 2011, 48(3): 501-510

[18] Shintani H, Nakamura A. Mechanism of degradation and crosslinking of polyurethane when irradiated by gamma-rays. Journal of Applied Polymer, 1991, 42(7): 1997-1987

[19] Gorna K, Gogolewski S. The effect of gamma radidtion on molecular stability and mechanical properties of biodegradable polyurethanes for medical applications. Polymer regradation and stability, 2003, 79(3): 415-474

[20] Roger A. Assink. Radidtion crosslinking of polyurethanes. Journal of Applied Polymer Science, 1985, 30(6): 30(6): 2701-2705

[21] Shintani H, Kikuchi H, Nakamuro A. Effects of gamma-ray irradidtion on the change of characteristics of polyurethane. Journal of Applied Polymer Science, 1990, 41(3-4): 661-675

[22] Beyer G, Steckenbiegler B. Radiation crosslinked thermoplastic polyurethane. Gummi Fasern Kunststoffe(Germany), 1991, 44(11): 614-617

[23] Hearon K, Gall K, Ware, T, et al. Post-polymerization crosslinked polyurethane shape memory polymers. Journal of Applied Polymer Science, 2011, 121(1): 144-153

[24] Haugen H J, Brunner M, Pellkofer F, et al, Effect of different γ-irradiation doses on cytotoxicity and material properties of porous polyether-urethane polymer. Journal of Biomedical Materials Research Part B: Applied Biomaterials, 2007, 80B(2): 415-423

[25] Jayabalan M, Lizymol P P. Effect of γ-radiation sterilization on the stability of polyurethane potting compounds based on castor oil/SMDI and caprolactone polyol/SMDI, used for hollow fibre haemodialyzer. Bulletin of materials sciences, 1997, 20(5): 727-735

[26] Keith H, Ken G, Tayler W, et al. Post-polymerization crosslinked polyurethane shape-memory polymers. Journal of Applied Polymer Science, 2011, 121(1): 144-153

[27] Shintani H, Akitada N. Degradation and cross-linking of polyurethane irradiated by gamma-rays. Polymer Regradation and Stability, 1991, 32(2): 191-208

[28] Milekhin Yu M, Sadavnichii D N, Mukhacher S V, et al. Effects of γ-irradiation on the properties of (polyester urethane) elastomer. High Energy Chemistry, 2008, 42(1): 18-22

[29] Azevedo E C, Chierice G O, Neto S C. Gamma radiation effects on mechanical properties and morphology of a polyurethane derivate from caster oil. Radiation Effects and Defects in Solide, 2011, 166(3): 208-214

[30] Nouh S A, Abutalib M M. A comparative study of the effect of gamma and electron beam irradiation on the optical and structural properties of polyurethane. Radiation Effects and Defects in Solids, 2011, 166(3): 165-177

[31] Yang Y S, Lee J L. Polymerization of polyurethane-polyester interpenetrating polymer network (IPN). Macromolecules, 1987, 20 (7): 1490–1495

[32] Saad Abouzahr, Garth L Wilkes. Structure property studies of polyester- and polyether-based MDI–BD segmented polyurethanes: Effect of one- vs. two-stage polymerization conditions. Journal of Applied Polymer Science, 1984, 29(9): 2695-2711

[33] Tetsuya Kogiso, Shin-Ichi Inoue. Synthesis and properties of elastic polyurethane-imide. Journal of Applied Polymer Science, 2010, 115(1): 242-248

[34] Kiyotsugu Asai, Shin-Ichi Inoue, Hiroshi Okamoto. Preparation and properties of imide-containing elastic polymers from elastic polyureas and pyromellitic dianhydride. Journal of Polymer Science Part A: Polymer Chemistry, 2000, 38(4): 715-723
[35] Chuang F S. Analysis of thermal degradation of diacetylene-containing polyurethane copolymers. Polymer Degradation and Stability， 2007, 92 (7): 1393-1407.
[36] Buckley L J, Hammond P T, Rubner M F. Morphological investigation of polyurethane/diacetylene segmented copolymers. Macromolecules, 1993, 26(9): 2380-2382.
[37] Hu Xiao, Stanford J I, Day R J, et al. Synthesis, characterization, and structure of glassy diacetylene-containing segmented block copolyurethanes. Macromolecules, 1992, 25(2): 672-683.
[38] Nallicheri R A, Rubner M F. Thermal and mechanical properties of polyurethane-diacetylene segmented copolymers. 2. Effects of diacetylene cross-polymerization. Macromolecules, 1990, 23(4): 1017-1025.
[39] Nallicheri R A, Rubner M F. Thermal and mechanical properties of polyurethane-diacetylene segmented copolymers. 1. Molecular weight and annealing effects. Macromolecules, 1990, 23 (4):1005-1016.
[40] Nitzsche S A, Hsu S L, Hammond P T, et al. Spectroscopic study of domain ordering in diacetylene-containing model polyurethanes. Macromolecules, 1992, 25(9): 2391-2396.
[41] Nallicheri R A, Rubner M F. Influence of cross-linking on the hysteresis behavior of poly(urethane-diacetylene) segmented copolymers. Macromolecules, 1991, 24(2): 526-529
[42] Hu X, Breach C D, Young R J. Elucidation of the hard segment transition in a diacetylene-containing copolyurethane using modulated differential scanning calorimetry. Polymer, 1997,38(4): 981-983
[43] Hammond P T, Nallicheri R A, Rubner M F. Examination of the strain-induced orientation of hard segment domains in 4,4-methylenebis(phenyl isocyanate)-based polyurethane-diacetylene segmented copolymers. Materials Science and Engineering A, 1990, A126(1-2): 281-287
[44] Nallicheri R A, Rubner M F. Investigations of the mechanochromic behavior of poly(urethane-diacetylene) segmented copolymers. Macromolecules, 1991, 24(2): 517-525
[45] Huang S J, Edelman P G. Polyetherurethaneureas containing diacetylene in the hard segments. Journal of Applied Polymer Science, 1990, 41(1-2): 3-11
[46] Estrada M R, Burillo G, Ogawa T. Diacetylene-containing polymers, II. Polyesters and polyurethanes containing m,m prime -butadiynylenedibenzyl groups. Polymers for Advanced Technologies, 1992,3(8):419
[47] Chengfei Zhou, Wei Cao, Tong Zhai, et al. Bulk preparation and γ-ray irradiation-induced effect of poly(urethane-imide) using 2-hydrooxyethyl methylacrylate as end capping reagent. Polymer Materials Science and Engineering, 2011, 27(6): 15-17,21

[48] Patel H S, Patel B P, Patel D B. Synthesis, Characterization and glass reinforcement of poly(ester amido imide)s. International Journal of Polymeric Materials, 2009, 58(12): 625-635

[49] Patel H S, Mathur A B, Bhardwaj I S. Synthesis and characterization of modified polyimides: poly(urethane-imide). Journal of Macromolecular Science, Part A: Pure and Applied Chemistry, 1995, 32(12): 2925-2034

[50] Patel J C, Patel K D, Daraji J M. Synthesis, characterization, and glass fiber reinforced composites of poly(urethane-imide)s.International Journal of Polymeric Materials, 2003, 52(5): 345-359

[51] Jiangxuan Song, Guangxin Chen, Yun Ding, et al. Preparation and characterization of epoxy resin modified with alkoxysilane- functionalized poly(urethane-imide) by the sol–gel process. Polymer International, 2011, 60(11): 1594-1599

[52] Hamid Yeganeh, Mohammad Atai, Pejman Hojati Talemi. Synthesis, characterization and properties of novel poly(urethane-imide) networks as electrical insulators with improved thermal stability. Macromolecular Materials and Engineering, 2006, 291(7): 883-894

[53] Hamid Yeganeh, Shahram Mehdipour-Ataei, Mehdi Ghaffari. Preparation and properties of novel poly(urethane-imide)s via blending of reactive polyimide and epoxy-terminated urethane prepolymers. High Performance Polymers. 2008, 20(2): 126-145

[54] Tetsuya Kogiso, Shin-Ichi Inoue. Synthesis and properties of elastic polyurethane-imide. Journal of Applied Polymer Science, 2010, 115(1): 242-248

[55] Jong-Young Jeon, Tae-Moon Tak. Synthesis and characterization of block copoly(urethane–imide). Journal of Applied Polymer Science, 1996, 62(5): 763-769

[56] Chia Zheng Cheng, Yeo Swee Chengl. Preparation of nanoporous polyimide films from poly(urethane-imide) by thermal treatment. .Macromolecular Materials and Engineering, 2003, 288(9): 730-736

[57] Jin Liu, Zhen Li, Xiaolie Luo, Dezhu Ma. Synthesis, structure, and properties of polyimide and polyurethane-urea-imide copolymers. Journal of Polymer Science Part B: Polymer Physics, 2004, 42(2): 216-225

[58] Gnanarajan T Philip, Nasser A Sultan , Iyer N. Padmanabha,et al.Synthesis of poly(urethane-imide) using aromatic secondary amine-blocked polyurethane prepolymer. Journal of Polymer Science Part A: Polymer Chemistry, 2000, 38(22): 4032-4037

[59] Mi-Hee Park, Wonbong Jang, Seung-Jin Yang,et al. crosslinked networks. Journal of Applied Polymer Science, 2006, 100(1): 113-123

[60] Gnanarajan T Philip, Iyer N. Padmanabha. Poly(urethane-imide)s from blocked polyurethane prepolymer and pyromellitic dianhydride: effect of alkali metal alkoxides and phenoxides and substituents on the blocking agent in the polymerization reaction. Journal of Macromolecular Science, Part A: Pure and Applied Chemistry, 2001, 38(8): 807-820

[61] Vladimír Šindelář, Petr Sysel, Vladimír Hynek, et al. Transport of gases and organic vapours through membranes made of poly(amide-imide)s crosslinked with poly(ethylene adipate). Collect. Czech. Chem. Commun. 2001, 66, 533-540
[62] Hossein Behniafar. Direct synthesis of new soluble and thermally stable poly(urethane-imide)s from an imide ring-containing dicarboxylic acid using diphenylphosphoryl azide. Journal of Applied Polymer Science, 2006, 101(2): 869-877
[63] Min Zuo, Tsutomu Takeichi. Novel method for the preparation of poly(urethane–imide)s and their properties. Journal of Polymer Science Part A: Polymer Chemistry, 1997, 35(17): 3745-3753
[64] XiuMin Qin, XiaoHui Yang, XinLing Wang. Synthesis and characterization of poly(imide-urethane) based on novel chain-extender containing both imide and sulphone functions. Journal of Polymer Science Part A: Polymer Chemistry, 2005, 43(19): 4469-4477
[65] Choonkeun Lee, Iyer N. Padmanabha, Kyungwook Min, et al. Synthesis and characterization of novel poly(amide-imide)s containing 1,3-diamino mesitylene moieties. Journal of Polymer Science Part A: Polymer Chemistry, 2004, 42(1): 137-143
[66] Pinggui Liu, Qingfang Zhang, Lihua He,et al. Thermal and mechanical properties of poly(urethane-imide)/epoxy/silica hybrids. Journal of Applied Polymer Science, 2010, 117(6): 3722-3728
[67] Chengfei Zhou, Wei Cao, Tong Zhai, et al. Preparation and characterization of radiation crosslinking polyurethane and its POSS-containing composites. Polyurethane Industry, 2011, 26(3): 9-12
[68] Efrat T,Dodiuk H,Kenig S,et al. Nanotailoring of polyurethane adhesive by polyhedral oligomeric silsesquioxane (POSS). Journzl of Adhesion Science and Technology, 2006, 20(12):1413-1430
[69] Madbouly S A, Otaigbe J U, Nanda A K, et al. Rheological behavior of POSS/polyurethane-urea nanocomposite films prepared by homogeneous solution polymerization in aqueous dispersions. Macromolecules, 2007,40 (14) 4982-4991
[70] Bliznyuk V N, Tereshchenko T A, Gumenna M A, et al. Structure of segmented poly(ether urethane)s containing amino and hydroxyl functionzlized polyhedral oligomeric silsesquioxanes (POSS). Polymer,2008,49(9):2298-2305
[71] Dnanasekaran D, Reddy B S R. Advanced materials, CNTs, Particles, Films and Composites, Chapter 4: Synthesis and characterization of poly(urethane-imide)-POSS nanocomposites. Nanotechnology,2011,1:445-448
[72] Jiangxuan Song, Guangxin Chen, Gang Wu, et al. Thermal and dynamic mechanical properties of epoxy resin/poly(urethane-imide)/polyhedral oligomeric silsesquioxane nanocomposites. Polymers for Advanced Technologies, 2011, 22(12):2069-2074
[73] Chengfei Zhou, Jianmei Guo, Tong Zhai. Research of radiation grafting polyurethane foam. China Rubber/Plastics Technology and Equipment, 2005, 31(11): 32-35

[74] Fenin A A, Ermakov V I, Revina A A. Radidtion-chemical synthesis of polymer polyols and related composites, Theoretical Foundations of chemical Engineering, 2008, 42 (5):662-665

[75] Chengfei Zhou. Advance in polymer polyols with high solid content. China Synthetic Rubber Industry, 2005, 28(5): 321-324

Chapter 10

Nitrogen-Rich Polymers as Candidates for Energetic Applications

Eric Pasquinet

Additional information is available at the end of the chapter

1. Introduction

The design of new energetic molecules is based on compounds exhibiting a high density and an elevated heat of formation. These fundamental properties, achieved through the presence of numerous nitrogen atoms and/or explosophoric groups, ensure high performance levels that can be useful in target applications such as explosives, propellants or gas generators.

The same basics also apply when considering the use of polymers, instead of single molecules, as energetic ingredients. However, the quest for high densities and heats of formation appears somewhat more challenging in the case of polymers, since the polymerization process often requires specific chemical functions that may be detrimental to the desired energetic properties. Currently, the most commonly used polymers in the field get their energetic content only from explosophoric groups: NO_2 for polyNIMMO (polynitratomethyl methyloxetane) or polyGLYN (polyglycidyl nitrate), N_3 for GAP (glycidyl azide polymer), to name a few. This survey gives an overview of research efforts that have been devoted to the synthesis of macromolecules with high nitrogen contents. Such a unique property is expected to produce materials with particularly elevated heats of formation, thus making them very valuable in the field of energetic compounds.

Azaheterocycles are obviously suitable scaffolds for achieving nitrogen-rich polymers. Tetrazole-, tetrazine-, triazole- and triazine-based materials are considered here. Three methodologies have been identified as the main routes for obtaining the target polymers: polymerization of vinylazaheterocycles, incorporation of the desired azaheterocycle in a preformed macromolecular architecture and polycondensation.

2. Tetrazole-based polymers

2.1. Polymerization of vinyltetrazoles

During the last decade, two major papers have reviewed the synthesis and polymerization of vinyltetrazoles, and together they present an extensive literature coverage [1,2].

2.1.1. *Synthesis of C-vinyltetrazoles*

Polymerization of the starting and/or target vinyl compound is a major concern in the reaction of acrylonitrile with sodium azide leading to *C*-vinyltetrazole (5-vinyltetrazole) **2**. Masking of the double bond of acrylonitrile **1** through addition of dimethylamine [3] or hydrogen chloride [4] circumvents this problem and renders the process more efficient (Figure 1).

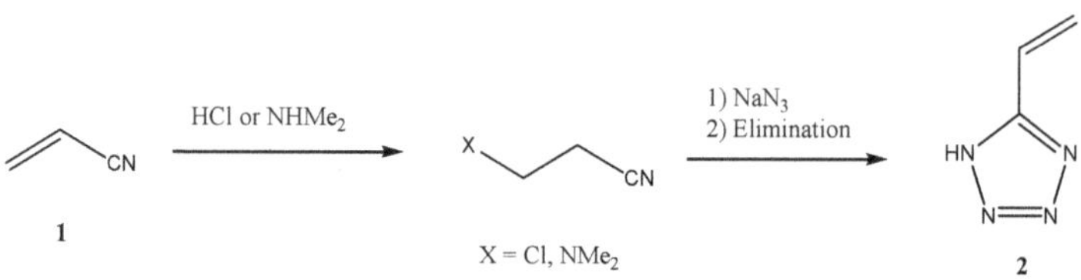

Figure 1.

2.1.2. *Synthesis of N-vinyltetrazoles*

Two main strategies have been used for the synthesis of *N*-vinyltetrazoles. The first one involves alkylation of the tetrazole N-H with a functionalized ethyl chain, followed by elimination. The second one is the catalyzed direct vinylation with vinylacetate or butylvinylether. Recently, a variant of the first approach, using dibromoethane as the alkylating agent, has been developed [5]. This method appears now as the method of choice, since it is a one-step, high-yielding procedure that only needs readily available, non-toxic chemicals. In the majority of cases, a mixture of 1- and 2-vinyltetrazole are produced and selectivity depends on the 5-substituent (Figure 2).

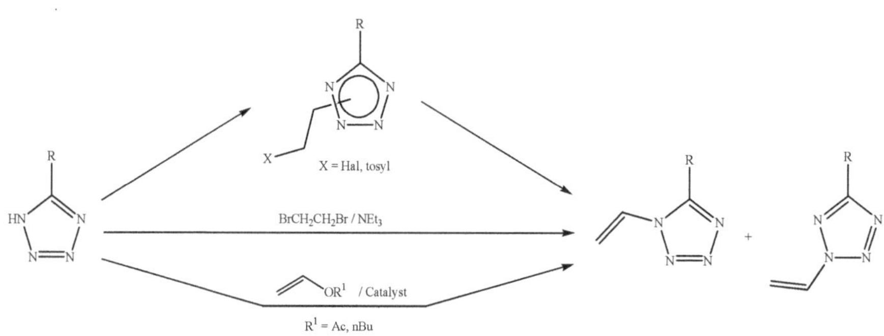

Figure 2.

2.1.3. *Synthesis of divinyltetrazoles*

Following the above-mentioned strategies, divinyltetrazole compounds have been obtained. Vinylation of 5-vinyltetrazole gave a mixture of 1,5- and 2,5-divinyltetrazole **3a-b** in a 1/2.3-2.4 ratio [5, 6]. In the vinylation of 5,5′-methylenebis(2-vinyltetrazole) **4**, only 2 of the 3 possible isomers were isolated, as the 1,1′-divinyl compound was not detected [5]. Such divinyl products could be useful as nitrogen-rich cross-linkers (Figure 3).

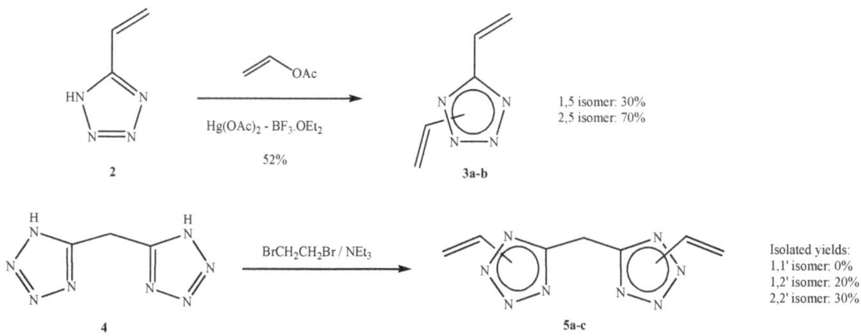

Figure 3.

2.1.4. *Polymerization of C-vinyltetrazoles*

5-Vinyltetrazoles are readily involved in free-radical polymerization (Figure 4). A comparison between *C*- and *N*-vinyltetrazoles shows that the former are more active in such processes. A number of reaction media have been used for the polymerization of 5-vinyltetrazoles, and it was shown that the activity was enhanced in more polar solvents, especially in water. Molecular weights also increased in aqueous systems. This trend prompted a polymerization study in ionic solvents. Of the various ionic liquids (ILs) tested, only 1-methyl-3-propylimidazolium bromide gave rise to decent yields. Other ILs gave low yields or even decomposition of the tetrazole ring [7].

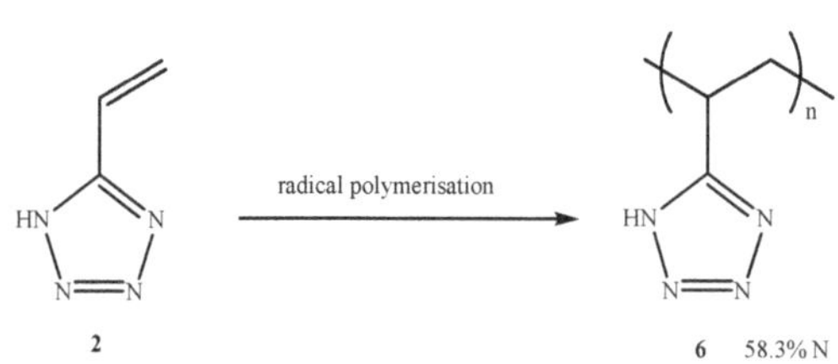

Figure 4.

Poly-5-vinyltetrazole **6** contains 58.3% nitrogen, which clearly makes it a high-nitrogen compound. However, it seems that its *N*-vinyl congener has been favored for industrial applications.

2.1.5. *Polymerization of N-vinyltetrazoles*

Figure 5 presents the most interesting polymers obtained in terms of nitrogen-content or expected energetic output. To the best of our knowledge, polymerization of 5-nitro-1-vinyltetrazole[8] is unknown. However, 1-vinyltetrazoles are generally readily polymerized in free-radical conditions, with water as the solvent of choice. The activity depends on the nature of the 5-substituent. For example, 5-amino-1-vinyltetrazole **7** is less active than the unsubstituted compound due to the electron-donating character of the amino group [[9],[10]]. Unusually, yields of poly(5-amino-1-vinyltetrazole) **8** are slightly higher in *N,N*-dimethylformamide (DMF) than in water. Recently, emulsion polymerization gave porous monoliths of the corresponding so-called 'polyHIPEs' of **8**, exhibiting remarkable mechanical properties [[11]].

The synthesis and polymerization of 5-(methyl)hydrazino-1-vinyltetrazole **9** have been described. It was necessary to add the initiator (AIBN) in several portions to enhance the conversion of the starting vinyl compound. However, while **9b** (R = Me) yielded 85% of the corresponding polymer **10b**, the yield of **10a** from **9a** (R = H) was limited to about 25-30% and the degree of polymerization was very low. The synthesis of the very nitrogen-rich azide-functionalized polymer **11** was attempted through diazotization of **10a**. However, elemental analyses were unsatisfactory because of partial decomposition of the tetrazolylhydrazine moieties [[12]].

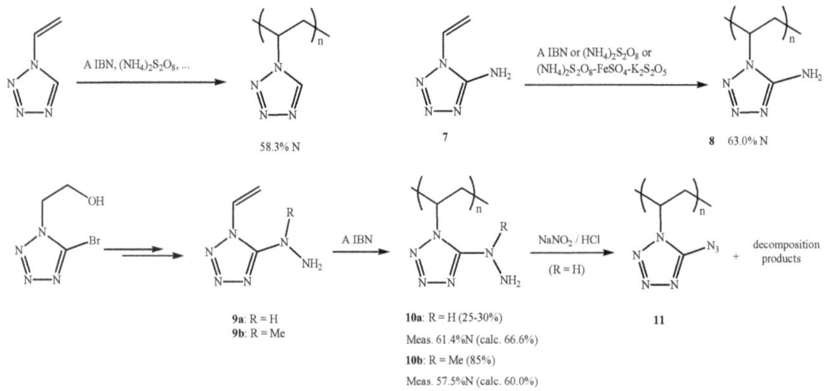

Figure 5.

The different examples cited herein highlight the high nitrogen content of polyvinyltetrazoles and their prospects as energetic compounds. In 2003, a pilot-plant production of polyvinyltetrazoles was started in two companies in Russia.

2.2. Incorporation of tetrazole via polymer-analogous transformation

It is well-known that tetrazoles can be obtained by addition of an azide onto a cyano derivative. This method was applied to polyacrylonitrile (PAN) in order to synthesize poly(5-vinyltetrazole) (Figure 6).

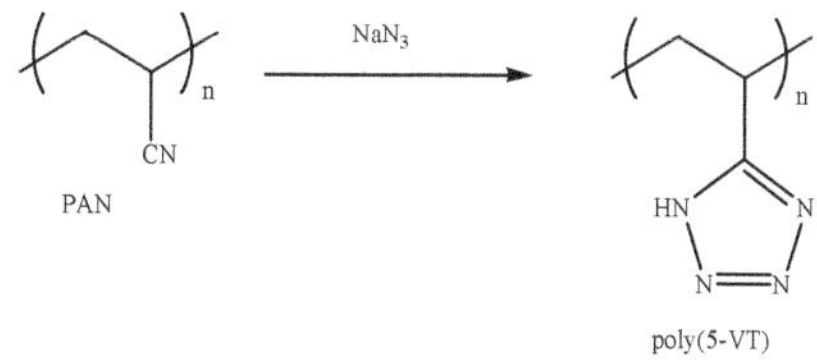

Figure 6.

When using NaN_3/NH_4Cl, it was shown that higher temperatures and molecular weights of the initial PAN resulted in a higher incorporation of tetrazole moieties in the polymer [13]. Thus, at 105°C, a PAN with a Mw of 180,000 g/mol was almost completely transformed in the corresponding poly(5-vinyltetrazole) (tetrazole units: 97.5%). The tetrazole content was estimated by 2 independent methods: weight measurements and acid-base titration. According to the authors, this poly(5-vinyltetrazole) synthesis using polymer-analogous transformation is advantageous since there is no commercial source of 5-vinyltetrazole and its synthesis is difficult.

Zinc chloride has also been used as a catalyst to carry out the desired transformation [14]. The best components ratio was $NaN_3/NH_4Cl/RCN$ 4/4/1. Infrared and NMR spectroscopies demonstrated the total conversion of nitrile functions into tetrazoles. In addition to PAN, other nitrile-containing polymers were succesfully tetrazolated.

Poly(5-vinyltetrazole) for gas generants was also synthesized upon the action of a zinc salt ($ZnBr_2$) but polymerization was carried out in emulsion in the presence of a surfactant [15]. A related patent subsequently extended this water-based synthesis [16]. At 115°C tetrazolation was limited to 70%, but reached 95% at 170°C. Further reaction of the resulting poly(5-vinyltetrazole) with ammonia yielded the corresponding ammonium salt **12** (Figure 7). The latter theoretically contains 61.9% of nitrogen and is therefore a promising material for gas generants. A related triaminoguanidinium salt **13** had been disclosed as early as in 1968 [17]. Taking into account the 6% of PAN units remaining in the starting poly-5-vinyltetrazole, a theoretical value of 69.2% of nitrogen was achieved in the salt (Figure 7).

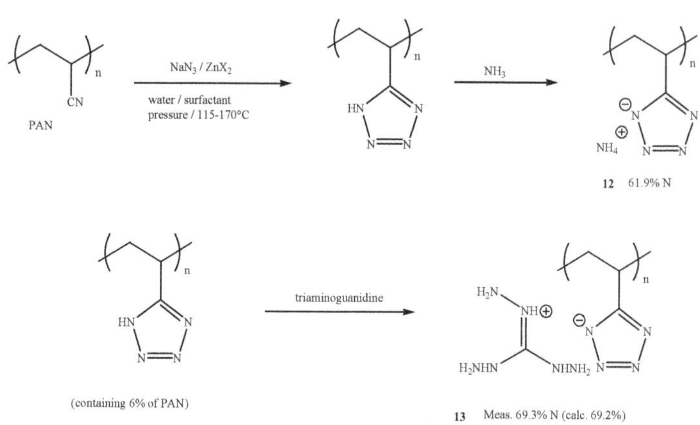

Figure 7.

All these conditions surpass earlier methods that yielded lower levels of tetrazole incorporation (see for example ref [18]).

Polyvinylchloride (PVC) can also be used as a polymer precursor. Upon reaction with a tetrazolate anion, the corresponding polyvinyltetrazole is formed (Figure 8). However, this method seems less practical since partial elimination of hydrogen chloride generates unsaturated fragments in the final product [1].

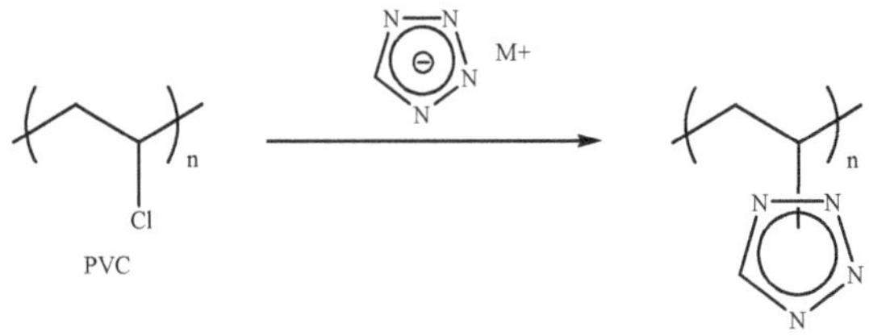

Figure 8.

2.3. Synthesis of tetrazole-based polymers via polycondensation

5-Chloromethyltetrazole **14** is an interesting substrate as it has both a nucleophilic (tetrazole N-H) and a highly electrophilic (CH_2-Cl) center. In the presence of triethylamine as a base, the tetrazole ring can be deprotonated which gives rise to poly(methylenetetrazole) **15** in good yields [19,20] (Figure 9).

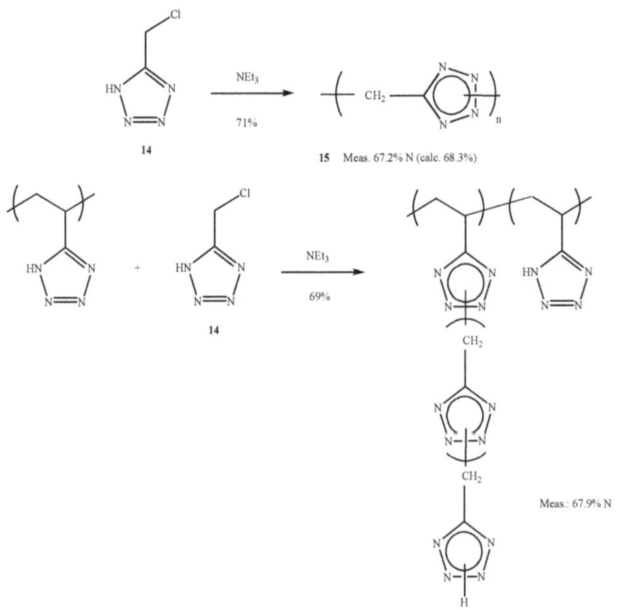

Figure 9.

This polyalkylation product is in fact an oligomer, with M_w below 2000 g/mol (ca 15-20 units). With a measured nitrogen content of up to 67.2%, it surpasses polyvinyltetrazole by almost 10%. In addition, its viscous state is attained at relatively low temperatures, thus making it a very promising candidate in gas-generating systems. It should be noted that 5-chloroethyltetrazole can also participate in a similar process. However, polymerization is slower and yields do not exceed 20%.

The high reactivity of 5-chloromethyltetrazole **14** has been used to graft poly(5-vinyltetrazole). Upon addition of an excess of **14** and triethylamine, a branched poly(5-vinyltetrazole) bearing poly(methylenetetrazole) bridges was obtained in yields around 70%. Ten equivalents of **14** produced the best results. Characterization of the polymer revealed a 6/1 ratio of methylenetetrazole vs vinyltetrazole moieties. The number of vinyltetrazole units between branching junctions was around 2. Its energetic prospects were highlighted by a very elevated density of 1.70 and a nitrogen content of 67.9% [19,20] (Figure 9).

A patent describes the polycondensation of various dinitriles with diazides to produce tetrazole containing polycondensates [21]. Of these, the most valuable polymer for energetic applications is the one obtained from dicyanofuroxane **15**. Indeed, despite the fact that the hexamethylene spacer lowers the N content, a high energetic output is expected from the presence of 2 tetrazole and 1 furoxane rings in the unit of polymer **16** (Figure 10).

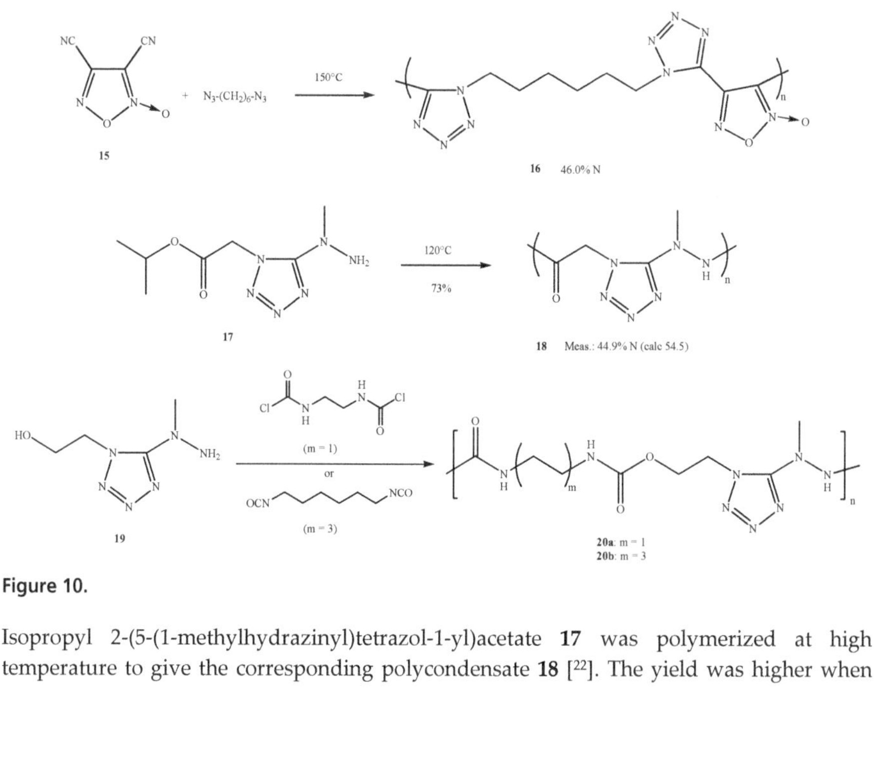

Figure 10.

Isopropyl 2-(5-(1-methylhydrazinyl)tetrazol-1-yl)acetate **17** was polymerized at high temperature to give the corresponding polycondensate **18** [22]. The yield was higher when working at 120°C rather than 140°C, however elemental analyses still showed a lower nitrogen content than expected (44.9 instead of 54.5%). The presence of low molecular weight polymers still bearing isopropoxy groups may account for this discrepancy.

The related monomer *N*-[1-(2-hydroxyethyl)tetrazol-5-yl]-*N*-methylhydrazine **19** was also involved in polycondensation processes. With succinyl chloride as the partner, only low molecular weights could be achieved, and results from elemental analysis were unsatisfactory. This was ascribed to the protonation of the hydrazino moieties during polycondensation. More positive results were obtained with hexamethylenediisocyanate but obviously the resulting polycondensate **20b** is less attractive for energetic applications due to its lower nitrogen content [23]. Another similar polymer was recently described by the same group [24] (Figure 10).

3. Tetrazine-based polymers

3.1. Polyvinyltetrazines

To the best of our knowledge, monovinyl-1,2,4,5-tetrazines are unknown, and the corresponding polyvinyltetrazines thus remain elusive compounds. However, the first member of the vinyltetrazine family, 3,6-divinyl-1,2,4,5-tetrazine **21** has been described (Figure 11).

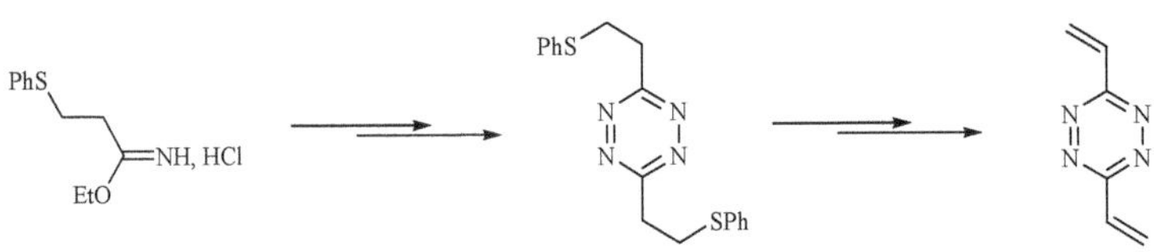

Figure 11.

The synthesis was achieved by using the phenylsulfanylethyl group as a masked vinyl moiety, the latter being the result of oxidation to the sulfone and elimination. Compound **21** was obtained as a volatile, pink oil that solidified in the freezer. Preliminary studies showed that **21** exhibited a limited stability at high temperatures and in the presence of radicals, which obviously precludes its use in polymer chemistry. However, 3,6-divinyl-1,2,4,5-tetrazine can be seen as a tetraaza analogue of the widely used 1,4-divinylbenzene and is therefore believed to be a useful building block in the synthesis of new tetrazine-based molecules and materials [25]. Further work is needed to establish the properties and reactivity of 21. These studies also open the way towards the synthesis of unsymmetrical vinyl-1,2,4,5-tetrazine compounds that may exhibit improved stability and reactivity in polymerization conditions.

3.2. Synthesis of tetrazine-based polymers via polycondensation

One of the most convenient ways to construct the 1,2,4,5-tetrazine (*s*-tetrazine) ring is by condensation of hydrazine with a diimidate to produce the corresponding 1,2-dihydro-

1,2,4,5-tetrazine, followed by oxidation. This strategy has been extended to the synthesis of diverse poly(phenylene-*s*-tetrazinylene)s by using diimidates. In order to increase the energetic prospects of such polymers, azahetaryl precursors were considered. However, the reaction of hydrazine monohydrate with pyridine-2,6-diimidate **22** did not yield the expected dihydrotetrazine polymer. Instead, the corresponding diamidrazone **23** was isolated in high yield. It is likely that the insolubility of **23** in THF is responsible for the uneffective polycondensation reaction. Diamidrazone **23** was used as the monomer in a microwave-assisted polycondensation to give the poly(2,6-pyridinediyl-dihydro-*s*-tetrazinylene) **24** which was further oxidized to the desired tetrazine polymer **25** using sodium nitrite/aq. AcOH [26] (Figure 12).

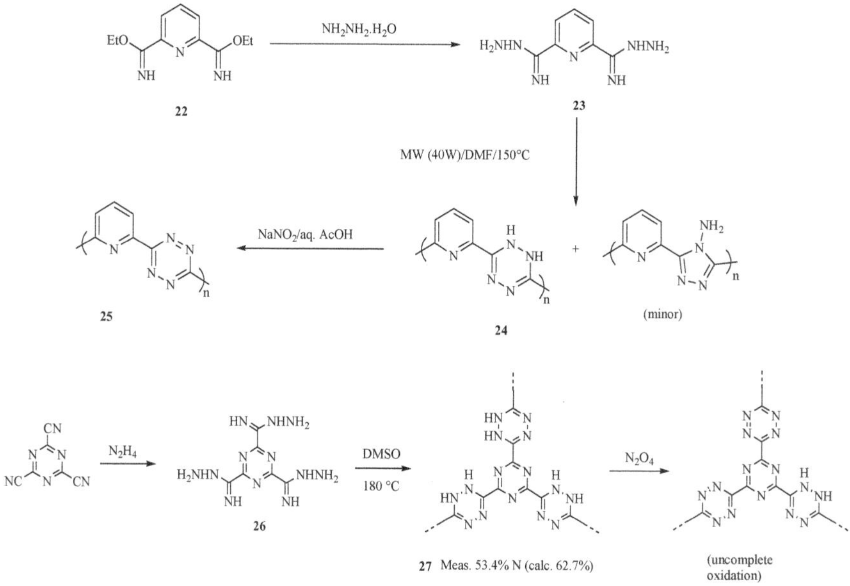

Figure 12.

Following the same strategy, the previously unknown 1,3,5-triazine-2,4,6-triamidrazone **26** was subjected to polycondensation to yield the hyperbranched polymer **27**. In this case, microwave activation was not necessary. Polymer **27** stands out as the most nitrogenated tetrazine-containing polymer, even if the measured N content was lower than expected, presumably due to water incorporation. A density of 1.56 was determined for **27**, which is significantly higher than the value reported for amorphous poly(*p*-phenylene) (1.11). This clearly shows how the introduction of nitrogen heterocycles in polyarylene structures is valuable in the field of energetic materials as it increases this crucial parameter. Unfortunately **27** was found unreactive to common oxidants, presumably because of its high insolubility. Only very strong oxidative conditions resulted in the formation of *s*-tetrazine rings, but oxidation remained uncomplete (Figure 12).

The structures of all polymers were confirmed using ^{13}C and ^{15}N solid state NMR [26,27]. High-resolution spectra also indicated low amounts of 4-amino-1,2,4-triazole moieties (coming from the known rearrangement of some 1,2-dihydro-1,2,4,5-tetrazine units).

4. Triazole-based polymers

Synthetic strategies for *C*-vinyl- and *N*-vinyltriazoles (1,2,3- or 1,2,4-triazoles) are essentially the same as the ones presented above for vinyltetrazoles. The preparation and polymerization of vinyltriazoles have been reviewed in 2008 [28], 2003 [29,30] and 1974 [31].

4.1. 1,2,4-Triazole-based polymers

4.1.1. *Poly(N-vinyl-1,2,4-triazole)s*

Figure 13 presents the most interesting polymers obtained in terms of nitrogen content or expected energetic output.

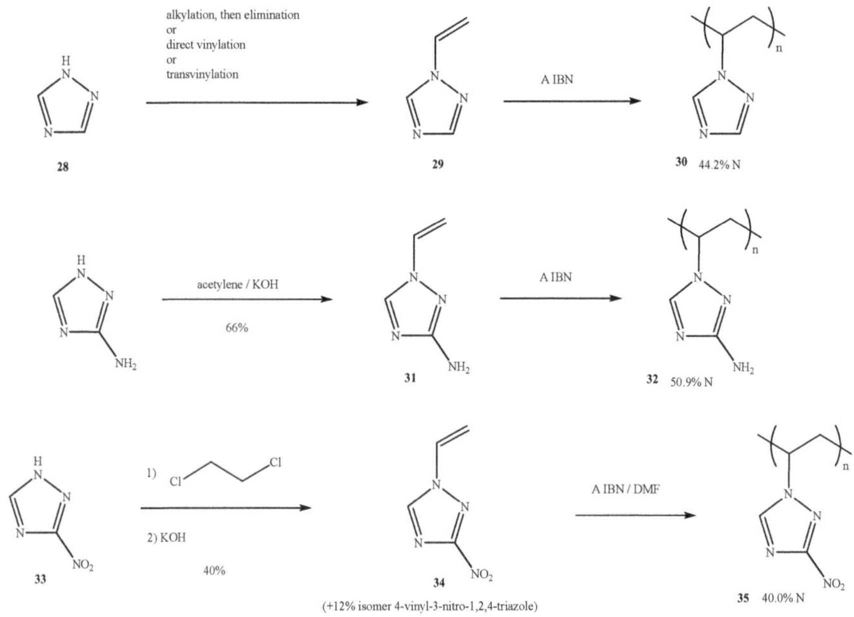

Figure 13.

1-Vinyl-1,2,4-triazole **29** was readily polymerized in free-radical conditions, and the process was more efficient in the presence of water. However, it was less active than 1-vinyl-1,2,3-triazole. This trend in activity was found to correlate with the corresponding Hammet constants [29]. Poly(1-vinyl-1,2,4-triazole) **30,** obtained in sufficiently aqueous media, was essentially insoluble in water or organic solvents. This polymer is a well-established compound that may be interesting in numerous applications. Recently, open-cell porous monoliths of poly(1-vinyl-1,2,4-triazole) cross-linked with a low amount of *N,N*-methylenebisacrylamide were prepared (Figure 14). They exhibited interesting mechanical properties due to the presence of water in the polymer structure, as evidenced by NMR [32].

Figure 14.

For the synthesis of 3-amino-1-vinyl-1,2,4-triazole **31**, the classical transvinylation using vinyl acetate in the presence of cuprous chloride was ineffective. The product was obtained through the reaction of acetylene in a basic medium at high temperature [33] (Figure 13). The presence of the amino group resulted in some changes compared to the unfunctionalized derivative: the polymerization activity was three times lower, and poly(3-amino-1-vinyl-1,2,4-triazole) **32** was soluble in water [34]. The nitrogen content was also significantly improved to 50.9%, with a density of 1.30-1.36 equalling that of polyglycidyl azide.

Poly(3-nitro-1-vinyl-1,2,4-triazole) **35** is also a useful target due to the expected output from the nitro group. Alkylation of 3-nitro-1,2,4-triazole **33** with dichloroethane followed by dehydrochlorination gave a mixture of 1- and 4-vinyltriazole. The former (**34**) was polymerized in DMF and gave a polymer that was soluble only in highly polar solvents [35] (Figure 13).

4.1.2. Poly(C-vinyl-1,2,4-triazole)s

The strategy involving the construction of the vinyl fragment through dehydrohalogenation, dehydration or deamination is also applicable for the synthesis of *C*-vinyl-1,2,4-triazoles [28,36]. Another approach is the use of the Wittig reaction, exemplified below for salts of poly(*C*-vinyl-1,2,4-triazole).

4.1.3. Energetic poly(vinyl-1,2,4-triazole)s salts

To further increase the potential of poly(1-vinyl-1,2,4-triazole), protonation with energetic inorganic or organic acids was investigated (Figure 15). Polymerization of protonated 1-vinyl-1,2,4-triazole **36** was successful only with the nitric or perchloric acid-protonated product, since the presence of nitro functionalities in the organic acids tended to inhibit polymerization. Moreover, only about 70% of the nitrate or perchlorate anions remained after polymerization. However, radical polymerization of 1-vinyl-1,2,4-triazole **29** in DMF followed by protonation produced a series of energetic salts of poly(vinyl-1,2,4-triazole) **38**. Elemental analyses established that the molar degree of substitution of anions was between

48 and 92% [37]. Thus this process represents a convenient way of improving the performances of the well-known poly(vinyl-1,2,4-triazole) **37**.

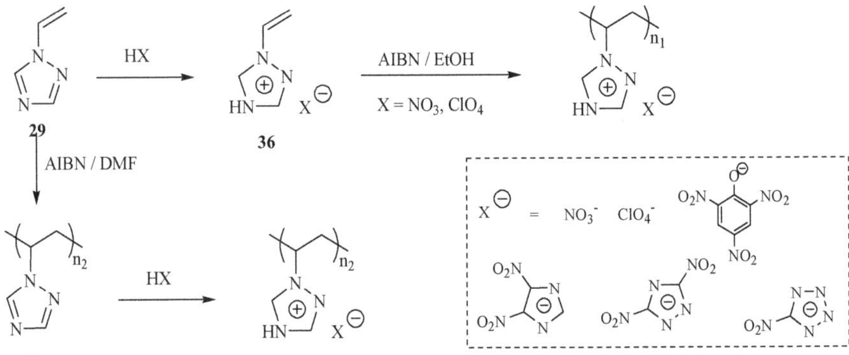

Figure 15.

The equivalent work in the poly(*C*-vinyl-1,2,4-triazole) series is also disclosed in a patent [38]. Salts (e.g., the ammonium salt) of 3(5)-bromomethyl-1,2,4-triazole **39** were converted to the 3(5)-vinyl derivatives **40** following a Wittig-type methodology. Free-radical or cationic polymerization produced the corresponding salt of poly(3(5*)*-vinyl-1,2,4-triazole) **41** (Figure 16). This compound has been used as a fuel in gas-generating compositions.

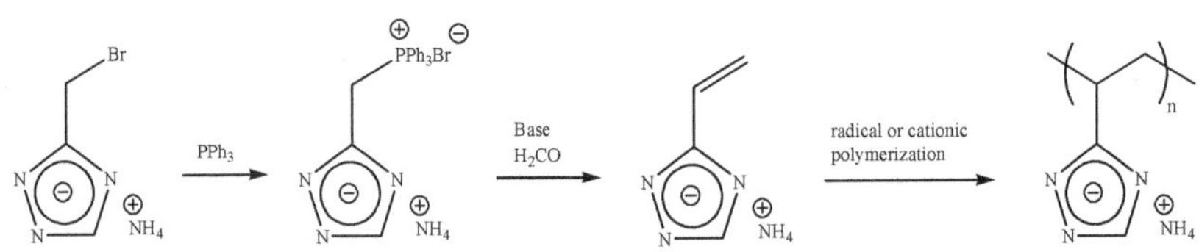

Figure 16.

4.1.4. Incorporation of 1,2,4-triazole via polymer-analogous transformation

Energetic heterocyclic analogues of the well-known polyglycidyl azide (GAP) have been prepared via substitution of the chlorine atom of polyepichlorhydrin **42** by the anion of 3(5)nitro-1,2,4-triazole **43-Na** [39]. The reaction proceeded in an almost quantitative fashion (residual chlorine: 0.1-0.4%) when carried out in high-boiling solvents at 100-130°C. Proton NMR enabled the determination of N-1 vs N-2 alkylation at the triazole ring, and a marked selectivity (9/1) towards the N-1 substituted polymer was found (Figure 17). The introduction of sodium azide together with the nitrotriazole salt resulted in a copolymer of glycidylazide and N-glycidyl-3-nitro-1,2,4-triazole.

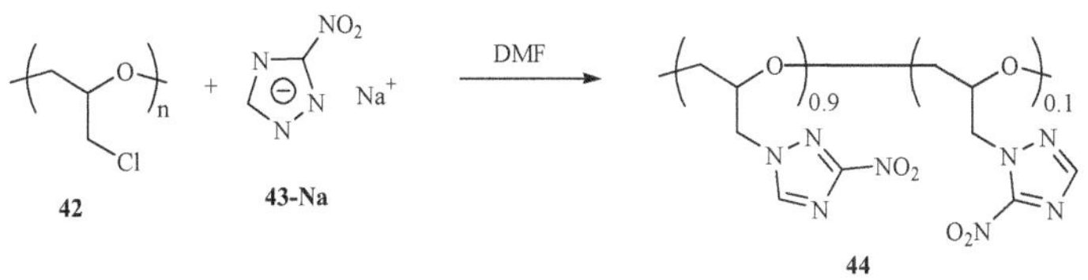

Figure 17.

4.1.5. Synthesis of 1,2,4-triazole-based polymers via polycondensation

Among the few 1,2,4-triazole-based condensation polymers known [40], the most promising as energetic materials are poly(4-amino-1,2,4-triazole)s [41]. The latter can be commonly obtained through isomerization of poly(dihydrotetrazine)s **45**, following the general scheme depicted in Figure 18. This process is closely related to the one mentioned in the tetrazine section of this chapter.

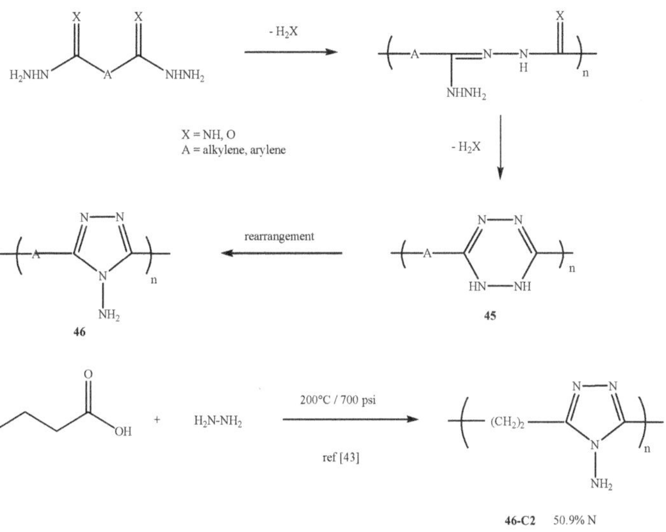

Figure 18.

Polymer **46-C2** stands out as the most nitrogenated member of this family with a calculated N content of 50.9%. Even though the first synthesis of **46-C2** dates back to 1954 [42], interest for this polymer or similar derivatives has been maintained for a long time for diverse applications [43, 44].

Another polycondensate incorporating 4-amino-1,2,4-triazole units has been mentioned in a patent. The synthesis is based on the simultaneous condensation of 3,5-dihydrazino-4-amino-1,2,4-triazole **47** and diaminoguanidinium azide **48** with formaldehyde/glyoxal (4/1) [45]. The exact structure of the resulting polymer **49** is not disclosed but has been claimed to provide a N/C ratio of 2.3 (Figure 19).

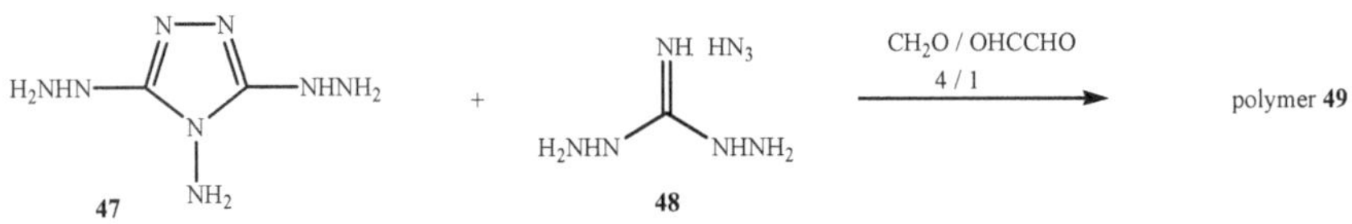

Figure 19.

It should also be noted that 3-amino-1,2,4-triazole can be subjected to electrooxidation to yield films of the corresponding polymer [46,47]. However, the use of these films seems to be limited to corrosion protection applications and they obviously cannot be implemented as energetic materials.

4.2. 1,2,3-Triazole-based polymers

4.2.1. Poly(N-vinyl-1,2,3-triazole)s

1-Vinyl-1,2,3-triazole **50** can be classically synthesized via reaction of 1,2,3-triazole with acetylene in the presence of a catalyst [48], or by transvinylation with vinyl acetate [49]. Another possibility is the construction of the 1,2,3-triazole ring through 'click' chemistry between a functionalized azide and (substituted) acetylene, followed by elimination. A recent synthesis [50] following this scheme is presented in Figure 20. By using vinylacetylene in this approach, 1,5-divinyl-1,2,3-triazole **52** can be obtained [51]. 1-Vinyl-1,2,3-triazole itself is highly active in classical free-radical polymerization [29] and can also be polymerized to **51** using reversible addition-fragmentation and transfer (RAFT) conditions [50], but 5-nitro-1-vinyl-1,2,3-triazole **53** [49] is inert in this process (see below).

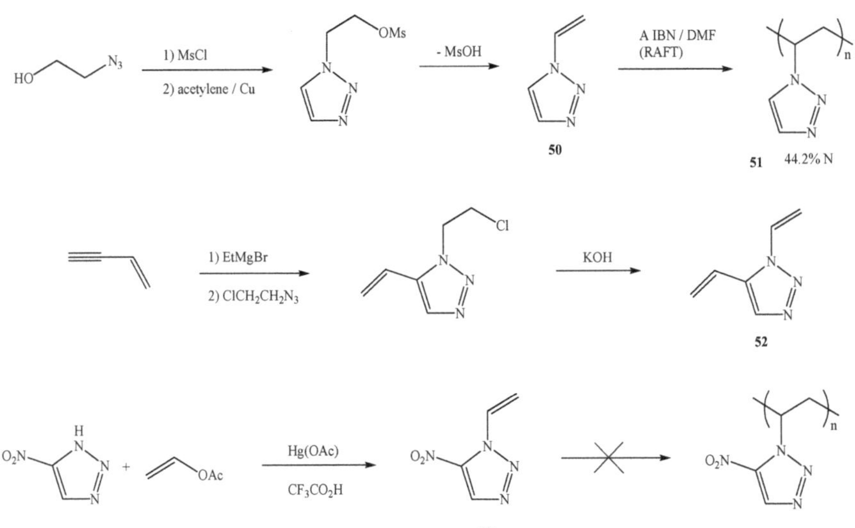

Figure 20.

4.2.2. *Poly(C-vinyl-1,2,4-triazole)s*

4-Vinyl-1,2,3-triazole **56** has been obtained through 'click' chemistry with different acetylene and azido derivatives. A recent example describes the use of vinylacetylene **54** and azidomethyl pivalate **55** as starting materials [52]. This monomer is also active in radical polymerization [53], but less than its *N*-vinyl congener. Poly(4-vinyl-1,2,3-triazole) **57** has also been synthesized by deprotection of poly[4-vinyl-1(4-methoxybenzyl)-1,2,3-triazole] **58** [54] (Figure 21).

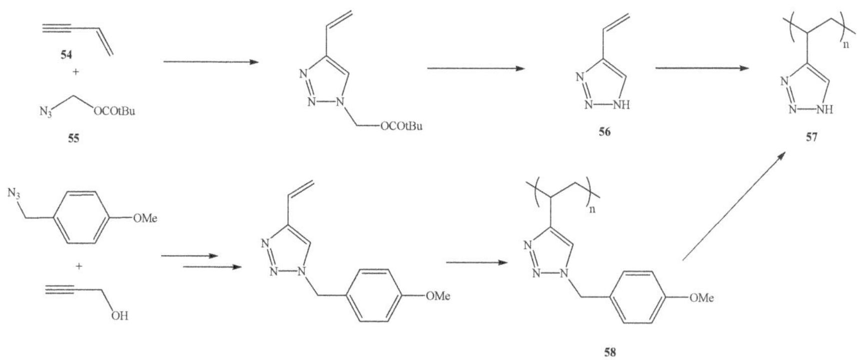

Figure 21.

4.2.3. *Incorporation of 1,2,3-triazole via polymer-analogous transformation*

Reaction of the sodium salt of 4(5)-nitro-1,2,3-triazole **59-Na** with polyvinylchloride (PVC) afforded poly(1-vinyl-4-nitro-1,2,3-triazole) **60**. By using at least 2 equivalents of **59-Na** at 120°C for 1 day, a complete substitution of the chloride functions could be achieved as demonstrated by elemental analysis showing 39.9% of N (calc. 40.0%) [55] (Figure 22).

This is the only method rendering it possible to obtain this nitrated polymer since it cannot be synthesized through polymerization of 1-vinyl-4-nitro-1,2,3-triazole **53**. As a matter of fact, in **53,** the nitro group acts as an inhibitor under radical and ionic conditions, which is apparently not the case for the isomeric 3-nitro-1-vinyl-1,2,4-triazole **34** (see above).

The same chemistry was applied to copolymers of vinylchloride and 2-methyl-5-vinyl-tetrazole to yield the corresponding nitro-1,2,3-triazole containing copolymers **61a-b**. Copolymer **61a** exhibited a measured N content of 47.2%.

Similarly, the salt **59-Na** was used to quantitatively substitute the chlorine atoms on polyepichlorhydrin **42.** As in the nitro-1,2,4-triazole series (see above), a mixture of isomers was obtained, but the selectivity was reversed: the N-2 isomer was the major one (60%) [56] (Figure 22).

4.2.4. *Synthesis of 1,2,3-triazole based polymers via polycondensation*

The well-established reaction of an azide with an alkyne, that has been extensively used for monomeric 1,2,3-triazole compounds, can be further extended to the synthesis of polymers.

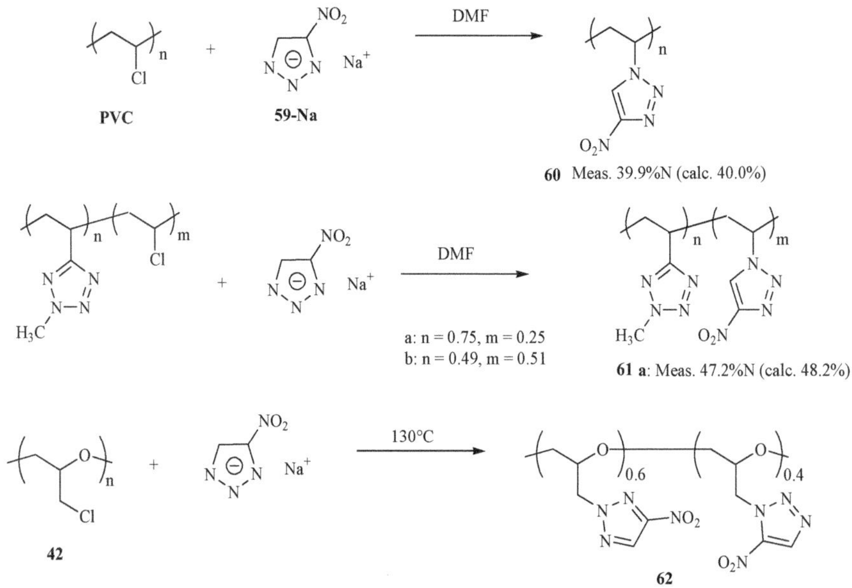

Figure 22.

Thus, (azidoalkyl)- or (azidoaryl) acetylenes were polymerized without catalyst to afford materials that were presumed to contain 1,4- and 1,5-substituted 1,2,3-triazole rings in a random distribution [57]. The most interesting compound for energetic applications was polymer **64,** obtained from 3-azido-1-butyne **63** (Figure 23).

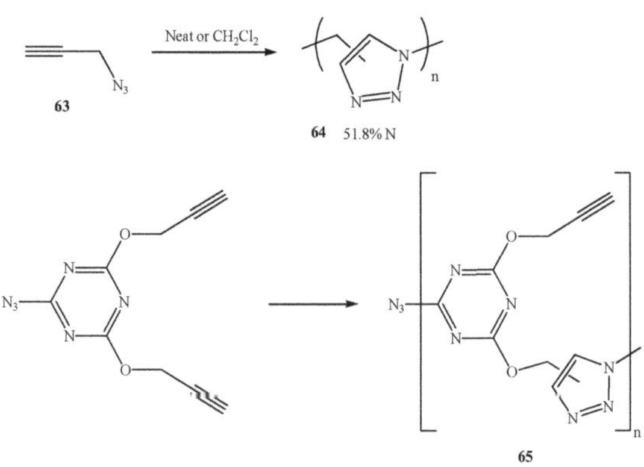

Figure 23.

Similarly, an azido-1,3,5-triazine containing propargyl ether was used as the monomer. Of the different conditions (bulk, solution, with or without catalyst or heating), polymerization in bulk without catalysts was found to be the best way to minimize decomposition reactions and formation of by-products [58]. The hyperbranched product **65** offers an original combination of both triazine and triazole rings, although the nitrogen content was obviously lowered by the remaining propargyl ethers moieties .

5. Triazine-based polymers

To the best of our knowledge, vinyl-1,2,4-triazines are unknown. Some poly-1,2,4-triazines have been obtained by polycondensation [59], however their nitrogen content is much too low for them to be considered as energetic materials. This section will thus focus on polymers containing 1,3,5-triazines (*s*-triazines), which have been studied for a long time. Ref [60] is a comprehensive review of earlier work.

5.1. Polyvinyl-1,3,5-triazines

2-Vinyl-1,3,5-triazine **67** [61] has been prepared from 1,3,5-triazine **66** but it seems that interest in it faded quickly due to the ease of hydrolysis. 2-Vinyl-4,6-disubstituted-1,3,5-triazines were claimed to be more stable [62].

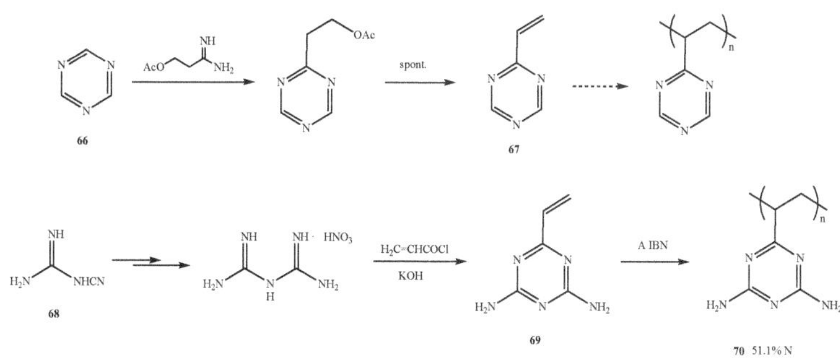

Figure 24.

2,4,6-Trivinyl-1,3,5-triazine would also be an interesting compound (as a nitrogen-containing cross-linker) but the reported yield for its synthesis is poor [61]. In fact, the most valuable vinyltriazine is 2,4-diamino-6-vinyl-1,3,5-triazine **69**, which can be prepared in 2 steps from dicyandiamide **68** [63]. Compound **69** has been polymerized in free-radical conditions, either by using potassium persulfate in water [63] or AIBN in DMSO or DMF [64]. The calculated N content of poly(2,4-diamino-6-vinyl-1,3,5-triazine) **70** exceeds 50% (Figure 24).

5.2. Synthesis of 1,3,5-triazine-based polymers via polycondensation

Amino-substituted triazines bearing a leaving group have been heated to obtain the corresponding homopolycondensates. A linear polytriazinylamine **72** was formed from 2,4-diamino-6-phenoxy-1,3,5-triazine **71** [65], whereas a hyperbranched structure **74** resulted from 2-amino-4,6-dichloro-1,3,5-triazine **73** since it was impossible to control the reaction in the sense that only one chlorine atom reacted [66] (Figure 25). Although no elemental analyses were reported to ascertain the structures, these products are believed to exhibit a high nitrogen content. Compound **74** is claimed to approach a cross-linked structure, which is corroborated by its limited degree of softening.

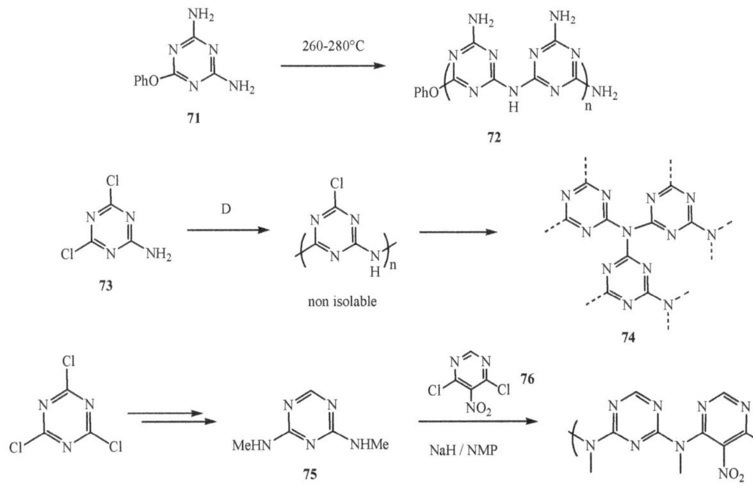

Figure 25.

In a similar fashion, 2,4-bis(methylamino)-1,3,5-triazine **75** was reacted with 4,6-dichloro-5-nitro-pyrimidine **76** in the presence of a base. Modest molecular masses were obtained (1000-10000 g/mol), as confirmed by the presence of end-chain NHMe signals in NMR spectra [67]. However, polymer **77** is still of interest thanks to the presence of the energetic nitro group in addition to a reasonable nitrogen content (Figure 25).

Aminoalkyl units can also serve as linkages between triazine rings, as exemplified by the commercially available polymer **78**, which has been used as a charring agent in flame-retardant compositions [68] (Figure 26).

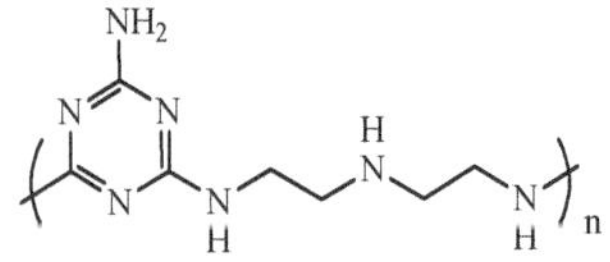

78

Figure 26.

Melamine-formaldehyde resins constitute a well-known class of compounds. Not surprisingly, related polymers can also be considered as derivatives with high nitrogen contents. For example, the reaction of trichloromelamine **79** with formaldehyde followed by dehydrochlorination has been claimed to give a structure in which triazine rings are linked together through carbodiimide or dihydrocarbodiimide units [69] (Figure 27).

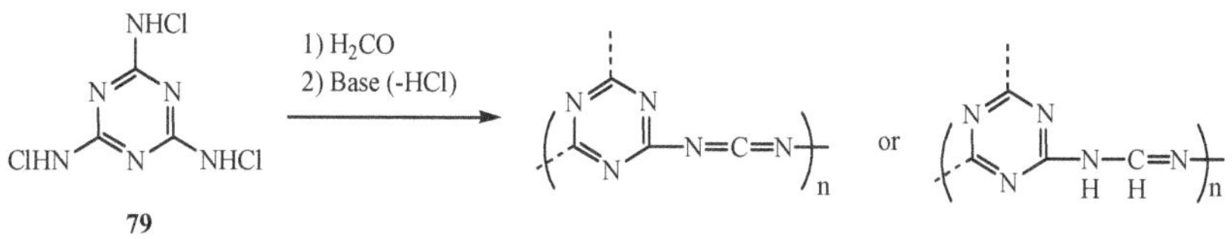

Figure 27.

The chemistry shown above in the case of polytriazinylamines has also been widely explored for the synthesis of polytriazinylethers. The latter are obviously less attractive in terms of nitrogen content. However, commercially available nitro-containing diols are valuable compounds when one wishes to incorporate explosophoric functions in the final material. Thus, 2-nitroresorcinol **80** was reacted with trichlorotriazine **81** (cyanuric chloride) in the presence of a base to produce a cross-linked polymer **82** with triazine units and nitro groups (Figure 28). The interfacial conditions readily employed for such polycondensations were naturally adapted for the synthesis of porous polyHIPEs of this material [70] (see tetrazole and triazole-based polymers for other examples of high-nitrogen polyHIPEs). The chemical structure should be similar to the one obtained by cyclotrimerization of 1,3-dicyanato-2-nitrobenzene [71].

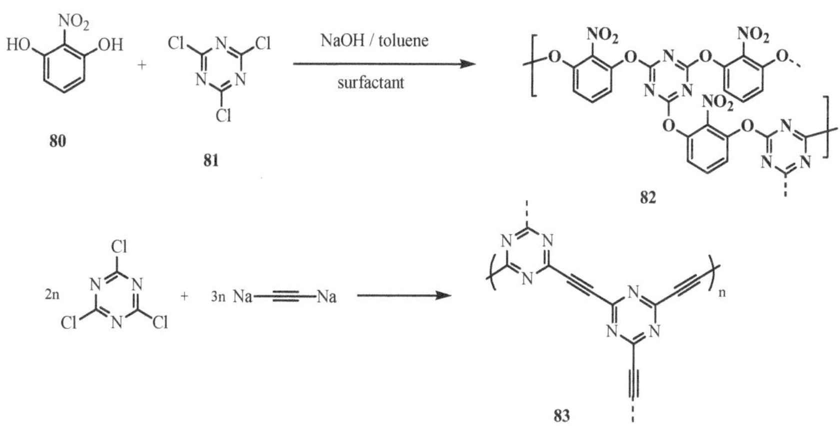

Figure 28.

It should also be noted that carbon dinucleophiles are effective in polycondensation reactions with cyanuric chloride. Thus, sodium carbide enabled the construction of cross-linked architectures **83** in which triazine rings were connected by ethynylene moieties [72] (Figure 28).

Another well-known way to obtain triazine compounds is the cyclotrimerization of nitriles. However, only scarce examples deal with azaheterocyclic nitriles. 2,6-dicyanopyridine **84** was converted at high temperatures, in the presence of zinc chloride, to a porous framework **85** with alternating pyridine and triazine nuclei [73]. The measured N content was slightly below that expected (27.7 vs 32.5%) (Figure 29)

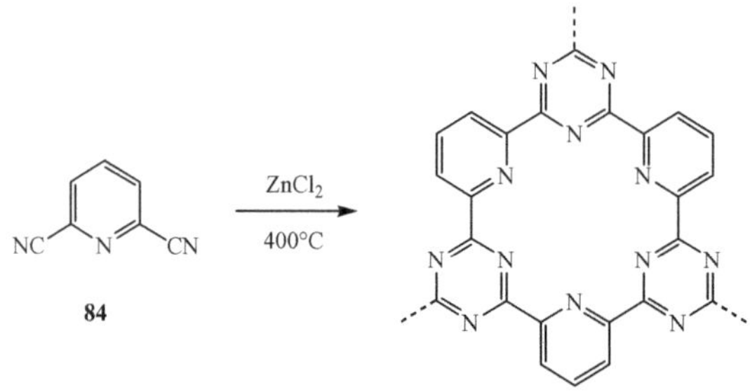

84

85

Figure 29.

The quest for graphitic forms of carbon nitride (g-C_3N_4) has stimulated a vast amount of research in order to find suitable molecular precursors. With a calculated nitrogen content of 60.9%, C_3N_4 surely has its place in this overview, although only a few precursors lead to materials approaching the theoretical N/C ratio due to the presence of hydrogen or oxygen. The structure of C_3N_4 is still under consideration, but surely involves 1,3,5-triazine moieties. Many useful references are provided to the reader in [74]. Theoretical as well as characterization results have indicated that tri-*s*-triazine structures **86** rather than *s*-triazines would be intermediates towards g-C_3N_4 [75] (Figure 30).

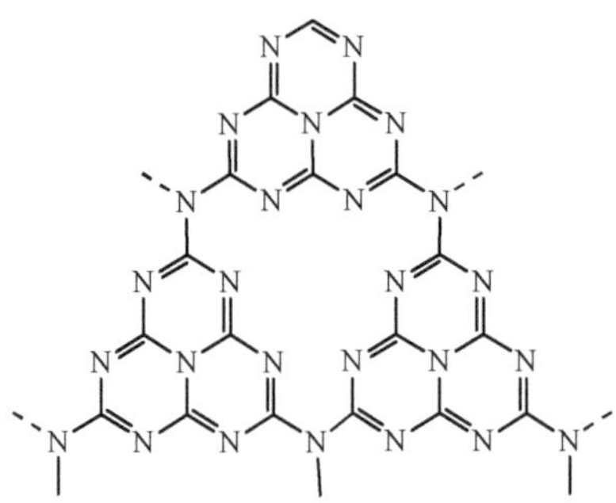

86

Figure 30.

6. Miscellaneous

Azide salts of bis(aminoguanidinium) compounds can be condensed with formaldehyde to produce the corresponding polymers **87a-b** (Figure 31). Thanks to the azide groups and the intrinsically nitrogen-rich aminoguadinium moieties, the nitrogen content of these polycondensates was remarkably high, up to 77% [45,76]. A related copolymer incorporating 4-amino-1,2,4-triazole units has already been described in the triazole section.

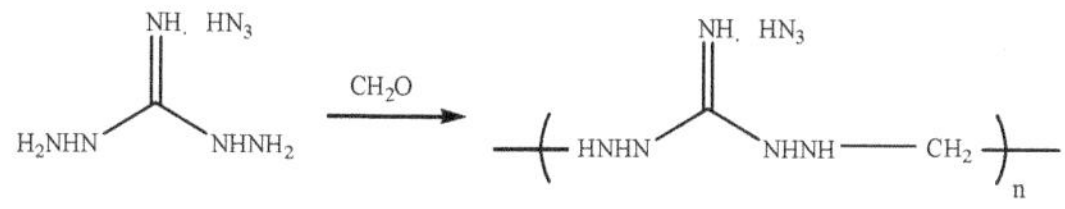

87a Meas. 77.0% N (calc. 77.7)

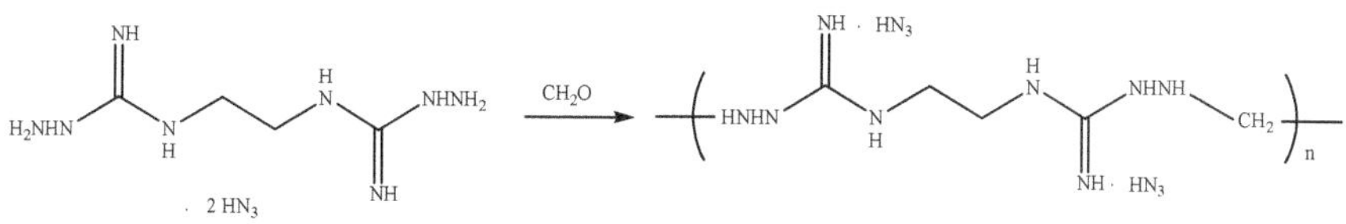

87b Meas. 71.8% N (calc. 72.0)

Figure 31.

Gaseous cyanogen (NC-CN) is a promising monomer as it shows a C/N ratio of 1/1. Studies have been devoted to its polymerization and given rise to so-called paracyanogen, either through chemical [77] or photochemical [78] methods. The determination of the exact structure of the polymer is a difficult task and a number of different hypotheses have been postulated (Figure 32). This work is made complicated by the incorporation of solvent or water/oxygen during the polymerization process, which leads to a significant presence of oxygen in the elemental analyses. Therefore, the theoretical figure of 50% of nitrogen can hardly be obtained. However, certain reaction conditions have enabled the synthesis of materials exceeding 40% of N.

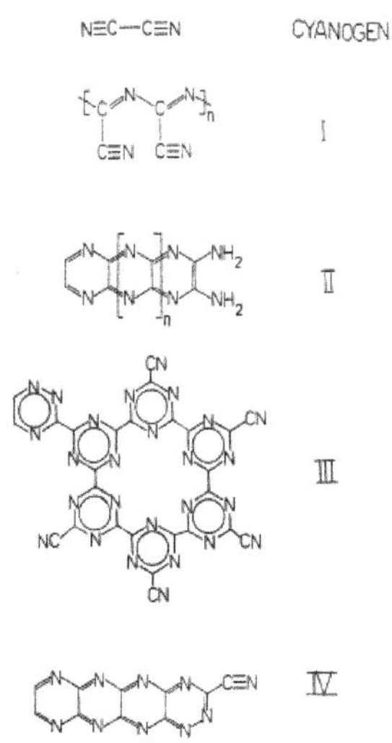

Figure 32. (from ref [78])

7. Conclusions

This review highlights a number of azaheterocycles-based polymer structures containing a high content of nitrogen, frequently around 50%. In some cases, over 60% and 70% N can be achieved, mainly in tetrazole-based materials. The presence of additional explosophoric groups such as nitro may also be encountered in such polymers which strengthens the interest in using them for energetic applications. The diversity of the described structures renders it possible to foresee a wide range of properties for these energetic materials. Consequently, a great opportunity is offered to select the appropriate material for a specific application in this field. Certain patents cited in this survey show that some of these polymers have already been exploited, and many other applied high-nitrogen polymers will surely see the light in the future.

Author details

Eric Pasquinet
CEA-DAM Le Ripault – BP16 – F-37260 Monts, France

8. References

[1] Kizhnyaev VN, Pokatilov FA, Vereshchagin LI (2008) Carbochain Polymers xith Oxadiazole, Triazole, and Tetrazole Cycles. Polym. sci. ser. C. 50:1-21.

[2] Kizhnyaev VN, Vereshchagin LI (2003) Vinyltetrazoles: Synthesis and Properties. Russ. chem. rev. 72: 143-164.

[3] Henry RA (1967) Process for Synthesis of Vinyltetrazole Monomers. US Pat. 3,383,389.

[4] Buzilova SR, Shulgina VM, Sakovich GV, Vereshchagin LI (1981) Synthesis of Vinyl Derivatives of Tetrazole. Chem. heterocycl. comp. 17:960-963.

[5] Aleshunin PA, Dmitrieva UN, Ostrovskii VA (2011) Vinyltetrazoles: II. Synthesis of 5-Substituted 1(2)-Vinyltetrazoles. Russ. j. org. chem. 47: 1882-1888

[6] Vereshchagin LI, Buzilova SR, Mityukova TK, Proidakov AG, Kizhnyaev VN, Il'ina VV, Sukhanov GT, Gareev GA, Bogens AK (1986) Synthesis of Functionally Substituted N-Vinyltetrazoles. Russ. j. org. chem. 22: 1777-1783.

[7] Vygodskii YS, Mel'nik OA, Kazakova EV, Shaplov AS, Komarova LI, Kizhnyaev VN (2008) Free-Radical Polymerization of C-Vinyltetrazoles: Effect of the Nature of Ionic Solvents. Polym. sci. ser. B. 50: 193-197.

[8] Koldobskii GI, Soldatenko DS, Gerasimova ES, Khokhryakova NR, Shcherbinin MB, Lebedev VP, Ostrovskii VA (1997) Tetrazoles: XXXVI. Synthesis, Structure, and Properties of 5-Nitrotetrazole. Russ. j. org. chem. 33:1771-1783.

[9] Kizhnyaev VN, Kruglova VA, Shivernovskaya OA, Ratovskii GV, Protasova LE, Vereshchagin LI (1991) Electronic Structure and Activity of Vinyltetrazoles in Radical Homopolymerization. Izv. akad. nauk. SSSR ser. khim. 2234-2238.

[10] Kizhnyaev VN, Kruglova VA, Ivanova NA, Ratovskii GV, Buzilova SR, Gareyev GA (1989) The Radical Polymerization of N-Vinyltetrazoles. Polym. sci. USSR. 31: 2728-2734.

[11] Youssef C, Backov R, Treguer M, Birot M, Deleuze H (2010) Preparation of Remarkably Tough PolyHIPE Materials via Polymerization of Oil-in-Water HIPEs Involving 1-Vinyl-5-Aminotetrazole. J. polym. sci. A: polym. chem. 48: 2942-2947.
[12] Klapötke TM, Sproll SM (2010) Nitrogen-Rich Polymers Based on 5-Bromo-1-vinyl-1*H*-tetrazole. Eur. j. org. chem. 1169-1175.
[13] Huang M-R, Li X-G, Li S-X, Zhang W (2004) Resultful Synthesis of Polyvinyltetrazole from Polyacrylonitrile. React. function. polym. 59: 53-61.
[14] Tsarevsky NV, Bernaerts KV, Dufour B, Du Prez FE, Matyjaszewski K (2004) Well-Defined (Co)polymers with 5-Vinyltetrazole Units via Combination of Atom Transfer Radical (Co)polymerization of Acrylonitrile and Click Chemistry-Type Postpolymerization Modification. Macromolecules 37: 9308-9313.
[15] Miller CG, Williams GK (2005) Gas Generant and Synthesis. PCT Appl. WO 2005/118715 A2.
[16] Miller CG, Williams GK (2006) Water-Based Synthesis of Poly(tetrazoles). PCT Appl. WO 2006/050444 A1.
[17] Torley RE, Sprague GS (1968) Triaminoguanidinium Salts of 5-Vinyltetrazole Polymers and a Method for Their Preparation. US Pat. 3,397,186.
[18] Gaponik PN, Ivashkevich OA, Karavai VP, Lesnikovic AI, Cheshavina NI, Sukhanov GT, Gareev GA (1994) Polymers and Copolymers Based on Vinyltetrazoles. 1. Synthesis of Poly(5-vinyltetrazole) by Polymer-Analogous Conversion of Polyacrylonitrile. Angew. makromol. chem. 219: 77-88.
[19] Kizhnyaev VN, Pokatilov FA, Vereshchagin LI (2007) Branched Tetrazole-Containing Polymers. Polym. sci. ser. A. 49: 28-34.
[20] Kizhnyaev VN, Pokatilov FA, Vereshchagin LI, Smirnov AI (2006) Method for Preparing Polymethylenetetrazoles. Pat. RU 2318003 C2.
[21] Carpenter WR (1968) Tetrazole Polymers. US Pat. 3,386,968.
[22] Klapötke TM, Sproll SM (2008) Synthesis and Characterization of Nitrogen Rich, Energetic Polymers Based on Tetrazoles. Proc. New trends in research of energetic materials, Pardubice, 796-804.
[23] Banert K, Klapötke TM, Sproll SM (2009) Synthesis of *N*-[1-(2-Hydroxyethyl)-1*H*-tetrazol-5-yl]-*N*-methylhydrazine as Polymeric Precursor. Eur. j. org. chem. 275-281.
[24] Klapötke TM, Sproll SM (2010) Investigation of Nitrogen-Rich Energetic Polymers Based on Alkylbridged Bis-(1-methyl-tetrazolylhydrazines). J. pol. sci. A: polym. chem. 48: 122-127.
[25 Pican S, Lapinte V, Pilard JF , Pasquinet E, Beller L, Fontaine L, Poullain D (2009) Synthesis of 3,6-Divinyl-1,2,4,5-Tetrazine, the First Member of the Elusive Vinyltetrazine Family. Synlett 5: 731-734.
[26] Sagot E, Le Roux A, Soulivet C, Pasquinet E, Poullain D, Girard E, Palmas P (2007), Synthesis of Linear and Hyperbranched Tetrazine-Based Polyhetarylenes Assemblies with High Nitrogen Content. Tetrahedron 63: 11189-11194.
[27] Palmas P, Girard E, Pasquinet E, Caron T, Poullain D (2007) Spectral Assignments for 1H, 13C and 15N Solution and Solid-State NMR Spectra of s-Tetrazines and Dihydro-s-Tetrazine Derivatives. Magn. reson. chem. 45: 65-71.

[28] Kizhnyaev VN, Pokatilov FA, Vereshchagin LI (2008) Carbochain Polymers xith Oxadiazole, Triazole, and Tetrazole Cycles. Polym. sci. ser. C. 50:1-21.

[29] Tsypina NA, Kizhnyaev VN, Pokatilov FA, Smirnov AI (2003) N-Vinyltriazoles in Radical Polymerization. Polym. sci. ser. B. 45:41-44.

[30] Kizhnyaev VN, Vereshchagin LI, Verkhozina ON, Pokatilov FA, Tsypina NA, Petrova TL, Sukhanov GT, Gareev GA, Smirnov AI (2003) Triazole and Tetrazole Containing Energetic Compounds. Proc. ICT conference, 75: 1-11.

[31] Smets GJ (1974) Some New Polymers, Their Syntheses and Properties. 24th Plenary main. sect. lect. int. congr. pure. appl. chemical. 1: 1-23. CA 1976:463755.

[32] Audouin F, Birot M, Pasquinet E, Besnard O, Palmas P, Poullain D, Deleuze H (2011) Preparation, Solid-State NMR and Physico-Chemical Characterization of Surprisingly Tough Open Cell PolyHIPEs Derived From 1-Vinyl-1,2,4-Triazole Oil-in-Water Emulsions. Macromolecules 44: 4879-4886.

[33] Skvortsova GG, Domnina ES, Makhno LP, Voronov VK, Taryashinova DD, Chipanina NN (1973) Structure and Properties of 3-Amino-1-vinyl-1,2,4-triazole. Khim. geterosikl. soedin. 11: 1566-1569.

[34] Kizhnyaev VN, Pokatilov FA, Adamova LV, Zelenkov LE, Smirnov AI (2008) Polymerization of 1-Vinyl-3-amino-1,2,4-triazole and Some Properties of Related Polymers. Polym. sci. ser. B 50: 16-19.

[35] Attarian OS, Asratian GV, Eliazian GA, Darbinian EG, Matsoian SG (1986) Synthesis and Polymerization of Vinyl Derivatives of 1,2,4-Triazole, 3-Nitro-1,2,4-triazole and Tetrazole. Arm. khim. zh. 39: 630-635.

[36] Overberger CG, Yuen PS (1970) Esterolytic Catalyses by Triazoles. J. am. chem. soc. 92: 1667-1671.

[37] Xue H, Gao H, Shreeve JM (2008) Energetic Polymer Salts from 1-Vinyl-1,2,4-Triazole Derivatives. J. polym. sci. A: polym. chem. 46: 2414-2421.

[38] Williams GK, Burns SP, Mishra IB (2005) Gas Generating Commpositions. PCT Appl. WO 2005/035466 A2.

[39] Sukhanova AG, Sakovich GV, Sukhanov GT (2010) Reaction of 3-Nitro-1,2,4-Triazoles with Alkylating Agents. 7. N-Monoalkylation by Monocyclic α-Oxide Polymers. Chem. heterocycl. comp. 46: 478-482.

[40] Korshak VV, Teplyakov MM (1969) Synthesis of Polymers with Azole Rings. Progr. polim. khim. 198-251. CA 1970:13074.

[41] Nishikubo T, Hasegawa M (1987) Polyaminotriazoles. In: Kroschwitz JI, editor. Encyclopedia of Polymer Science in English. New York: Wiley. pp. 507-514.

[42] Fisher JW (1954) Polyaminotriazoles as Fiber-Forming Materials. J. appl. chem. 4: 212-219.

[43] Jorgensen B, Liepins R (1983) Fluorescers and Filters of Unusual Chemical Composition for Low-Energy X-Ray Measurements. J. vac. sci. technol. A. 1: 894-896.

[44] Suzuki T, Yamamoto K, Hino Y, Harada Y, Nagura H (1995) Secondary Batteries with Nonaqueous electrolytes. Jap. Pat. JP07169459.

[45] Oja PD, Creek W, Niles ET (1968) High Nitrogen Polymers Prepared by Reacting Aldehydes with Aminoguanidines, Tetrazoles, or Triazoles. US Pat. 3,375,230.

[46] Qafsaoui W, Takenouti H (2010) Corrosion Protection of 2024-T3 Aluminium Alloy by Electro-Polymerized 3-Amino-1,2,4-triazole in Sulphate Solution Containing Chloride. Corros. sci. 52: 3667-3676.
[47] Trachli B, Keddam M, Takenouti H, Srhiri A (2002). Protective Effect of Electropolymerized 3-Amino 1,2,4-triazole Towards Corrosion of Copper in 0.5 M NaCl. Corros. sci. 44: 997-1008.
[48] Voychtcheva OV, Galkine VD, Mikhantyev BI, Chatalov GV (1973) Vinylation of 1,2,3-Triazole. Izv. Vyssh. Ucheb. Zaved, Khim. Tekhnol. 16: 1913-1914.
[49] Kizhnyaev VN, Pokatilov FA, Tsypina NA, Ratovskii GV, Vereshchagin LI, Smirnov AI (2002) Synthesis of N-Vinyl-1,2,3-triazole Derivatives. Russ. J. org. chem. 38: 1056-1059.
[50] Nulwala H, Burke DJ, Khan A, Serrano A, Hawker CJ (2010) N-Vinyltriazoles: a New Functional Monomer Family through Click Chemistry. Macromolecules 43: 5474-5477.
[51] Vereshchagin LI, Tikhonova LG, Maksikova AV, Gavrilov LD, Gareev GA (1979) Synthesis of Acyl- and Vinyl-substituted 1,2,3-Triazoles. 15: 544-549.
[52] Sanghi S, Fassbender B, Tuominen M, Brunklaus G, Spiess HW, Coughlin EB (2010) Proton Conducting Properties of Poly(4-vinyl-1H-1,2,3-triazole). Polym. Preprints 51: 76-77.
[53] Wouters G, Smets G (1982) Copolymerization of C-Vinyltriazoles and C-Vinyltetrazole with Vinyl Monomers. Makromol. chem. 183: 1861-1868.
[54] Zhou Z, Li S, Zhang Y, Liu M (2006) 1H-1,2,3-Triazole Based Polymer Materials for Proton Exchange Membranes. Polym. mater. sci. eng. 94: 609.
[55] Petrova TL, Kizhnyaev VN, Tsypina NA, Vereshchagin LI, Smirnov AI (2002) Method for Preparing Vinylnitrotriazole Polymers. Pat. RU 2261873 C2.
[56] Sukhanov GT, Sukhanova AG, Filippova YV (2004) Selectivity of N-Monoalkylation and Exhaustive Alkylation in Synthesis of Polymers of N-Glycidyl Derivatives of 3-Nitro-5-R-1,2,4- and 4-nitro-1,2,3-triazoles. CA 2006:9757.
[57] Baldwin MG, Johnson KE, Lovinger JA, Parker CO (1967) 1,3-Dipolar Cycloaddition Polymerization of Compounds Containing Both Azido and Acetylene Groups. Polym. lett. 5: 803-806.
[58] Malkov GV, Shastin AV, Estrin YI, Badamshina ER, Mikhailov YM (2008) Synthesis and Characterization of the Nitrogen-Rich Hyperbranched Polymers – Poly([1,2,3]-Triazole-[1,3,5]-Triazine)s. Propellants explos. pyrotech. 33: 431-436.
[59] Hergenrother PM (1975) Poly-as-triazines. J. macromol. sci. revs. macromol. chem. C13: 189-218.
[60] Pankratov VA, Vinogradova SV (1972) Polytriazines. Russ. chem. rev. 41: 66-82.
[61] American Cyanamid Co (1963) Triazine Derivatives and Polymers Thereof. US Pat. 935,787.
[62] American Cyanamid Co (1958) Substituted s-Triazine and Method of Preparation. US Pat. 2,845,422.
[63] Overberger CG, Michelotti FW (1958) The Preparation of Polymers and Copolymers from Vinylpyrimidines and Triazines. J. am. chem. soc. 80: 988-991.
[64] Yuki Y, Hiramatsu N, Kakurai T, Noguchi T (1969) Polymerization of 2,4-Diamino-6-Vinyl-s-Triazine. Kobunshi Kagaku 26: 134-140.

[65] Thurston JT, Schaefer FC, Dudley JR, Holm-Hansen D (1951) Cyanuric Chloride Derivatives. V. Reaction of Alkoxy-s-Triazines and Aryloxy-s-Triazines with Amines. J. am. chem. soc. 73: 2992-2996.
[66] Ehlers GFL, Ray JD (1964) Synthesis and Characterization of Poly-s-Triazinyleneimides. J. polym. sci. A. 2: 4989-5003.
[67] Courvoisier C, Mercier R, Pasquinet E, Poullain D, Palmas P (2004) Synthèse de Polymères Hétérocycliques Azotés à Partir de Précurseurs Triaziniques. Journées de chimie organique, Palaiseau.
[68] Liu Y, Deng C-L, Zhao J, Wang J-S, Chen L, Wang Y-Z (2011) An Efficiently Halogen-Free Flame-Retardant Long-Glass-Fiber-Reinforced Polypropylene System. Polym. degrad. stab. 96: 363-370.
[69] Komatsu T (2000) Heat-Resistant (Dihydro)Carbodiimide Group-Containing Triazine Polymers. JP Pat. 2000-63467.
[70]Audouin F, Birot M, Pasquinet E, Deleuze H, Besnard O, Poullain D (2008) Synthesis of Porous Materials by 2–Nitroresorcinol/Cyanuric Chloride Thermal Polycondensation in Emulsion. J. appl. polym. sci. 108: 2808-2813.
[71] Grigat E, Schminke H-D, Pütter R (1965) Verfahren zur Herstellung von Hochmolekularen Polytriazinen. BRD Pat. 1190184.
[72] Paushkin YM, Aleksandrova VA, Lunin AF, Pisarenko TA (1969) Polycondensation of Cyanuric Chloride with Disodium Acetylide. Doklady. akademii. nauk SSSR 184:863-866. CA 1969:106920.
[73] Kuhn P, Antonietti M, Thomas A. (2008) Porous, Covalent Triazine-Based Frameworks Prepared by Ionothermal Synthesis. Angew. chem. int. ed. 47: 3450-3453.
[74] Schmidt CL, Jansen M (2010) New Directions in Carbonitride Research: Synthesis of Resin-Like Dense-Packed C3N4 using a Hydrogen-Free Precursor. J. mater. chem. 20:4183-4192.
[75] Jürgens B, Irran E, Senker J, Kroll P, Müller H, Schnick W (2003) Melem (2,5,8-Triamino-tri-s-triazine), an Important Intermediate during Condensation of Melamine Rings to Graphitic Carbon Nitride: Synthesis, Structure Determination by X-ray Powder Diffractometry, Solid-State NMR, and Theoretical Studies. J.am. chem. soc. 125: 10288-10300.
[76] Niles ET, Oja PD, Pannell CE (1969) High Nitrogen Containing Monomeric 1,2-Ethylene Bis(aminoguanidine) Compound-Aldehyde Condensation Polymer. US Pat. 3,408,331.
[77] Jenneskens LW, Mahy JWG, Vliestra EJ, Goede SJ, Bickelhaupt F (1994) Structural Studies on Paracyanogen and Paraisocyanogen. J. chem. soc. faraday trans. 90: 327-332.
[78] Cataldo F (1999) On Cyanogen Photopolymerization. Eur. polym. j. 35: 571-579.

Chapter 11

Electroreductive Synthesis of Polysilanes with Ordered Sequences

Manabu Ishifune

Additional information is available at the end of the chapter

1. Introduction

Polysilanes (**Scheme 1**) [1] have attracted considerable attention due to their usefulness as the precursors for thermally stable ceramics [2, 3] or a material for microlithography [4, 5], and also due to their potentiality in the preparation of new types of material showing semiconducting, photoconducting, or nonlinear optical property [6-8].

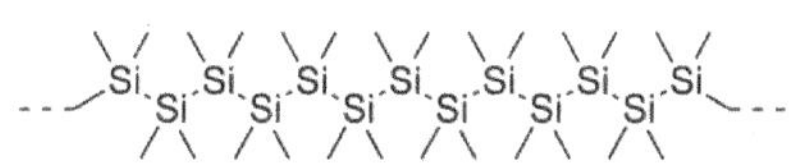

Scheme 1.

In contrast to the growing interest with the polysilane, the method of preparation hitherto known is highly limited. So far, the almost only practical method is the condensation of organodichlorosilane with alkali metal (Wurtz type condensation). This method, however, requires drastic reaction conditions and hence, is very much limited in the type of substituent that is allowed to be located on the monomer. Although several modified or alternative methods such as sonochemical coupling of dichlorosilane promoted by alkali metal [9-11], transition metal catalyzed reaction of hydrosilane [12, 13], anionic polymerization of masked disilene [14, 15], and ring opening polymerization of cyclic organosilane [16, 17] have been proposed, they are not always extensively effective as preparative methods.

The electroreductive coupling of dichlorosilanes with mercury electrode has been reported by Hengge in 1976 as a method to form disilanes [18], though this method was not effective in the preparation of polysilanes [19, 20].

On the other hand, we have recently found that the electroreduction of organic compounds with Mg electrode promotes a variety of unique reactions which can not be attained without using the Mg electrode. The use of Mg electrodes was highly effective to the formation of Si-Si bond and the synthesis of high molecular weight polysilanes [21, 22].

In this chapter, we describe the details of the electroreductive synthesis of high molecular weight polysilane and some types of functionalized polysilanes and also polygermanes, including the additional information about the effects of electrode material and monomer concentration. We also demonstrate that our electroreduction system is successfully applied for the synthesis of the sequence-ordered oligosilanes and polysilanes.

2. Formation of Si-Si bonds by electroreductive coupling of chlorosilanes [21, 22]

The electroreduction of chlorodimethylphenylsilane (**1a**) was studied as the model reaction (**Scheme 2**) and carried out under a variety of reaction conditions. In the first place, the cathodic reduction was performed in a divided cell since Si-Si bond is electrochemically oxidized at the potential range 0.7-1.6V vs. SCE. The yield of 1,1,2,2-tetramethyl-1,2-diphenyldisilane (**2a**) was, however, unexpectedly low under this reaction conditions. In the next place, the idea of using a sacrificial electrode was studied in order to avoid the undesirable anodic oxidation of Si-Si bond in an undivided cell and it was found that the electroreduction of **1a** with Mg electrode was highly effective for the formation of Si-Si bond and **2a** was obtained in an excellent yield.

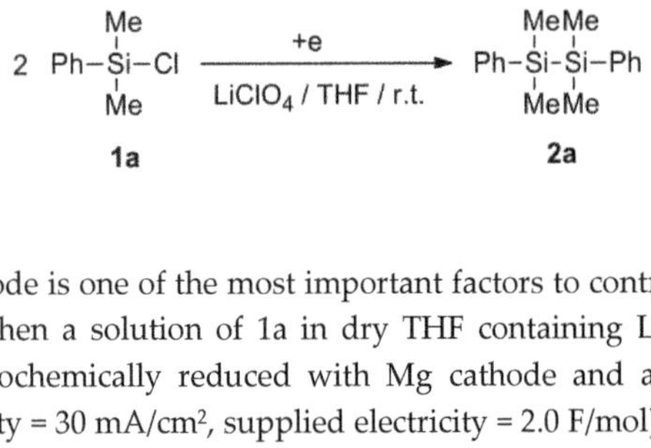

1a **2a**

Scheme 2.

The material of electrode is one of the most important factors to control the formation of Si-Si bond (**Table 1**). When a solution of 1a in dry THF containing $LiClO_4$ as a supporting electrolyte was electrochemically reduced with Mg cathode and anode with a constant current (current density = 30 mA/cm^2, supplied electricity = 2.0 F/mol), the coupling product 2a was obtained in 92% isolated yield (entry 1). The results that Pt, carbon, or Zn is not effective electrode in the formation of Si-Si bond (entries 2-4) clearly indicate that Mg plays some important roles in the formation of Si-Si bond. Although details of the role of Mg in the mechanism of formation of Si-Si bond is not always clear at present, the unique reactivity of Mg electrode is undoubtedly shown in this reaction.

The cathodic coupling of other organochlorosilanes was carried out under the optimized reaction condition, that is, Mg cathode and anode were alternated with the interval of 1 min., supporting electrolyte was $LiClO_4$, solvent was THF, and the electricity passed was 2.0 F/mol (**Scheme 3**). The results summarized in **Table 2** show the high potentiality of this method in the synthesis of a variety of disilanes. Moreover, it is remarkable that the extent of contamination with siloxane (Si-O-Si) was less than 2%.

entry	anode	cathode	alternation[b]	yield of **2a**, %[c]
1	Mg	Mg	yes	92
2	Pt	Pt	yes	0
3	C	C	yes	0
4	Zn	Zn	yes	trace
5	Pt	Mg	no	0
6	Mg	Pt	no	93[d]

[a]The electroreduction was carried out under the constant current conditions (current density = 30 mA/cm^2, supplied electricity = 2.0 F/mol). [b]The anode and cathode were alternated with an interval of 1 min. [c]Material yield based on **1a**. [d]Ultrasound (47 kHz) was applied during the reaction.

Table 1. Effect of electrode materials in the electroreductive formation of 1,1,2,2-tetramethyl-1,2-diphenyldisilane (**2a**)[a]

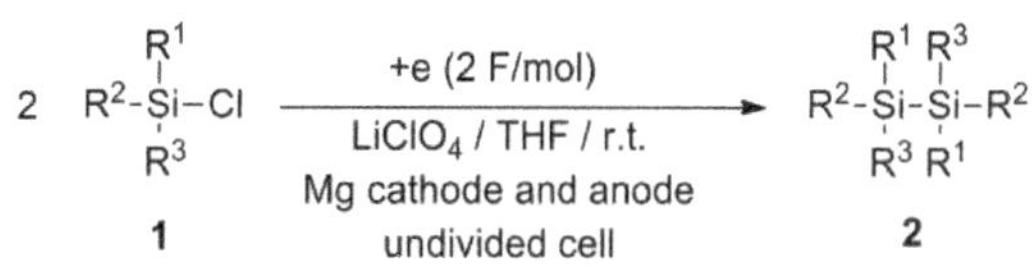

Scheme 3.

entry	chlorosilanes **1**				yield of **2**, %[a]	
		R[1]	R[2]	R[3]		
1	**1b**	Me	Me	Me	**2b**	82
2	**1c**	Me	Ph	Ph	**2c**	77
3	**1d**	Ph	Ph	Ph	**2d**	85

[a]Isolated yield based on **1**.

Table 2. Electroreductive synthesis of disilanes **2**

Two types of mechanism may be proposed to this electroreductive Si-Si bond forming reaction. The first prosible mechanism is a radical coupling in which a silyl radical formed by one electron reduction of the starting chlorosilane couples with another silyl radical to give the disilane. In the second mechanism, two-electron reduction of the chlorosilane yields an active species equivalent to silyl anion which reacts with chlorosilane to give a dimer. In order to have an insight into the mechanism, the products obtained in the mixed system of chlorotrimethylsilane (**1b**) and chlorotriphenylsilane (**1d**) (**1b** : **1d** = 1 : 1) were studied in detail (**Scheme 4**). The resulting products were 1 : 1 mixture of the mixed coupling product (**2e**) and the homocoupling product of **1d** (hexaphenyldisilane (**2d**)), whereas the homocoupling product of 1b, hexamethyldisilane (**2b**) was not found in the products at all. This result seems to agree with the anionic mechanism. That is, **1d** is first reduced to triphenylsilyl anion that reacts with **1b** and **1d** to afford disilanes **2e** and **2d** respectively. The electrophilicity of chlorosilanes **1b** and **1d** are high enough to be attacked equally by the triphenylsilyl anion at the half conversion (supplied electricity = 1 F/mol based on the total amount of **1b** and **1d**). In the radical mechanism, however, if only **1d** is reduced to yield a radical, the formation of **2e** is not reasonable, whereas if two types of radical are formed by the reduction of both **1b** and **1d**, the absence of **2b** in the products is unreasonable. Thus, the anionic mechanism is the most reasonable in this coupling reaction. The fact that the reduction potential of **1d** is much more positive than that of 1b also supports the above mentioned reaction mechanism.

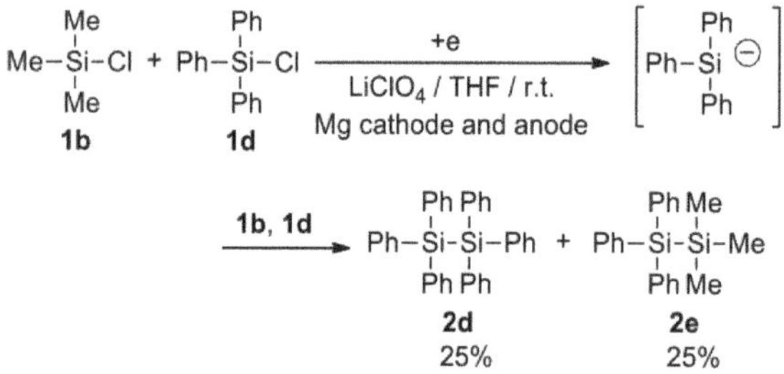

Scheme 4.

3. Stepwise synthesis of oligosilanes [22, 23]

This method is also applicable to the synthesis of trisilanes and tetrasilanes. For example, the electroreductive cross coupling of organodichlorosilanes (**3**) with chlorotrimethylsilane (**1b**) (5 equivalent to **3**) gave the corresponding trisilanes **4** in moderate to good material yields (**Scheme 5**) and that of 1,2-dichloro-1,1,2-trimethyl-2-phenyldisilane (**5**) and **1b** (5 equivalent to **5**) gave tetrasilane **6** in 55% yield (**Scheme 6**). Trisilane **4c** is a key intermediate for the photochemical synthesis of tetramesityldisilene which is known as an isolable disilene.

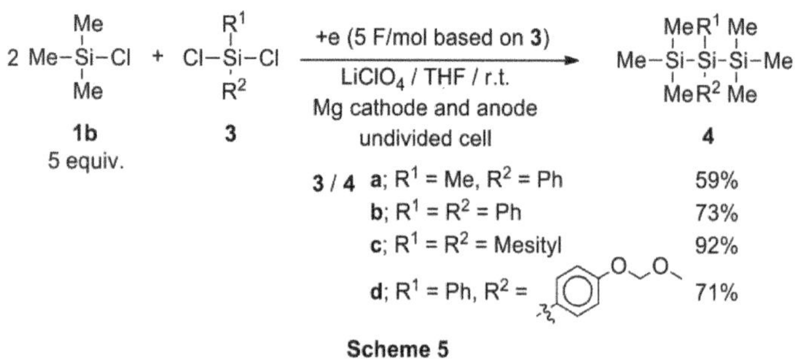

Scheme 5.

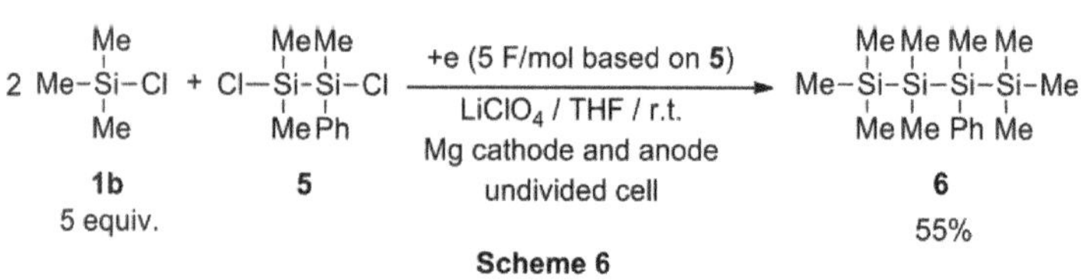

Scheme 6.

The mildness of the reaction conditions of this electroreductive method is favorable for the synthesis of oligosilanes having various functionalities such as Si-H bonds which are known to be reactive under radical or anionic condition. The electroreductive cross-coupling reaction of chlorodimethylsilane (7) with dichlorodiphenylsilane (**3b**), in fact, gave the corresponding trisilane **8** (**Scheme** 7). The Si-H bond was readily transformed to the Si-Cl bond by the treatment with catalytic amount of benzoylperoxide in carbontetrachloride. Using this method 1,3-dihydro-1,1,3,3-tetramethyl-2,2-diphenyltrisilane (**8**) was transformed to the corresponding chloride **9** (**Scheme 7**). The further electroreduction of 1,3-dichloro-1,1,3,3-tetramethyl-2,2-diphenyltrisilane (**9**) with chlorodimethylsilane (**7**) gave the corresponding pentasilane **10** (**Scheme 8**), and these sequences were utilized for the synthesis of the odd-numbered oligosilanes. In the same manner, the even-numbered oligosilanes can be prepared. For instance, 1,4-dichloro-1,1,2,3,3,4,4-heptamethyl-2-phenyltetrasilane (**11**) can be prepared by the reaction between **7** and 1,2-dichloro-1,1,2-trimethyl-1-phenyldisilane (**5**) (**Scheme 9**). Accordingly, the electroreductive cross-coupling reactions followed by the chlorination provided a powerful method for the stepwise elongation of Si-Si bonds and synthesis of sequence-controlled oligosilanes.

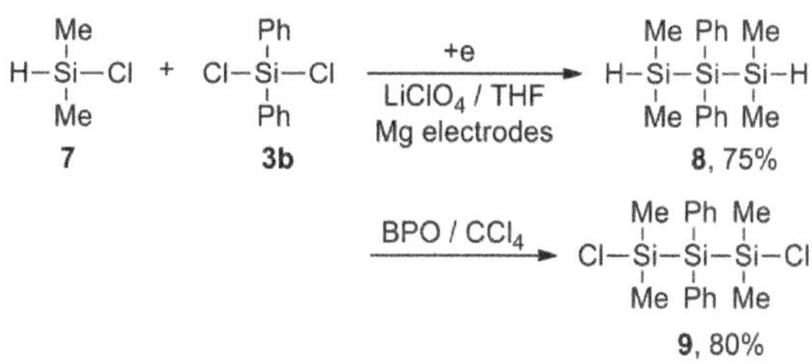

Scheme 7.

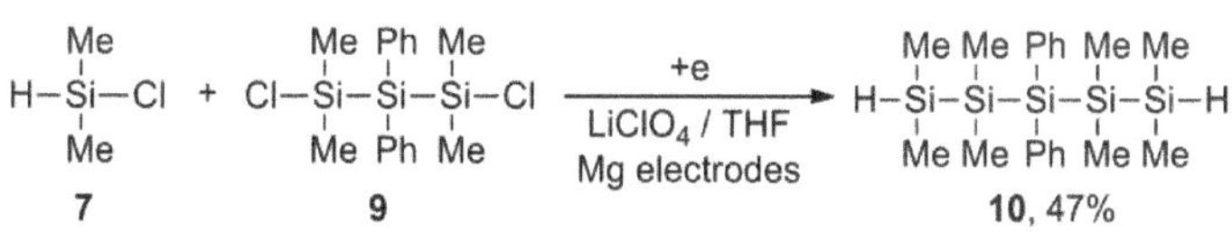

Scheme 8.

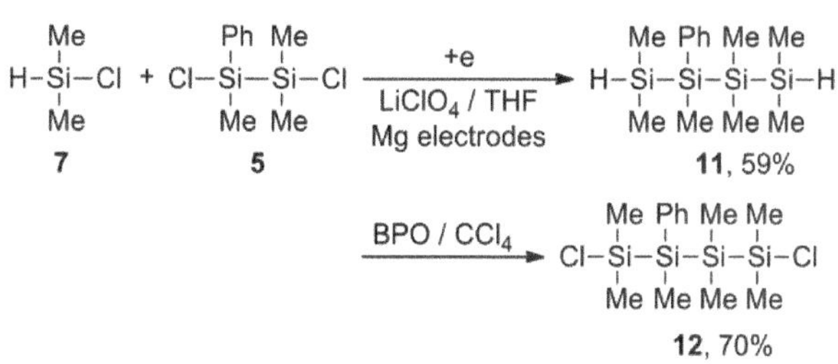

Scheme 9.

4. Electroreductive polymerization of dichloromethylphenylsilane [22]

Electroreduction of dichloromethylphenylsilane (**3a**) (**Scheme 10**) carried out under the above mentioned reaction conditions gave polymethylphenylsilane (**13**) in low yield (**Table 3**, entries 1, 2). The sonication of ultrasound was found to be necessary for the electroreductive polymerization of **3a** (entry 3). The low yield of **13** may be explained by the difficulty of keeping the electric current in a suitable level due to the increase of the terminal voltage with progress of the reaction. This difficulty was overcome by the alternation of anode and cathode with a suitable interval (15 sec., **Table 3**, entry 4) and the material yield of **13** was remarkably improved.

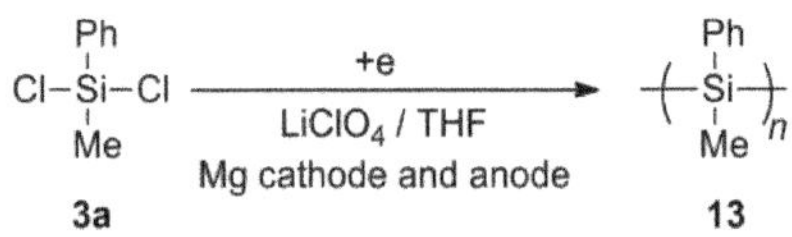

Scheme 10.

entry	alternation[c]	sonication[d]	*Mn*[e]	*Mw/Mn*[e]	yield of **13**, %[f, g]
1	no	no	-	-	-
2	yes	no	4000	1.4	7
3	no	yes	3900	1.4	17
4	yes	yes	5200	1.5	43

[a]Concentration of monomer **3a** is 0.33 mol/L. [b]Supplied electricity is 4 F/mol. [c]The anode and cathode were alternated with an interval of 15 sec. [d]The ultrasound (47 kHz) was applied during the electroreduction. [e]Determined by GPC based on polystyrene standard. [f]Purified by reprecipitation from benzene-EtOH. [g]Material yield based on **3a**.

Table 3. Electroreductive synthesis of poly(methylphenylsilane) (**13**)[a, b]

Mg is a remarkably effective material of electrode also for the formation of **13** (**Table 4**), whereas Al gave low yield (entry 4) and other materials such as Cu and Ni were rather ineffective (entries 2, 3). The electroreduction systems using Al or Cu anode in other electrolytes (Al anode/Bu_4NCl/DME [24], Al anode/LiCl/THF-HMPT [25], or Cu/Bu_4NClO_4/DME [26]) have been also reported, however, the molecular weight of the resulting polysilanes is relatively low. Other than Mg electrodes, the use of Ag anode and Pt cathode in DME containing Bu_4NCLO_4 is also reported to be effective to obtain high molecular weight polysilane [27, 28].

entry	electrode materials	*Mn*[b]	*Mw/Mn*[b]	yield of **13**, %[c, d]
1	Mg	5200	1.5	43
2	Cu	700	1.1	-[e]
3	Ni	640	1.1	-[e]
4	Al	4700	1.5	15

[a]Conditions: [monomer **3a**] = 0.33 mol/L; supplied electricity = 4 F/mol; The ultrasound (47 kHz) was applied during the electroreduction. The anode and cathode were alternated with an interval of 15 sec. [b]Determined by GPC based on polystyrene standard. [c]Purified by reprecipitation from benzene-EtOH. [d]Material yield based on **3a**. [e]No precipitate was obtained after usual reprecipitation procedure.

Table 4. Effect of electrode materials in the electroreductive synthesis of poly(methylphenylsilane) (**13**)

The effect of monomer concentration was investigated in order to obtain high molecular weight polysilane (**Table 5**). The molecular weight of **13** becomes higher with the increase in the concentration of **3a**. The molecular weight (*Mn*) of **13** was, for instance, 31,000 when the electroreduction of **3a** was carried out under high concentration condition (1.2 mol/L) at 0.5 F/mol of supplied electricity though the material yield of **13** decreased.

The most satisfactory result, in which material yield was 79 % and molecular weight (*Mn*) was 9900, was obtained when the concentration of 3a was 0.67 mol/L (entry 2). The polysilane **13** obtained here showed relatively sharp monomodal distribution of molecular weight in the elution profile of GPC, whereas the polysilanes prepared by the alkali metal condensation method usually showed broad bimodal distribution.

entry	monomer **3a**, mol/L	supplied electricity, F/mol	*Mn*[b]	*Mw/Mn*[b]	yield of **13**, %[c, d]
1	0.33	4.0	5200	1.5	43
2	0.67	4.0	9900	2.1	79
3	2.5	2.2	18000	2.1	43
4	6.3	0.8	19000	2.8	15
5	12	0.5	31000	1.8	8

[a]The electroreduction was carried out by using Mg electrodes under sonication of ultrasound (47 kHz). The anode and cathode were alternated with an interval of 15 sec. [b]Determined by GPC based on polystyrene standard. [c]Purified by reprecipitation from benzene-EtOH. [d]Material yield based on **3a**.

Table 5. Effect of monomer concentration in the electroreductive synthesis of **13a**

The mechanism of electroreductive formation of polysilane is not always perfectly clear, though the initial step of reaction is obviously the reduction of **3a** to a silyl anion species **14**. Two types of reaction patterns may be proposable to the propagation step. In the first case, the reaction of **14** with **3a** gives a dimer that yields trimer, tetramer, and finally polymer upon repeated reaction with **14** (**Scheme 11**). In another pattern of the propagation, the oligomers such as dimer, trimer, and the like **15** are reduced to give the oligomeric active species, which then react with oligomer **15** or **3a** to give finally polymer. (**Scheme 12**). Although it is not always possible to specify the extent of contribution of each pattern to the propagation step, the former reaction probably proceeds mainly, since the electrochemical reduction of oligomeric silyl chloride **15** may be rather difficult when it is analogized with the electroreduction of long chain alkyl chlorides.

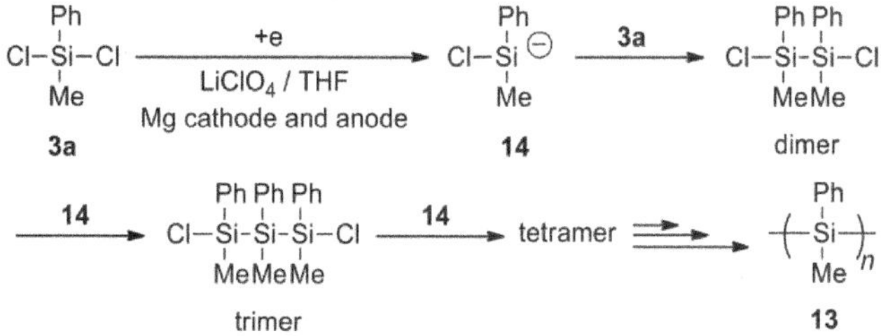

Scheme 11.

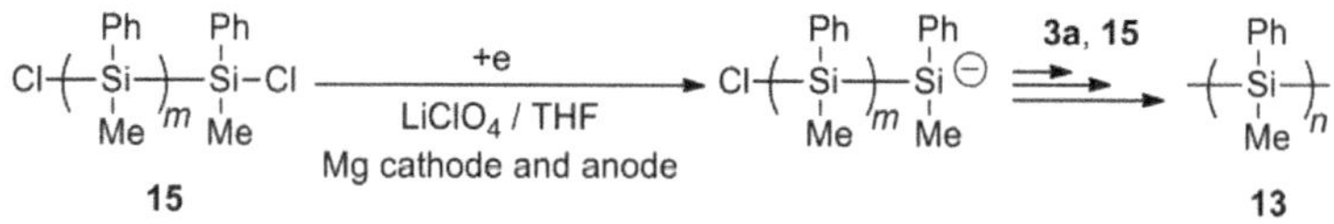

Scheme 12.

5. Electroreductive synthesis of functionalized polysilanes [29]

The mildness of the polymerization conditions of the electroreductive method is favorable for the synthesis of the polysilanes having a variety of hydroxyl-related functional groups. The electroreduction of a mixture of **3a** and the dichlorosilanes having protected hydoxyphenyl groups (**3d-g**) with Mg electrode afforded the corresponding copolymers **16** (**16d-g**, **Scheme 13**), and the deprotection of the resulting copolymers gave the polysilanes having hydroxyl groups. The reactivity of **3d-g** highly depends on the type of protecting group (**Table 6**). Homopolymerization of **3g**, for example, gave **16g** (entry 10), whereas other monomers did not afford polymers but oligomeric compounds (entries 3, 5, 8).

The modification or the property of polysilane must be achieved by using the hydroxyl group located on the polymer **16d** as a key functional group. The deprotection of the methoxymethyl group of **16d** (**Table 6**, entry 1) with 10%HCl aqueous solution followed by the reaction with hexamethylene diisocyanate resulted in a remarkable increase in the molecular weight with indicating the linkage of the polymer chain (**Scheme 14**).

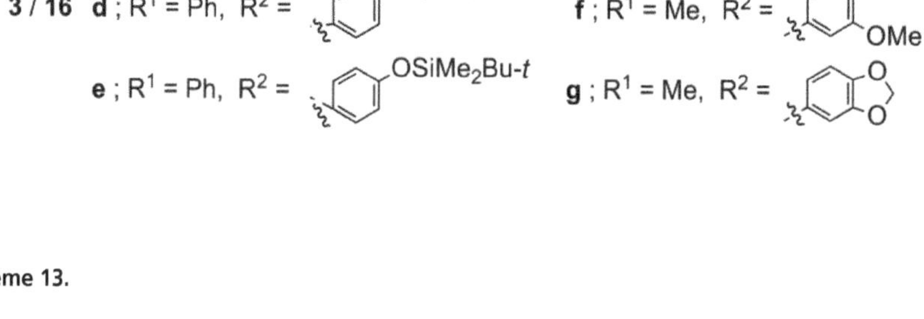

Scheme 13.

entry	charged mol% of **3d-g**[b]	yield of **16**, %[c, d]	Mn[e]	Mw/Mn[e]
1	7 (**3d**)	79 (7)	9900	1.9
2	10 (**3d**)	57 (12)	6900	1.7
3	100 (**3d**)	28 (100)	1100	1.2
4	10 (**3e**)	36 (11)	6100	1.5
5	100 (**3e**)	-[f] (100)	1100	1.2
6	10 (**3f**)	50 (6)	4500	1.3
7	50 (**3f**)	22 (46)	4600	1.3
8	100 (**3f**)	-[f] (100)	1700	1.3
9	10 (**3g**)	56 (17)	4600	1.3
10	100 (**3g**)	57 (100)	4000	1.1

[a]The electroreduction was carried out by using Mg electrodes under sonocation (47 kHz), and the anode and cathode were alternated with an interval of 15 sec. Total monomer concentration, 0.67 mol/L; supplied electricity, 4 F/mol. [b]**3d-g**/(**3d-g**+**3a**) x 100. [c]Purified by reprecipitaion from benzene-EtOH. [d]The values in parentheses indicate the mol% of **3d-g** units in the resulting copolymers determined by 1H NMR. [e]Determined by GPC based on polystyrene standard. [f]Polymer was not obtained by reprecipitation.

Table 6. Electroreductive synthesis of functionalized polysilanes[a]

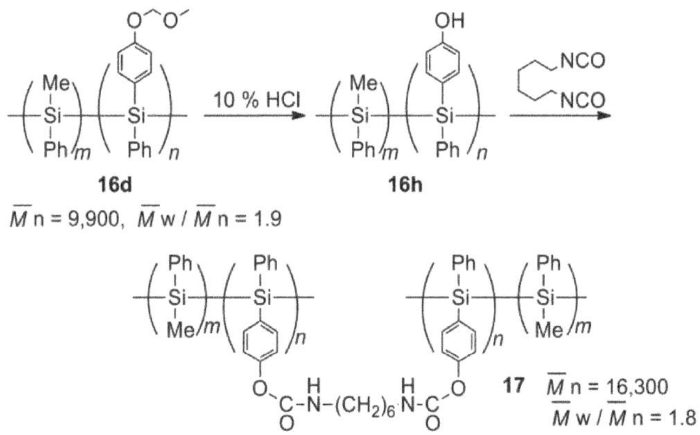

Scheme 14.

6. Electroreductive Polymerization of Dichlorooligosilanes [23]

The electroreductive polymerization of the dichlorooligosilanes is highly promising for the synthesis of sequence-ordered polysilanes. The electroreduction of dicholodisilane **5** was found to give the corresponding polysilane **18** consisting of disilane units (**Scheme 15**). The electroreductive polymerization was carried out under a variety of reaction conditions, however, the yield of the resulting polymer was very low (**Table 7**). It is probably due to high reactivity of the disilene intermediate formed by the electroreduction of **5**. In fact, the addition of naphthalene, which could make a masked disilene intermediate, into the reaction system slightly increased the yield of the polysilane **18** (entry 6).

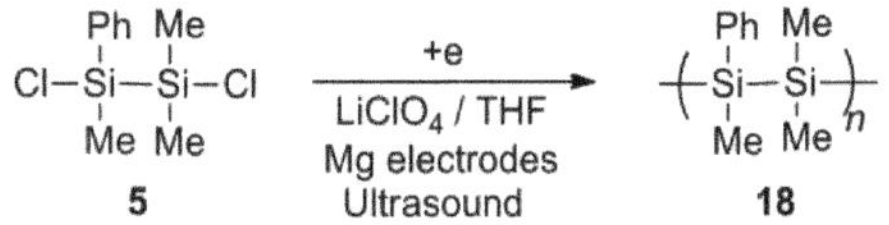

Scheme 15.

entry	dichlorosilane **5**, mol/L	supplied electricity, F/mol	polysilane **18**		
			Mn[b]	*Mw/Mn*[b]	yield, %[c]
1	0.11	4	2900	2.7	2.7
2	0.33	4	3600	2.3	2.7
3	0.67	4	2100	2.9	1.0
4	0.33	2	2800	2.7	3.9
5	0.33	6	3000	3.0	1.0
6[c]	0.33	4	2500	1.8	13.0

[a]The electroreduction was carried out by using Mg electrodes under sonocation (47 kHz), and the anode and cathode were alternated with an interval of 15 sec. Total monomer concentration, 0.67 mol/L; supplied electricity, 4 F/mol. [b]Determined by GPC based on polystyrene standard. [c]The electroreduction was carried out in the presence of naphthalene.

Table 7. Electroreduction polymerization of dichlorosilane **5**[a]

Dichlorooligosilanes, such as dichlorotrisilane 9 was found to be good monomers for the electroreductive synthesis of the polysilanes having longer sequence units (Schemes 16). The temperature control is found to be very important in the electroreductive polymerization of 9 (Table 8). The reaction at higher temperature, the backbiting reaction of the propagating polymer proceeded forming cyclohexasilane as a by-product (entry 1). This side reaction was successfully suppressed when the reaction was carried out below 0°C, and polysilanes 19 having relatively high molecular weight were obtained (entries 3, 4). In the optimized reaction conditions, the electroreduction of dichlorotetrasilane 12 gave the corresponding polysilane 20, units of which were ordered in four sequences in satisfactory yield (Scheme 17). The polymerizability of dichlorooligosilanes under the electroreduction conditions seems to be mainly affected by the substituents on the chlorinated terminal silicon atom, and this fact provides a wide possibility to design the oligosilane sequences of the inner silicon atoms.

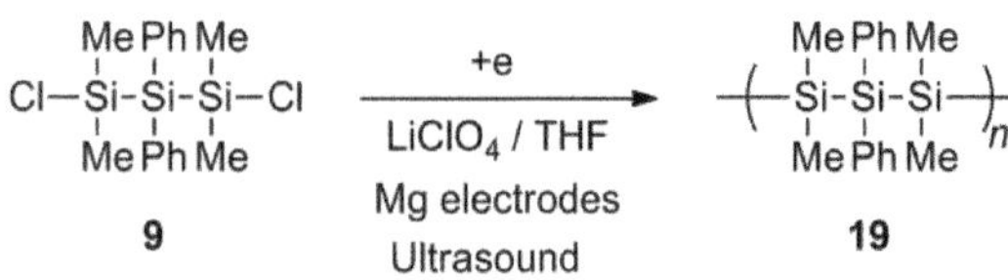

Scheme 16.

entry	polymerization temperature, °C	*Mn*[b]	*Mw/Mn*[b]	yield of **19**, %[c]
1	18	3800	1.44	(42)[d]
2	0	4700	1.87	50
3	-10	5500	1.54	35
4	-15	4400	1.42	16

[a]The electroreduction was carried out by using Mg electrodes under sonocation (47 kHz), and the anode and cathode were alternated with an interval of 15 sec. [Dichlorosilane **9**] = 0.27 mol/L; [$LiClO_4$] = 0.35 mol/L; supplied electricity = 4 F/mol. [b]Determined by GPC based on polystyrene standard. [c]Purified by reprecipitation from benzene-EtOH. [d]Contaminated by 1,1,2,2,4,4,5,5-octamethyl-3,3,6,6-tetraphenylcylcohexasilane.

Table 8. Electroreductive polymerization of dichlorosilane **9**[a]

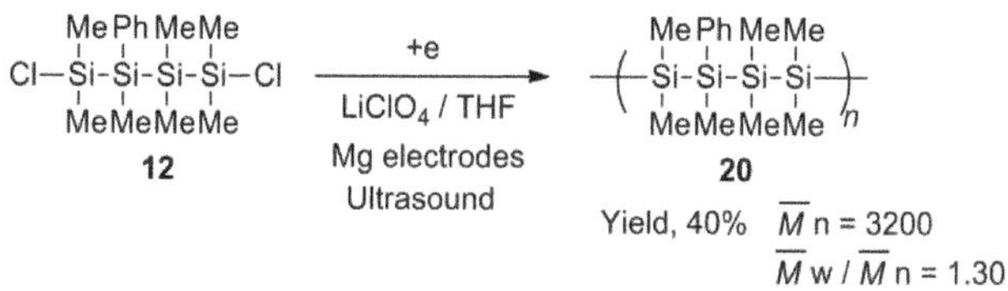

Scheme 17.

7. Electroreductive polymerization of dichlorosilanes in the presence of disilane additives [30]

The disilane additives, which are generated *in situ* in electroreductive coupling of the corresponding chlorosilanes, were found to be effective to the promotion of the electroreductive polymerization of dichloromethylphenylsilane (**3a**) and the control of the molecular weight distribution of the resulting polysilanes (**Scheme 18**, **Table 9**). The electroreduction of dichlorosilane **3a** in the presence of 1,1,1-trimethyl-2,2,2-triphenyldisilane gives the corresponding polysilane **13** in 56% yield, and the number average molecular weight and the molecular weight distribution are determined by GPC to be 3000, and 1.10 respectively (**Table 9**, entry 3). The reduction of dichloromethylphenylsilane (**3a**) by Wurtz type condensation using metal lithium in the presence of catalytic amount 1,1,1-trimethyl-2,2,2-triphenyldisilane affords five- and six-members ring products [31]. On the other hand, the cyclosilanes are not detected under the electroreductive conditions. The use of 1,1,12,2,2-hexaphenyldisilane affords the polysilane **13** in 59% yield, and the *Mw/Mn* is 1.08 (entry 4). Thus, the polysilanes prepared in the presence of the disilane additive containing triphenylsilyl group show narrower molecular weight distributions than the polysilane prepared without the disilane additive.

The mechanism for the triphenylsilyl substituted disilane to control the electroreductive polymerization suggested is as follows (**Scheme 19**). The electroreductively generated reactive silyl anion species at the terminus of the propagating polymer is trapped with the disilane forming a relatively stable triphenylsilyl anion and inhibits the undesirable side reactions such as backbiting reaction. The triphenylsilyl anion attacks as a nucleophile to the chlorinated silicon atom at the terminal of the propagating polymer to give the

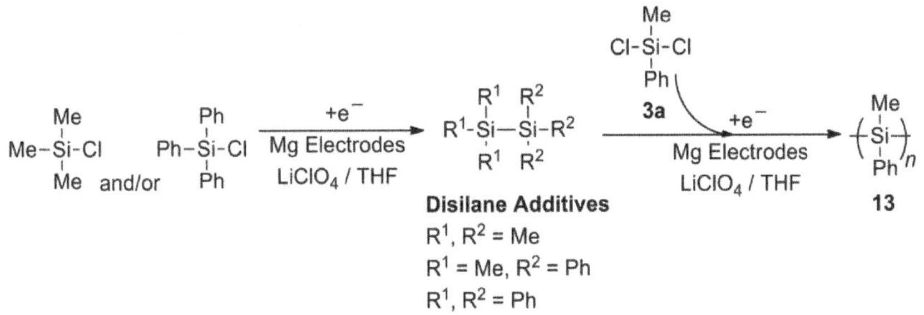

Scheme 18.

entry	disilane additives[b]	Mn^c	Mw/Mn^c	yield of **2a**, %[d]
1	-	3200	1.65	38
2	$Me_3SiSiMe_3$	3700	1.39	54
3	$Me_3SiSiPh_3$	3000	1.10	56
4	$Ph_3SiSiPh_3$	2600	1.08	59

[a]The electroreduction was carried out by using Mg electrodes, and the ultrasound (47 kHz) was applied during the reaction. The anode and cathode were alternated with the interval of 15 sec. [Dichlorosilane **3a**] = 0.50 mol/L; [LiClO4] = 0.35 mol/L; supplied electricity, 4 F/mol. [b]The disilane additives were prepared in situ by electroreductive coupoling of the corresponding chlorosilanes before electroreductive polymerization of **3a**. Total concentration of chlorosilanes was 0.04 mol/L. [c]Determined by GPC based on polystyrene standard. [d]Purified by reprecipitation from benzene-EtOH.

Table 9. Electroreductive polymerizartion of dichlomethylphenylsilane (**3a**) in the presence of disilane additives[a]

triphenylsilyl group terminated polysilane. The resulting polysilane is isolable but does not lose its polymerizability completely since the triphenylsilyl group at the terminal position acts as an activator, that is, it probably reacts as a macroinitiator.

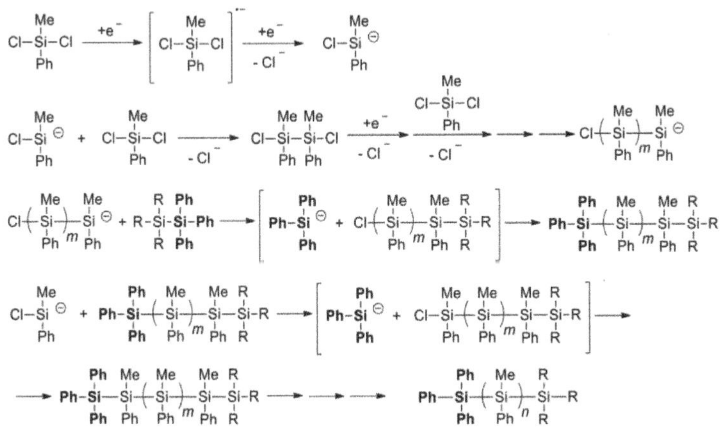

Scheme 19.

8. Electroreductive block copolymerization using triphenylsilyl group-terminated polysilane [32]

The triphenylsilyl group terminated polysilanes have been synthesized by the electroreductive polymerization of dichloromethylphenylsilane (**3a**) in the presence of the electroreductively prepared disilane additives **2**. The electroreductive termination with chlorotriphenylsilane (**1d**) was carried out to ensure the terminus of the resulting polysilane for triphenylsilyl group. The polysilanes were obtained as white powders in 15-32% yields after reprecipitation from benzene-ethanol, and the number average molecular weights were estimated by GPC to be 3000-3740 (**Table 10**). By using the isolated triphenylsilyl group-terminated poly(methylphenylsilane) **13** as a macroinitiator, the electroreductive polymerization of dibutyldichlorosilane (**3h**) was carried out (**Scheme 20, Table 10**). Under these conditions dichlorsilane **3h** was first electroreduced to form the corresponding oligomeric silyl anion. Electroreductive copolymerization was found to proceed by the attack of the oligomeric silyl anion to triphenylsilyl group-terminated polysilane **13** and further electroreductive condensation with dichlorosilane **3h** affording the corresponding copolymer, polymethylphenylsilane-block-polydibutylsilane (**21**), in 16-38% yields depending on the disilane additives (entries 1-3). The molecular weight of the copolymer obtained from the polysilane **13** (*Mn* = 3350) was 4730 (entry 3). The GPC profiles of the resulting copolymers **21** were monomodal and the polydispersity index values (*Mw*/*Mn*) were 1.2-1.4. The repeat unit ratio (-Si(Me)Ph- : -SiBu2-) of the resulting copolymer **21** (entry 3) was 75 : 25, which showed a good agreement with the calculated ratio (74 : 26).

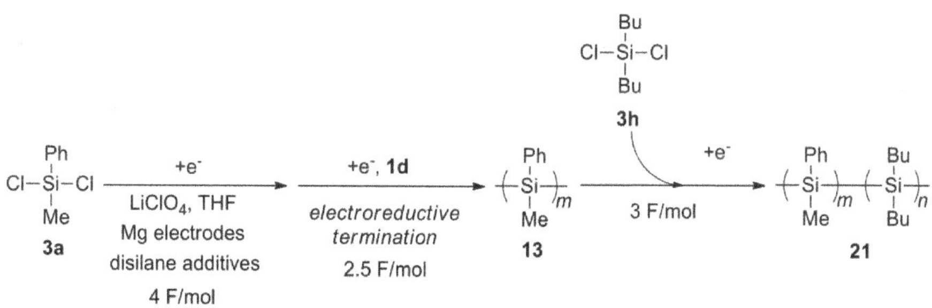

Scheme 20.

Polydibutylsilane-block-polymethylphenylsilane (**21′**) was also obtained by using triphenylsilyl group-terminated polydibutylsilane (**22**) as a macroinitiator (**Scheme 21**). The electroreductive polymerization of dibutyldichlorosilane (**3h**) in the presence of the disilane **2d** followed by electroreductive termination with chlorotriphenylsilane (**1d**) afforded the macroinitiator **22** (*Mn* = 3950, *Mw*/*Mn* = 1.7). The electroreductive polymerization of dichloromethylphenylsilane (**3a**) was found to initiate from **22** producing the corresponding copolymer **21′** in 25% yield, and the molecular weight of **21′** was 4390 (**Scheme 21**). The polydispersity index values (*Mw*/*Mn*) was 1.3, and the repeat unit ratio (-SiBu2- : -Si(Me)Ph-) was 61 : 39.

entry	preparation of the macroinitiator 13[a]			polymethylphenylsilane-*block*-polydibutylsilane (21)[b]				
	disilane additives	*Mn*[c]	*Mw/Mn*[c]	*Mn*[c]	*Mw/Mn*[c]	*m* : *n* observed[d]	*m* : *n* calculated[e]	yield, %[f]
1	$Me_3SiSiMe_3$	3740	1.9	5530	1.4	83 : 17	70 : 30	16
2	$Me_3SiSiPh_3$	3000	1.3	4080	1.4	66 : 34	77 : 23	28
3	$Ph_3SiSiPh_3$	3350	1.4	4730	1.2	75 : 25	74 : 26	38

[a]The reaction conditions of the preparation of the macroinitiator **13** are as follows: The electroreduction was carried out by using magnesium electrodes under sonication (50 kHz) and the polarity of the electrodes was alternated with an interval of 15 sec. The concentration of **3a** was 0.67 mol/L. Polymerization was terminated by electroreduction with triphenylchlorosilane (**1d**) (0.5 mol% of **3a**). [b]Electroreductive copolymerization was carried out without disilane additives. [c]Determined by GPC based on polystyrene standard. [d]Determined by ^{1}H NMR. [e]Calculated from the observed molecular weight of the copolymer **21** as the assumption that the molecular weight of the macroinitiator **13** was not changed. [f]Purified by reprecipitation from benzene-ethanol.

Table 10. Electroreductive block copolymerization with dibutyldichlorosilane (**3h**) using triphenylsilyl group-terminated poly(methylphenylsilane) (**13**) as a macroinitiator

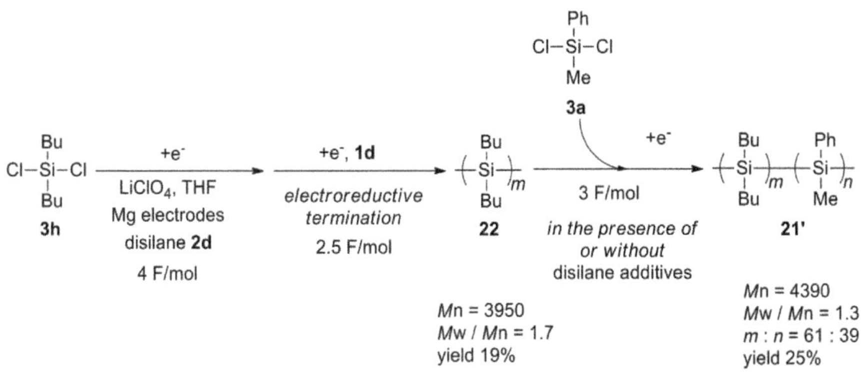

Scheme 21.

The UV absorption spectra of the resulting polysilane **21** (entry 3, **Table 10**) was compared with those of poly(methylphenylsilane) (*Mn* = 3350, *Mw*/*Mn* = 1.4), poly(dibutylsilane) (*Mn* = 3950, *Mw*/*Mn* = 1.7), and poly(methylphenylsilane-co-dibutylsilane) (random copolymer) (**Figure**). Poly(methylphenylsilane) showed a π-π* band at 271 nm (ε_{max} = 4831) and a σ-σ* band at 326 nm (ε_{max} = 4766) at 4°C (**Figure (a)**). Poly(methylphenylsilane) (*Mw* = 9000) prepared by Wurtz coupling under sonication was reported to show a σ-σ* band at 332 nm (ε_{max} = 4100) [33]. The λmax value of the electroreductively prepared poly(methyphenylsilane) was a little shorter and it is probably due to relatively lower molecular weight. In fact Wurtz coupling-synthesized poly(methylphenylsilane) having 5000 of *Mw* showed a σ-σ* band at 328 nm [34]. Poly(dibutylsilane) showed only a σ-σ* band at 309 nm, and the ε_{max} value (9611) was two times higher than that of poly(methylphenylsilane) (**Figure (b)**). The UV absorption spectrum of the helical form of poly(di-*n*-alkylsilane)s is typically centered near 315 nm, while the more planar conformation is typically centered at 375 nm [35, 36]. Poly(dibutylsilane) obtained in this study showed the λmax value corresponding to helical backbone conformation. The ε_{max} of the σ-σ* band of poly(methylphenylsilane) was observed to decrease with an increase in temperature [5]. The UV spectra of copolymer **5** showed a π-π* band at 273 nm (ε_{max} = 2723) and a σ-σ* band at 306 nm (ε_{max} = 2949) (**Figure (d)**), while the random copolymer showed a π-π* band at 272 nm (ε_{max} = 2438) and a σ-σ* band at 309 nm (ε_{max} = 2065) (**Figure (c)**). The ε_{max} value of the σ-σ* band of copolymer **21** was higher than that of the random copolymer, and less temperature dependence of ε_{max} was observed in the spectrum of copolymer **21**. These results indicate that copolymer **21** has long sequences of dibutylsilylene units, that is, block structure.

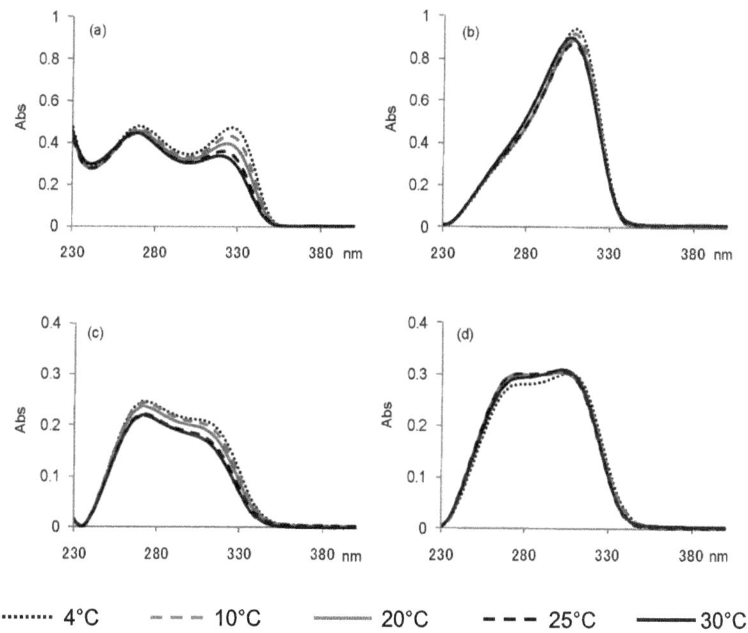

Figure 1. UV absorption spectra of (a) poly(methylphenylsilane) (Mn = 3350, Mw/Mn = 1.4), (b) poly(dibutylsilane) (Mn = 3950, Mw/Mn = 1.7), (c) poly(methylphenylsilane-*co*-dibutylsilane) (Mn = 4340, Mw/Mn = 1.5, -Si(Me)Ph- : $-SiBu_2-$ = 67 : 33), and (d) poly(methylphenylsilane)-*block*-poly(dibutylsilane) (Mn = 4080, Mw/Mn = 1.4, -Si(Me)Ph- : $-SiBu_2-$ = 66 : 34) in THF at 0, 10, 20, 25, and 30°C.

9. Conclusion

The formation of Si-Si bonds was achievable by the electroreductive condensation of organochlorosilanes with Mg sacrificial electrode. Disilanes, trisilanes, and tetrasilanes were readily obtained in good to moderate yield. Moreover, this method was also remarkably effective to the synthesis of polysilanes. The molecular weight and yield of the polymers was controlled by the concentration of monomers and the supplied electricity. The mildness of the reaction conditions allowed to use a wide variety of monomers, and enabled the synthesis of the functionalized polysilanes and the structure-controlled polysilanes. The electroreductive polymerization of the dichlorooligosilanes was highly useful for the synthesis of sequence-ordered polysilanes. Moreover, this electroreductive method also provided a new procedure to synthesize well-controlled di-block polysilane copolymers. Since the present electroreductive polymerization requires only a single compartment cell, it is undoubtedly one of the simplest and most powerful tools for synthesis of polysilanes.

Author details

Manabu Ishifune
Department of Applied Chemistry, Faculty of Science and Engeneering,
Kinki University, Higashi-Osaka, Osaka, Japan

10. References

[1] West R (1986) The polysilane high polymers. J. Organomet. Chem. 300(1-2): 327-346.

[2] Yajima S, Hasegawa Y, Hayashi J, Iimura M (1978) Synthesis of continuous silicon carbide fiber with high tensile strength and high Young's modulus. Part 1. Synthesis of polycarbosilane as precursor. J. Mater. Sci. 13(12): 2569-2576.

[3] Hasegawa Y, Okamura K (1985) Silicon carbide-carbon composite materials synthesized by pyrolysis of polycarbosilane. J. Mater. Sci. Lett. 4(3): 356-8.

[4] Miller RD, Willson CG, Wallroff GM, Clecak N, Sooriyakumaran R, Michl J, Karatsu T, McKinley AJ, Klingensmith KA, Downing J (1989) Polysilanes: photochemistry and deep-UV lithography. Polym Eng Sci 29(13): 882-886

[5] Miller RD, Michl J (1989) Polysilane high polymers. Chem. Rev. 89(6): 1359-1410.

[6] West R, David LD, Djurovich PI, Stearley KL, Srinivasan KSV, Yu H (1981) Phenylmethylpolysilanes: formable silane copolymers with potential semiconducting properties. J. Am. Chem. Soc. 103(24): 7352-7354.

[7] Kepler RG, Zeigler JM, Harrah LA, Kurtz SR (1987) Photocarrier generation and transport in σ-bonded polysilanes. Phys. Rev. B 35(6): 2818-2822.

[8] Baumert JC, Bjorklund GC, Jundt DH, Jurich MC, Looser H, Miller RD, Rabolt J, Soorijakumaran R, Swalen JD, Twing RJ (1988) Temperature dependence of the third-order nonlinear optical susceptibilities in polysilanes and polygermanes. Appl. Phys. Lett. 53(13): 1147-1149.

[9] Matyjaszewski K, Greszta D, Hrkach JS, Kim HK (1995) Sonochemical synthesis of polysilylenes by reductive coupling of disubstituted dichlorosilanes with alkali metals. Macromolecules 28(1): 59-72.

[10] Jones RD, Holder SJ (2006) High-yield controlled syntheses of polysilanes by the Wurtz-type reductive coupling reaction. Polym. Int. 55(7): 711-718.

[11] Koe J (2008) Contemporary polysilane synthesis and functionalisation. Polym. Int. 58(3): 255-260.

[12] Tilley TD (1993) The coordination polymerization of silanes to polysilanes by a " σ-bond metathesis" mechanism. Implications for linear chain growth. Acc. Chem. Res. 26(1): 22-29.

[13] Minato M, Matsumoto T, Ichikawa M, Ito T (2003) Dehydropolymerization of arylsilanes catalyzed by a novel silylmolybdenum complex. Chem. Commun. (24): 2968-2969.

[14] Sanji T, Kawabata K, Sakurai H (2000) Alkoxide initiation of anionic polymerization of masked disilenes to polysilanes. J. Organomet. Chem. 611(1-2): 32-35.

[15] Sanji T, Isozaki S, Yoshida M, Sakamoto K, Sakurai H (2003) Functional transformation of poly(dialkylaminotrimethyldisilene) prepared by anionic polymeriztion of the masked disilenes. The preparation of a true polysilastyrene. J. Organomet. Chem. 685 (1-2): 65-69.

[16] Cypryk M, Gupta Y, Matyjaszewski K (1991) Anionic ring-opening polymerization of 1,2,3,4-tetramethyl-1,2,3,4-tetraphenylcyclotetrasilane. J. Am. Chem. Soc. 113(3): 1046-7.

[17] Suzuki M, Kotani J, Gyobu S, Kaneko T, Saegusa T (1994) Synthesis of sequence-ordered polysilane by anionic ring-opening polymerization of phenylnonamethylcyclopentasilane. Macromolecules 27(8): 2360-2363.

[18] Hengge E, Litscher, G (1976) A new electrochemical method for the formation of silicon-silicon bonds. Angew] Chem] 88(12): 414.

[19] Hengge E, Litscher G (1978) Electrochemical formation of di-, oligo- and polysilanes. Monatsh. Chem. 109(5): 1217-1225.

[20] Hengge E, Firgo H (1981) An electrochemical method for the synthesis of silicon-silicon bonds. J. Organomet. Chem. 212(2): 155-161.

[21] Shono T, Kashimura S, Ishifune M, Nishida R (1990) Electroreductive formation of polysilanes. J. Chem. Soc. Chem. Commun. (17): 1160-1161.

[22] Kashimura S, Ishifune M, Yamashita N, Bu HB, Takebayashi M, Kitajima S, Yoshihara D, Kataoka Y, Nishida R, Kawasaki S, Murase H, Shono T (1999) Electroreductive synthesis of polysilanes, polygermanes, and related polymers with magnesium electrodes. J. Org. Chem. 64(18): 6615-6621.

[23] Ishifune M, Kashimura S, Kogai Y, Fukuhara Y, Kato T, Bu HB, Yamashita N, Murai Y, Murase H, Nishida R (2000) Electroreductive synthesis of oligosilanes and polysilanes with ordered sequences. J. Organomet. Chem. 611(1-2): 26-31.

[24] Umezawa M, Takeda M, Ichikawa H, Ishikawa T, Koizumi T, Nonaka, T (1991) Electroreductive polymerization of mixtures of chloromonosilanes. Electrochim. Acta 36(3-4): 621-624.

[25] Biran C, Bordeau M, Pons P, Leger MP, Dunogues J (1990) Electrosynthesis, a convenient route to di- and polysilanes. J. Organomet. Chem. 382 (3): C17-C20.

[26] Kunai A, Kawakami T, Toyoda E, Ishikawa M (1991) Electrochemistry of organosilicon compounds. 2. Synthesis of polysilane oligomers by a copper electrode system. Organometallics 10(6): 2001-2003.

[27] Okano M, Takeda K, Toriumi T, Hamano H (1998) Electrochemical synthesis of polygermanes. Electrochim. Acta 44(4): 659-666.

[28] Yamada K, Okano M (2006) Electrochemical synthesis of poly(cyclotetramethylene-silylene) Electrochemistry 74(8): 668-671.

[29] Kashimura S, Ishifune M, Bu HB, Takebayashi M, Kitajima S, Yoshihara D, Nishida R, Kawasaki S, Murase H, Shono T (1997) Electroorganic chemistry. 153. Electroreductive synthesis of some functionalized polysilanes and related polymers. Tetrahedron Lett. 38(26): 4607-4610.

[30] Ishifune M, Kogai Y, Uchida K (2005) Effect of disilane additives on the electroreductive polymerization of organodichlorosilanes. J. Macromol. Sci. Part A Pure and Appl. Chem. 42(7): 921-929.

[31] Chen SM, David LD, Haller KJ, Wadsworth CL, West R (1983) Isomers of $(PhMeSi)_6$ and $(PhMeSi)_5$ Organometallics 2(3): 409-414.

[32] Ishifune M, Sana C, Ando M, Tsuyama Y (2011) Electroreductive block copolymerization of dichlorosilanes in the presence of disilane additives. Polym. Int. 60(8): 1208-1214.

[33] Terunuma D, Nagumo K, Kamata N, Matsuoka K, Kuzuhara H (2000) Preparation and characterization of water-soluble polysilanes bearing chiral pendant ammonium moieties. Polymer Journal 32: 113-117.

[34] Herzog U, West R (1999) Hererosubstituted polysilanes. Macromolecules 32: 2210-2214.

[35] Hu Z, Zhang F, Huang H, Zhang M, He T (2004) Morphology and structure of poly(di-*n*-butylsilane) single crystals prepared by controlling kinetic process of solvent evaporation. Macromolecules 37: 3310-3318.

[36] Chunwachirasiri W, Kanaglekar I, Winokur MJ, Koe JC, West R (2001) Structure and chain conformation in poly(methyl-*n*-alkyl)silanes. Macromolecules 34: 6719-6726.

Permissions

The contributors of this book come from diverse backgrounds, making this book a truly international effort. This book will bring forth new frontiers with its revolutionizing research information and detailed analysis of the nascent developments around the world.

We would like to thank Dr. Ailton de Souza Gomes, for lending his expertise to make the book truly unique. He has played a crucial role in the development of this book. Without his invaluable contribution this book wouldn't have been possible. He has made vital efforts to compile up to date information on the varied aspects of this subject to make this book a valuable addition to the collection of many professionals and students.

This book was conceptualized with the vision of imparting up-to-date information and advanced data in this field. To ensure the same, a matchless editorial board was set up. Every individual on the board went through rigorous rounds of assessment to prove their worth. After which they invested a large part of their time researching and compiling the most relevant data for our readers. Conferences and sessions were held from time to time between the editorial board and the contributing authors to present the data in the most comprehensible form. The editorial team has worked tirelessly to provide valuable and valid information to help people across the globe.

Every chapter published in this book has been scrutinized by our experts. Their significance has been extensively debated. The topics covered herein carry significant findings which will fuel the growth of the discipline. They may even be implemented as practical applications or may be referred to as a beginning point for another development.

The editorial board has been involved in producing this book since its inception. They have spent rigorous hours researching and exploring the diverse topics which have resulted in the successful publishing of this book. They have passed on their knowledge of decades through this book. To expedite this challenging task, the publisher supported the team at every step. A small team of assistant editors was also appointed to further simplify the editing procedure and attain best results for the readers.

Our editorial team has been hand-picked from every corner of the world. Their multi-ethnicity adds dynamic inputs to the discussions which result in innovative outcomes. These outcomes are then further discussed with the researchers and contributors who give their valuable feedback and opinion regarding the same. The feedback is then collaborated with the researches and they are edited in a comprehensive manner to aid the understanding of the subject.

Apart from the editorial board, the designing team has also invested a significant amount of their time in understanding the subject and creating the most relevant covers. They scrutinized every image to scout for the most suitable representation of the subject and create an appropriate cover for the book.

The publishing team has been involved in this book since its early stages. They were actively engaged in every process, be it collecting the data, connecting with the contributors or procuring relevant information. The team has been an ardent support to the editorial, designing and production team. Their endless efforts to recruit the best for this project, has resulted in the accomplishment of this book. They are a veteran in the field of academics and their pool of knowledge is as vast as their experience in printing. Their expertise and guidance has proved useful at every step. Their uncompromising quality standards have made this book an exceptional effort. Their encouragement from time to time has been an inspiration for everyone.

The publisher and the editorial board hope that this book will prove to be a valuable piece of knowledge for researchers, students, practitioners and scholars across the globe.

List of Contributors

Parveen Saini and Manju Arora
National Physical Laboratory, New Delhi, India

Markus Woehrmann
Technische Universität Berlin (TUB), Berlin, Germany

Michael Toepper
Fraunhofer Institute for Reliability and Microintegration (IZM), Berlin, Germany

Michael R. Gleeson
National University of Ireland Maynooth, Department of Computer Science, Faculty of Science and Engineering, Maynooth, Co. Kildare, Republic of Ireland

Jinxin Guo and John T. Sheridan
UCD School of Electrical, Electronic and Communications Engineering, College of Engineering, Mathematical and Physical Sciences, Communications and Optoelectronic Research Centre, SFI Strategic Research Centre in Solar Energy Conversion, University College Dublin, Belfield, Republic of Ireland

Kareema Majeed Ziadan
University of Basrah, College of Science, Department of Physics, Basrah, Iraq

Sergey V. Kazakov
Department of Chemistry & Physical Sciences, Pace University, Pleasantville, NY, USA

Ana Isabel Mouquinho, Krasimira Petrova, Maria Teresa Barros and João Sotomayor
REQUIMTE, CQFB, Departamento de Química, Faculdade de Ciências e Tecnologia, Universidade Nova de Lisboa, Caparica, Portugal

Karen Segala
Department of Bioprocesses and Materials Engineering (DEMBio), State University of Campinas, UNICAMP, Brazil

Angela S. Pereira
CICECO, University of Aveiro, Portugal

I.Yu. Sapurina
Institute of Macromolecular Compounds RAS, Sankt-Petersburg, Russia

M.A. Shishov
St.-Petersburg State Electro-technical University "LETI" Saint-Petersburg, Russia

Chengfei Zhou
Beijing Research Center for Radiation Application, Key Laboratory of Radiation Technology and Advanced Materials of Beijing Academy of Science and Technology, Beijing, P.R. China

Eric Pasquinet
CEA-DAM Le Ripault - BP16 - F-37260 Monts, France

Manabu Ishifune
Department of Applied Chemistry, Faculty of Science and Engeneering, Kinki University, Higashi-Osaka, Osaka, Japan